HB 539 .Z34 2002
OCLC:49326927
Interest rate management

P9-CAM-145

THE RICHARD STOCKTON COLLEGE
OF NEW JERSEY LIBRARY
POMONA, NEW JERSEY 08240-0195

Springer Finance

Editorial Board

M. Avellaneda
G. Barone-Adesi
M. Broadie
M.H.A. Davis
C. Klüppelberg
E. Kopp
W. Schachermayer

Springer
Berlin
Heidelberg
New York
Barcelona
Hong Kong
London
Milan
Paris
Tokyo

Springer Finance

Springer Finance is a programme of books aimed at students, academics and practitioners working on increasingly technical approaches to the analysis of financial markets. It aims to cover a variety of topics, not only mathematical finance but foreign exchanges, term structure, risk management, portfolio theory, equity derivatives, and financial economics.

M. Ammann, Credit Risk Valuation: Methods, Models, and Application (2001)

N.H. Bingham and R. Kiesel, Risk-Neutral Valuation: Pricing and Hedging of Financial Derivatives (1998)

T.R. Bielecki and M. Rutkowski, Credit Risk: Modeling, Valuation and Hedging (2001)

D. Brigo and F. Mercurio, Interest Rate Models: Theory and Practice (2001)

R. Buff, Uncertain Volatility Models-Theory and Application (2002)

G. Deboeck and T. Kohonen (Editors), Visual Explorations in Finance with Self-Organizing Maps (1998)

R. J. Elliott and P. E. Kopp, Mathematics of Financial Markets (1999)

H. Geman, D. Madan, S.R. Pliska and T. Vorst (Editors), Mathematical Finance-Bachelier Congress 2000 (2001)

Y.-K. Kwok, Mathematical Models of Financial Derivatives (1998)

A. Pelsser, Efficient Methods for Valuing Interest Rate Derivatives (2000)

M. Yor, Exponential Functionals of Brownian Motion and Related Processes (2001)

R. Zagst, Interest-Rate Management (2002)

Rudi Zagst

Interest-Rate Management

 Springer

THE RICHARD STOCKTON COLLEGE
OF NEW JERSEY LIBRARY
POMONA, NEW JERSEY 08240-0195

Rudi Zagst
RiskLab GmbH
Arabellastraße 4
D-81925 München
e-mail: zagst@risklab.de
and
Zentrum Mathematik
Technische Universität München
D-80290 München
e-mail: zagst@ma.tum.de

Mathematics Subject Classification (2000): 60HXX, 60G15, 60G44, 62P05, 90B50, 90C11, 90C20
JEL Classification: G11, G12, G13, E43, E44, E47, D81

Cataloging-in-Publication Data applied for

Die Deutsche Bibliothek - CIP Einheitsaufnahme

Zagst, Rudi:
Interest, rate, management/Rudi Zagst.-Berlin; Heidelberg; New York;
Barcelona; Hong Kong; London; Milan; Paris; Tokyo: Springer, 2002
  (Springer finance)
  ISBN 3-540-67594-9

ISBN 3-540-67594-9   Springer-Verlag Berlin Heidelberg New York

This work is subject to copyright. All rights are reserved, whether the whole or part of the material is concerned, specifically the rights of translation, reprinting, reuse of illustrations, recitation, broadcasting, reproduction on microfilm or in any other way, and storage in data banks. Duplication of this publication or parts thereof is permitted only under the provisions of the German Copyright Law of September 9, 1965, in its current version, and permission for use must always be obtained from Springer-Verlag. Violations are liable for prosecution under the German Copyright Law.

Springer-Verlag Berlin Heidelberg New York
a member of BertelsmannSpringer Science+Business Media GmbH

http://www.springer.de

© Springer-Verlag Berlin Heidelberg 2002
Printed in Germany

The use of general descriptive names, registered names, trademarks etc. in this publication does not imply, even in the absence of a specific statement, that such names are exempt from the relevant protective laws and regulations and therefore free for general use.

Cover design: *design & production*, Heidelberg

Typesetting by the author using a Springer LATEX macro package
Printed on acid-free paper   SPIN 10755144   41/3142db-5 4 3 2 1 0

Meinen Eltern !

Als ich klein und hilflos war,
gabt Ihr mir Geborgenheit und Schutz.
Als ich größer wurde und Weisheit suchte,
halft Ihr mir, sie zu finden.
Als ich meinen Weg suchen mußte,
habt Ihr mich gehen lassen.
Als ich ihn fand,
halft Ihr mir, meine Ziele zu erreichen.
Als ich dieses Buch schrieb,
habe ich gemerkt, daß ich mich viel zu selten bei Euch bedankt habe.

Vielen Dank dafür,
daß ich mich stets auf Eure Liebe und Treue verlassen konnte.
Es war und ist ein Glück, daß es Euch gibt !

Preface

Who gains all his ends did set the level too low.

Although the history of trading on financial markets started a long and possibly not exactly definable time ago, most financial analysts agree that the core of mathematical finance dates back to the year 1973. Not only did the world's first option exchange open its doors in Chicago in that year but Black and Scholes published their pioneering paper [BS73] on the pricing and hedging of contingent claims. Since then their explicit pricing formula has become the market standard for pricing European stock options and related financial derivatives. In contrast to the equity market, no comparable model is accepted as standard for the interest-rate market as a whole. One of the reasons is that interest-rate derivatives usually depend on the change of a complete yield curve rather than only one single interest rate. This complicates the pricing of these products as well as the process of managing their market risk in an essential way. Consequently, a large number of interest-rate models have appeared in the literature using one or more factors to explain the potential changes of the yield curve. Beside the Black ([Bla76]) and the Heath-Jarrow-Morton model ([HJM92]) which are widely used in practice, the LIBOR and swap market models introduced by Brace, Gątarek, and Musiela [BGM97], Miltersen, Sandmann, and Sondermann [MSS97], and Jamshidian [Jam98] are among the most promising ones. However, up to now, none of the existing models can be considered as more than a standard for a sub-market such as the cap or swap market.

Inconsistencies usually appear once these models are to be used for pricing other interest-rate derivatives jointly.

To understand all the different interest-rate models, and to be able to develop new models, one needs a thorough background in stochastic calculus and financial mathematics. Excellent books for the advanced reader in this field are, e.g., Lamberton and Lapeyre [LL97], Musiela and Rutkowski [MR97], or Øksendal [Øks98]. On the other hand, there are also books written for a more economics oriented readership. Very good representatives, e.g., are Hull [Hul00] or Baxter and Rennie [BR96]. Books aiming for a middle way between these two species are, for instance, the excellent texts of Bingham and Kiesel [BK98] or Korn and Korn [KK99]. However, none of these books addresses the complete financial engineering process, i.e. modelling, pricing, hedging as well as medium and long-term risk and asset management. And indeed, this is the main reason

...why I have written this book.

In many discussions with my students at the universities of Ulm, Augsburg and Munich, as well as during my courses and consulting activities for banks, insurance companies, and other financial institutions, the question appeared of whether there is a book describing the whole process - from mathematical modelling and pricing to the risk and asset management of a complete portfolio or trading book. A list of different books has been the best advice I could give. Then some years ago, when we were discussing this very topic during a car ride from Munich to Ulm, a good friend of mine, Dr. Gerhard Scheuenstuhl, encouraged me to close this gap by writing a book about both sides of the coin, the mathematical modelling and the risk management. Of course, covering the whole story would have been a daunting task and would have resulted in many more pages than you hold in your hand. The background material of stochastic calculus had to be restricted, as well as the number of models and derivatives being discussed and the topics covering risk management issues. However, it was the aim of the author to give an insight into the long road of modelling an interest-rate market, mark-to-market a selection of interest-rate derivatives and simulate their future value using the market model (mark-to-future), as well as deriving valuable risk numbers applied within a reliable risk management process. So after all this,

...what is this book about?

We begin with an overview of the most important mathematical tools for describing financial markets, i.e. stochastic processes and martingales. These methods are applied to modelling a financial market and, in particular, to modelling an interest-rate market. We will learn about different interest-rate models driving the prices of financial assets, as well as different methods for pricing interest-rate derivatives. These are a pure application of the martingale theory, an application of the theory of Green's functions, and an application of the important change-of-numéraire technique. Each of these methods is applied within a specific model showing the wide spectrum of possibilities we have in the evaluation of financial products. However, it was not possible to cover all or even most of the existing interest-rate models available in the literature. Also it was not the intention of the author to compete with the books of Lamberton and Lapeyre, Musiela and Rutkowski, or Øksendal which go much further into the mathematical details than this book. Rather, we follow the books of Bingham and Kiesel or Korn and Korn on their middle way through mathematical finance before we leave their path to aim for the measuring and management of market risk. Short- and long-term risk measures will be discussed, as well as a selection of optimization problems, which are solved to maximize the performance of a portfolio under limited downside risk. Now you may ask

... for whom have I written this book?

This book is written for students, researchers, and practitioners who want to get an insight into the modelling of interest-rate markets as well as the pricing and management of interest-rate derivatives. Chapters 2 and 3 of the book give a rigorous overview of the mathematics of financial markets. They present the most important tools needed to describe the movement of market prices and define the theoretical framework for the pricing and hedging of contingent claims. A basic knowledge of probability theory and a certain quantitative background are recommended as prerequisites. Those looking for a crash course in stochastic processes and the modelling of financial markets will hopefully find this part to be a valuable source. However, if you are already familiar with the Itô calculus and the methods for pricing and hedging financial derivatives, you could immediately start with Chapter 4. Here and in Section 5 we focus on the modelling and pricing in interest-rate markets. If you already know most of the specific interest-rate models you may just glance over Section 4, which shows how an interest-rate market can be embedded in the financial market framework of Section 3. In Chapter 5 we describe the most popular interest-rate derivatives, and show how they can be priced using a specific interest-rate model. Because the trading and risk management of derivatives are dominated mainly by the application of specific models and techniques, the style of the book will also change gradually to a more economics oriented one. Real-world applications have to take care of market conventions such as daycounts or special rates which are, from a mathematical point of view, not too much of a deal but which may vary between different markets and result in misleading risk numbers and prices once they are ignored. Chapters 6 and 7 are intended to give an insight into the practical application of interest-rate models to the risk and portfolio management of interest-rate derivatives. They cover a selection of short- and long-term-oriented risk measures as well as comprehensive case studies based on real market data. We hope those interested in mark-to-future simulations, specific risk numbers, and their use for risk management will enjoy reading these chapters. Should you be more practice-oriented, you will not need a full understanding of stochastic calculus and martingale theory for Sections 5 to 7. A basic understanding of the main results of Chapters 2 to 4 will be sufficient. However, those getting fascinated with the potential of financial modelling may look for the "math necessities" within these chapters. Since all parts of the book have been used in teaching mathematical finance, financial engineering and risk management at different universities, this textbook may also serve as a valuable source for graduate and PhD students in mathematics or finance who want to acquire some knowledge of financial markets and risk management.

A final word!

Satisfying the needs of both practitioners and researchers is always a hard and sometimes too hard a problem to solve. A gap still exists between these two worlds, and accordingly, there remains a gap between the corresponding parts of the book. I have tried to make this gap as small as possible. I also had to restrict myself to a brief overview of stochastic calculus, where a lot more could have been said and proved. On account of the idea and limited size of the book, I had to select a small variety of interest-rate models and discuss their pricing effects rather than show for which market which model works best. The reader interested in this question may, for instance, refer to Brigo and Mercurio [BM01]. I also could not cover all risk-management topics, since this is a boundless field in its own right. Therefore I surely didn't gain all my intended ends. Nevertheless, I hope that I have succeeded in finding a good middle way to describe the process from mathematical modelling and pricing to the risk and asset management of interest-rate derivatives portfolios. Whether I have reached this target will be for the reader to judge.

Acknowledgements

It is a pleasure to thank all those who have influenced this work. I am especially thankful to Dr. Rüdiger Kiesel for his excellent and useful suggestions, and his never-ending patience in discussing mathematical problems, even on some week-ends when he surely had more practical things in mind. I also owe a special debt to Professor Nick H. Bingham for his helpful comments and suggestions.

I thank Professor Stefan R. Mayer, Dr. Frank Proske, and Dr. Thomas Zausinger for their valuable comments on an earlier draft of this book. I would also like to thank Reinhold Hafner and Dr. Bernd Schmid for their patience in explaining the secrets of Scientific Word. Special thanks to Katrin Schöttle and Jan Kehrbaum for calculating many of the examples.

I am particularly grateful to Dr. Gerhard Scheuenstuhl. Not only did he help to create the first idea of the book but, as a friend and colleague, always found the right, encouraging words I needed to go on.

It is a pleasure to thank Professor Ulrich Rieder for his support and helpful comments and suggestions on the concept and content of this book. I would also like to thank Professor Dieter Kalin, Professor Frank Stehling and Professor Hans Wolff for their advice and support.

I thank Algorithmics Incorporated for the possibility of using their powerful software tools, especially Risk++ and RiskWatch, and GAMS Software GmbH for making it possible to work with their optimization tools. Most of the case studies were carried out using the software tool Risk *Advisor* from risklab *germany*.

I am especially grateful to HypoVereinsbank Asset Management GmbH and the people from risklab *germany* for their support and patience during the time when I wrote this book.

Johann Goldbrunner, Thomas Neisse, and Horst Schmidt deserve a special acknowledgment for their support and confidence, and for making it possible to create an environment where I could realize my ideas.

I would also like to thank Dr. Catriona Byrne at Springer-Verlag for her helpful comments as well as Susanne Denskus and the staff at Springer-Verlag for their support and help throughout this project.

And, last and most, I thank my wife Edith for her love, support, and forbearance while this book was being written.

Munich, January 2002   Rudi Zagst

Contents

II   Modelling and Pricing in Interest-Rate Markets

III   Measuring and Managing Interest-Rate Risk

1
Introduction

Financial markets at all times fascinated people trying to predict price movements and making money out of it. Fortune-tellers and self-made market prophets often tried to influence the market participants by their forecasts, very often not to their own disadvantage. However, following such a forecast and putting all one's eggs in one basket is a very risky thing to do. Of course, if prices move in the investor's favour, he might get rich. But unfortunately, bad things happen. On Black Friday, October 25, 1929 a tremendous market crash finished ten bullish years of increasing stock prices, leaving millions of people with empty pockets and starting one of the most significant depressions of the modern age. On Black Monday, October 19, 1987 the Dow Jones Index lost 23% or 500 points within minutes while it increased by only 1700 points over the previous five years. Could traders have known before about the forthcoming worst-case events?

A lot of people came up with more or less sophisticated statistical examinations and correlations, trying to find some regularities in the market. For example, they found empirical evidence for the hypothesis that *the shorter the skirts, the higher the stock prices.* While, from a real life perspective, this seems to be a little implausible, rules of thumb were set up such as *buy on bad news, sell on good news* or *sell in May and go away*, considering the flow of money or the nervousness of the people. Other rules which are very popular at the trading desks are *the trend is your friend* or *never catch a falling knife*, telling traders not to invest in a falling market. But even the famous Sir Isaac Newton (1643-1727), who lost a fortune at the time of the South Sea Bubble in 1720, sighed: *"I can measure the motion of bodies, but I cannot measure human folly."*

Nevertheless, researchers tried to model financial markets and build a theoretical framework to explain market price behaviour. In 1829, the botanist R. Brown watched pollen particles under a microscope, observing that they may move according to a so-called Brownian motion. In 1900, L. Bachelier [Bac00] was the first to consider Brownian motion as a tool to describe the behaviour of stock prices. But not until 1923 was this process rigorously defined and constructed by N. Wiener [Wie23]. We will follow his way and give a brief overview of the basic *mathematical concepts* used to describe market price movements in Chapter 2. The definition and characteristics of stochastic processes are given in Sections 2.1 and 2.2. One of the central tools in stochastic calculus is Itô's lemma, which is described in Section 2.4 and can be applied to determine the stochastic differential equation for the prices of financial derivatives. This concept is closely related to the stochastic integral, which is defined in Section 2.3. Since martingales are one of the most important elements for evaluating the prices of financial instruments, they are defined in Section 2.5 and used to describe the prices of financial derivatives in terms of conditional expectations. The Feynman-Kac formula provides the gateway between such conditional expectations and the differential equations which can be solved numerically. This is discussed in Section 2.6.

In Chapter 3 we introduce the basic building blocks and assumptions for setting up a consistent framework to describe a *financial market* using stochastic processes. The primary traded assets and the basic trading principles are defined in Section 3.1. We also give conditions under which the normalized or discounted market prices can be described by martingales. In Section 3.2 we show under which conditions there are no arbitrage opportunities in the financial market. Another important characteristic of a financial market is its completeness, which will be discussed in Section 3.3. In Section 3.4 we show that financial derivatives can be uniquely priced if the financial market is complete. One of the most famous complete market models, the Black-Scholes model, is discussed in Section 3.5. It is used to derive the prices for (European) options on contingent claims. To price financial derivatives it might sometimes be more comfortable to use a numéraire other than the cash account. We could, for example, change the numéraire and use the so-called T−forward measure for describing the discounted market prices. This technique is described in Sections 3.6 and 3.7.

Chapter 4 deals with *interest-rate markets* and the zero-coupon bonds as primary traded assets. Because this market is a specific financial market, we will apply the results of Chapter 3 to embed it in the general framework of Section 3.1. We start by defining the general interest-rate market model in Section 4.1. No-arbitrage and completeness conditions in the interest-rate market model are given in Section 4.2, while Section 4.3 deals with the pricing of interest rate related contingent claims. Since there are infinitely many zero-coupon bonds corresponding to different maturity dates it has been one of the main challenges in interest rate theory to find the driving factors of the zero-coupon bond prices. One of the most general models or frameworks was introduced by Heath, Jarrow, and Morton [HJM92] and will be discussed in Section 4.4. One-factor models, such as the famous short-rate models and the Gaussian models, are discussed in Sections 4.5. A brief overview of multi-factor models can be found in Section 4.6. The LIBOR market models, a new class of models describing the behaviour of market rates rather than that of the short or forward short rates, is presented in Section 4.7. None of these models deal with the possibility of zero-rate changes because of defaults in the financial market. Therefore, we give an overview of credit risk models in Section 4.8.

Many years were to pass between the first steps of Bachelier and Wiener and the introduction of standardized *financial products* and financial markets ready to support the exchange of these instruments on a common platform. Describing all of the existing financial products would have been a heroic task, and was not in the intention of this book. Nevertheless, Chapter 5 is dedicated to describing and pricing at least the most important financial instruments, which are all built on zero-coupon bonds as primary traded assets and may be enriched by additional conditions such as optionality or agreements with respect to future points in time. We start with a brief discussion on how the financial market defines the time between two specific dates in Section 5.1. Probably the simplest financial instrument derived from the primary traded assets is a portfolio of zero-coupon bonds which is, under special assumptions on their notional amount, called a coupon bond and described in Section 5.2. Coupon and zero-coupon bonds are the underlying instruments for the forward agreements and futures discussed in Sections 5.3 and 5.4. Zero bonds are also the main building block for another family of interest-rate instruments, the interest-rate swaps, which are presented in Section 5.5. Probably the easiest option in interest-rate markets is an option on a zero-coupon bond, which will be priced in Section 5.6.1. It is the basic tool for evaluating a great variety of optional interest-rate instruments. Examples are the caps and floors of Section 5.6.2, as well as the coupon bond options of Section 5.6.3. In Section 5.6.4 we show how options on interest-rate swaps can be priced. All the previous interest-rate options are considered to be market standard. Contingent claims with payoffs more complicated than that of standard (European) interest-rate call and put options are called exotic interest-rate options. An overview of some of these products is given in Section 5.7. All financial instruments are described from a practical point of view and priced according to the theoretical framework of the previous chapters. But the quality of pricing of all these instruments pretty much depends on the availability of good market or price information, especially with respect to yields and volatilities. Each model we use for pricing interest-rate derivatives has to be fitted to market data. Section 5.8 gives an overview of different sources of interest-rate information expressed by yield or zero-rate curves, market prices or volatilities. The latter are always quoted with respect to a benchmark model which is, most of the time, a version of the Black model. We will give a little more detail in Sections 5.8.2 and 5.8.3. A practical case study on how Black volatility information can be transformed to an implied volatility curve for the Hull-White model is shown in Section 5.8.4.

In 1991, the Basle Comittee on Banking Supervision set up directions for the risk management of derivatives and defined *market risk* to be the risk of a negative impact of changing market prices for the financial situation of an institution. For interest-rate instruments this risk is due to changes of the yield or zero-rate curve. Traditionally, it was measured by a parallel shift of the yield curve, but many risk and portfolio managers had to learn about the possibility of non-parallel movements of the yield curve, which is sometimes called shape risk. It is in the focus of this book to give an overview of how market risk can be measured and managed. In Section 6.1 we define the different risk measures based on small market movements and small periods of time, such as the first- and second-order sensitivities. Examples are the Black or key-rate deltas and gammas as well as duration and convexity. Beside this short-term sensitivity risk portfolio managers will change their portfolio if the risk of the portfolio return falling below a given benchmark is too high. This considers large market movements as well as longer time horizons and is also known as downside risk. It is usually carried out by a scenario analysis and involves calculating the portfolio's profit or loss over a specified period of time under a variety of different scenarios. The scenarios can either be chosen by the management or generated by a specific interest-rate model as it was discussed in Chapter 4. Using these scenarios, different measures of downside risk, such as the lower partial moments or the value at risk, can be calculated as discussed in Section 6.2. An interesting question is what properties a somehow "good" risk measure should have. We try to answer this question in Section 6.3. Because simulations are often very time-consuming, it may make sense to concentrate on just a few explanatory risk factors. This idea is discussed in Section 6.4.

Having defined the different possibilities for calculating market risk, what kind of risk does a fortune-teller consider? Well, she may look into her cards and see what will happen giving this scenario a probability of one. Unfortunately, we will have to work a little harder to find an adequate probability distribution or even a representative set of market scenarios. One special method, tailor-made for the Hull-White model, is described in the appendix but there is a whole variety of simulation methods available on modern computer systems and programs. So what should we do with these simulations and risk numbers? The answer is *risk management*.

Basically, risk management deals with the problem of protecting a portfolio or trading book against unexpected losses. It therefore expresses the desire of a portfolio manager or trader to guarantee a minimum holding period return or to create a portfolio which helps to cover specific liabilities over time. Done right, risk management may avoid extreme movements of the portfolio value, reduce the tracking error or even the trading costs. However, there are different possibilities to set up a risk-management or hedging process which are adressed in Chapter 7. If we are interested in controlling short-term risk or if we like to hedge against small movements in market prices, we may choose a sensitivity-based risk management. This method is described in Section 7.1. If we are dealing with a longer time horizon, or want to be safe against large market movements, we may prefer a downside risk management, as introduced and discussed in Section 7.2. We explicitly state the corresponding optimization programs and prove that they solve the given practical problems. The differences and possibilities of the methods described are documented by extensive practical case studies, showing that risk management is much more than setting some parameters to zero. These case studies may also help to emphasize that high quality risk management is one of the key requirements for a modern portfolio management.

Part I

Mathematical Finance Background

2

Stochastic Processes and Martingales

In this chapter we give a brief overview of the basic mathematical concepts used for pricing financial products. Modern pricing theory is mainly based on describing the ups and downs of market prices via stochastic processes. Therefore, we start with the definition and characteristics of stochastic processes in Section 2.1. Usually the price-changes of the most basic financial instruments are modelled in terms of a so-called stochastic differential equation (SDE). The price of a financial derivative is considered to be a function of the basic financial instruments' prices. One of the central tools in stochastic calculus which can be applied to determine the stochastic differential equation for the prices of these derivatives, Itô's lemma, is described in Section 2.4. This concept is closely related to the stochastic integral, which is defined in Section 2.3 to be the mean square limit of some random sums, carefully put together so that the resulting limit is a (local) martingale. And in fact, martingales are one of the most important elements needed for the evaluation of financial instruments. They are defined in Section 2.5 and are used to describe the prices of financial derivatives in terms of conditional expectations. The Feynman-Kac formula provides a partial differential equation (PDE) that corresponds to such conditional expectations as described in Section 2.6. The advantage of dealing with partial differential equations is that they can be solved numerically. Also, PDE methods come naturally to applied mathematicians and physicists. This approach led Black and Scholes [BS73] to their famous equation for pricing European options. Their model will be discussed in Section 3.5. By contrast, martingales and Itô (stochastic) calculus come naturally to probabilists, and this approach led Merton [Mer74] to his option pricing formula.

If you are already familiar with stochastic processes and Itô calculus you may immediately switch to Chapter 3. If you are, however, interested in more details on this topic you may refer to Bingham and Kiesel [BK98], Karatzas and Shreve [KS91], Korn and Korn [KK99], or Musiela and Rutkowski [MR97]. To simplify the search for more information, at least one reference including the exact location of the respective statement or proof is listed wherever it seemed to be appropriate and helpful.

2.1 Stochastic Processes

Let us begin this brief overview of probability theory and stochastic processes by defining the basic items. A measurable space $(\Omega, \mathcal{F})$ is given by a non-empty set Ω, called the sample space, and a sigma-algebra $\mathcal{F}$ on Ω, i.e. a collection of subsets of Ω with $\Omega \in \mathcal{F}$, $A^c := \Omega - A \in \mathcal{F}$ for all $A \in \mathcal{F}$, and $\cup_{n=1}^{\infty} A_n \in \mathcal{F}$ for any sequence $(A_n)_{n \in I\!N}$ with $A_n \in \mathcal{F}$ for all $n \in I\!N$, where $I\!N$ is the set of all natural numbers. Each $A \in \mathcal{F}$ is called a measurable set. An important example of such a sigma-algebra is the so-called Borel sigma-algebra $\mathcal{B}\left(I\!R^k\right)$ in $I\!R^k$, with $I\!R$ denoting the real numbers, which is the smallest sigma-algebra that contains all open sets in $I\!R^k$. A k-dimensional function $f : \Omega \rightarrow I\!R^k$ is called $\left(\mathcal{F} - \mathcal{B}\left(I\!R^k\right) -\right)$ measurable or simply $(\mathcal{F}-)$ measurable if $f^{-1}(B) = \{\omega \in \Omega : f(\omega) \in B\} \in \mathcal{F}$ for all $B \in \mathcal{B}\left(I\!R^k\right)$. A set function $Q : \mathcal{F} \rightarrow [0, \infty]$ on the measurable space $(\Omega, \mathcal{F})$ is called a measure on $(\Omega, \mathcal{F})$ if Q is countably additive, i.e. for any sequence $(A_n)_{n \in I\!N}$ of disjoint measurable sets $A_n \in \mathcal{F}, n \in I\!N$, we have $Q\left(\cup_{n=1}^{\infty} A_n\right) = \sum_{n=1}^{\infty} Q(A_n)$. The triple $(\Omega, \mathcal{F}, Q)$ is called a measure space. Q is called sigma-finite if there exists an non-decreasing sequence of measurable sets $(A_n)_{n \in I\!N}$, that is $A_n \subset A_{n+1}$ for all $n \in I\!N$, with $\cup_{n=1}^{\infty} A_n = \Omega$ such that $Q(A_n) < \infty$ for all $n \in I\!N$. Q is called a probability measure if $Q(\Omega) = 1$. In this case $(\Omega, \mathcal{F}, Q)$ is called a probability space. Note that each probability measure is sigma-finite. A set $A \in \mathcal{F}$ with $Q(A) = 0$ is called a $(Q-)$ null set. A measure space $(\Omega, \mathcal{F}, Q)$ is called complete if every subset of a $(Q-)$ null set is also a measurable set, i.e. a member of $\mathcal{F}$. A measure space $(\Omega, \mathcal{F}, Q)$ that is not already complete can be easily completed by adjoining the set $\mathcal{N}_Q$ of all subsets of the $(Q-)$ null sets. To do this we extend $\mathcal{F}$ to $\mathcal{F}_{\mathcal{N}_Q}$ which is defined to be the smallest sigma-algebra that contains all sets of the form $A \cup N$ with $A \in \mathcal{F}$ and $N \in \mathcal{N}_Q$, i.e. $\mathcal{F}_{\mathcal{N}_Q} = \{A \cup N : A \in \mathcal{F} \text{ and } N \in \mathcal{N}_Q\}$, and we extend the measure Q to the measure $Q_{\mathcal{N}_Q}$ by setting $Q_{\mathcal{N}_Q}(A \cup N) = Q(A)$ for all $A \in \mathcal{F}, N \in \mathcal{N}_Q$. Then the measure space $\left(\Omega, \mathcal{F}_{\mathcal{N}_Q}, Q_{\mathcal{N}_Q}\right)$ is complete.

Definition 2.1 (Equivalent Measures) *Let Q and $\widetilde{Q}$ be two measures defined on the same measurable space $(\Omega, \mathcal{F})$. We say that $\widetilde{Q}$ is absolutely continuous with respect to Q, written $\widetilde{Q} \ll Q$, if $\widetilde{Q}(A) = 0$ whenever $Q(A) = 0$, $A \in \mathcal{F}$. If both $\widetilde{Q} \ll Q$ and $Q \ll \widetilde{Q}$, we call Q and $\widetilde{Q}$ equivalent measures and denote this by $\widetilde{Q} \sim Q$.*

If Q and $\widetilde{Q}$ are equivalent measures, we have $\widetilde{Q}(A) = 0$ if and only if $Q(A) = 0$ for all sets $A \in \mathcal{F}$, which means that Q and $\widetilde{Q}$ have the same null sets. This is equivalent to the statement that Q and $\widetilde{Q}$ have the same sets of positive measure, and also equivalent to the statement that Q and $\widetilde{Q}$ have the same sets of probability 1, also called the almost sure (a.s.) sets, if Q and $\widetilde{Q}$ are probability measures. A measurable function $f : \Omega \to I\!R$ is called (Q−quasi) integrable if $\int_\Omega f^+ \, dQ < \infty$ and (or) $\int_\Omega f^- \, dQ < \infty$ with $f^+ = \max\{f, 0\}$ and $f^- = \max\{-f, 0\}$ where $\int_\Omega f \, dQ$ denotes the so-called (Lebesgue) integral of f with respect to Q. Using these definitions the following theorem holds:

Theorem 2.2 (Radon-Nikodým) *Let Q be a sigma-finite measure and $\widetilde{Q}$ be a measure on the measurable space $(\Omega, \mathcal{F})$ with $\widetilde{Q} < \infty$. Then $\widetilde{Q} \ll Q$ if and only if there exists an integrable function $f \geq 0$ $Q - a.s.$ such that*

$$\widetilde{Q}(A) = \int_A f \, dQ \quad \text{for all } A \in \mathcal{F}.$$

f is called the Radon-Nikodým derivative of $\widetilde{Q}$ with respect to Q and is also written as $f = d\widetilde{Q}/dQ$.

From now on we assume that we are working on a *complete probability space* $(\Omega, \mathcal{F}, Q)$. In this case a k-dimensional measurable function $X : \Omega \to I\!R^k$, $k \in I\!N$ is called a random vector. For $k = 1$ we call X a random variable. The smallest sigma-algebra containing all sets $X^{-1}(B) = \{\omega \in \Omega : X(\omega) \in B\}$, where B runs through the Borel sigma-algebra $\mathcal{B}(I\!R^k)$, is called the sigma-algebra generated by X, and will be denoted by $\mathcal{F}(X)$. For a Q−quasi integrable random variable $X : \Omega \to I\!R$ we call $E_Q[X] = \int_\Omega X \, dQ$ the expected value or expectation of X under the measure Q. A random variable X is called square-integrable, or of order two, if $E_Q[X^2] < \infty$. We say that two random variables X and Y are equivalent, if $X = Y$ $Q - a.s.$ If we identify equivalent random variables, the set of all random variables of order two spans a linear vector space V. The inner product on this vector space is defined by

$$< X, Y > := E_Q[X \cdot Y], \; X, Y \in V.$$

Because of Schwarz' inequality, we know that $| < X, Y > | < \infty$ for all $X, Y \in V$. Furthermore, it is easy to see that

$$\|X\|_2 := < X, X >^{\frac{1}{2}} = \sqrt{E_Q[X^2]}, \; X \in V,$$

defines a norm on V. We call this norm the L_2-norm and use it to define a metric on V by $d(X,Y) := ||X-Y||_2$. The space $(V, || \ ||_2)$ or (V, d) is called the L_2-space. We say that a sequence of random variables converges in mean square (m.s.) to a random variable X for $n \to \infty$, if $\lim_{n \to \infty} ||X_n - X||_2 = 0$. We also write $X_n \overset{m.s.}{\longrightarrow} X$ or $\lim_{\substack{n \to \infty \\ m.s.}} X_n = X$ and call this convergence a convergence of order two or mean square convergence.

Theorem 2.3 (Completeness)

a) *The L_2-space is complete. That is, for each sequence $(X_n)_{n \in \mathbb{N}}$ of random variables in V the following statements are equivalent:*

  (i) $d(X_n, X_m) = ||X_n - X_m||_2 \to 0$ for $m, n \to \infty$.
  (ii) *There is a random variable X in V with $d(X_n, X) \to 0$ for* $n \to \infty$.

b) *Let $(X_n)_{n \in \mathbb{N}}$ be a sequence of random variables of order two with $X_n \overset{m.s.}{\longrightarrow} X$ and $X_n \overset{m.s.}{\longrightarrow} X^*$. Then the two random variables X and X^* are equivalent, i.e. $X = X^*$ (in V).*

We thus know that the L_2-space is a special Hilbert space. Also interesting to note is the following theorem.

Theorem 2.4   *Let $(X_n)_{n \in \mathbb{N}}$ be a sequence of random variables of order two with $X_n \overset{m.s.}{\longrightarrow} X$. Then $\lim_{n \to \infty} E_Q[X_n] = E_Q[X]$.*

Now suppose that $\mathcal{G} \subset \mathcal{F}$ is a sub-sigma-algebra of $\mathcal{F}$, i.e. $\mathcal{G}$ is a sigma-algebra and also a subset of $\mathcal{F}$. $\mathcal{F}$ is sometimes called the finer, $\mathcal{G}$ the coarser sigma-algebra. Furthermore, let X be a non-negative integrable random variable on $(\Omega, \mathcal{F}, Q)$. It is easy to show that the set-function $\widetilde{Q}(A) := \int_A X \, dQ$ defined on all $A \in \mathcal{G}$ is a measure on $(\Omega, \mathcal{G})$ with $\widetilde{Q} \ll Q$. Therefore, applying Theorem 2.2, there exists an integrable ($\mathcal{G}$ – measurable) random variable $f_{\mathcal{G}}$ such that $\widetilde{Q}(A) = \int_A f_{\mathcal{G}} \, dQ$ for all $A \in \mathcal{G}$. We call $f_{\mathcal{G}}$ the conditional expectation of X given (or conditional on) $\mathcal{G}$ and denote it by $f_{\mathcal{G}} = E_Q[X \mid \mathcal{G}]$. If X may also be negative we decompose X as $X = X^+ - X^-$ and define $E_Q[X \mid \mathcal{G}]$ by $E_Q[X \mid \mathcal{G}] = E_Q[X^+ \mid \mathcal{G}] - E_Q[X^- \mid \mathcal{G}]$. If the sigma-algebra $\mathcal{G}$ is generated by the random variable Y, i.e. $\mathcal{G} = \mathcal{G}(Y)$, we simply denote the conditional expectation $E_Q[X \mid \mathcal{G}(Y)]$ by $E_Q[X \mid Y]$.

Definition 2.5 (Conditional Expectation) *Let X be an integrable random variable on the probability space $(\Omega, \mathcal{F}, Q)$ and $\mathcal{G} \subset \mathcal{F}$ be a sub-sigma-algebra of $\mathcal{F}$. The conditional expectation of X given (or conditional on, or knowing) $\mathcal{G}$ is implicitly defined to be the $(Q - a.s.$ unique) $\mathcal{G}$-measurable function $E_Q[X \mid \mathcal{G}]$ with*

$$\int_A X \, dQ = \int_A E_Q[X \mid \mathcal{G}] \, dQ \quad \text{for all } A \in \mathcal{G}.$$

The following lemma summarizes the main properties of the conditional expectation[1]. For a proof see, e.g. Bingham and Kiesel [BK98], p. 51ff.

Lemma 2.6 *Let X be an integrable random variable on the probability space $(\Omega, \mathcal{F}, Q)$ and $\mathcal{G} \subset \mathcal{F}$ a sub-sigma-algebra of $\mathcal{F}$. Then we have*

a) for the smallest possible sigma-algebra $\mathcal{G} = \{\emptyset, \Omega\}$:

$$E_Q[X \mid \{\emptyset, \Omega\}] = E_Q[X],$$

b) for the largest possible sigma-algebra $\mathcal{G} = \mathcal{F}$: $E_Q[X \mid \mathcal{F}] = X$ Q–a.s.,

c) if X is $(\mathcal{G})$-measurable: $E_Q[X \mid \mathcal{G}] = X$ Q – a.s.,

d) for any integrable random variable Z, if X is $(\mathcal{G})$-measurable:

$$E_Q[X \cdot Z \mid \mathcal{G}] = X \cdot E_Q[Z \mid \mathcal{G}] \quad Q - a.s.,$$

e) for any sub-sigma-algebra $\mathcal{G}_0 \subset \mathcal{G}$:

$$E_Q[E_Q[X \mid \mathcal{G}_0] \mid \mathcal{G}] = E_Q[X \mid \mathcal{G}_0] = E_Q[E_Q[X \mid \mathcal{G}] \mid \mathcal{G}_0] \quad Q - a.s.,$$

f) always: $E_Q[E_Q[X \mid \mathcal{G}]] = E_Q[X]$ Q – a.s.,

g) if X is independent of $\mathcal{G}$, i.e. X is independent of $1_A : \Omega \to \mathbb{R}$ for every $A \in \mathcal{G}$ with $1_A(\omega) := 1$ if $\omega \in A$ and $1_A(\omega) := 0$ else:

$$E_Q[X \mid \mathcal{G}] = E_Q[X] \quad Q - a.s.,$$

h) for any integrable random variable Z and any real numbers a, b:

$$E_Q[a \cdot X + b \cdot Z \mid \mathcal{G}] = a \cdot E_Q[X \mid \mathcal{G}] + b \cdot E_Q[Z \mid \mathcal{G}] \quad Q - a.s.,$$

i) for any random variable Z with $X \leq Z$: $E_Q[X \mid \mathcal{G}] \leq E_Q[Z \mid \mathcal{G}]$ $Q - a.s.

We could interpret a sigma-algebra as a state of information, where the sigma-algebra $\mathcal{F}$ stands for full information and the sub-sigma-algebra $\mathcal{G} \subset \mathcal{F}$ stands for partial information. Following this interpretation $\mathcal{G} = \{\emptyset, \Omega\}$ means that we have no information and conditioning on nothing doesn´t change the expectation (see Lemma 2.6a)). On the other hand, full information means that we know X and thus, conditioning on knowing everything leaves us with the variable X (see Lemma 2.6b)). This argument

---

[1] It should be mentioned that Definition 2.5 as well as Lemma 2.6 can be extended to Q–quasi integrable random variables X as long as all appearing integrals exist. For a proof and more details see, e.g., Hinderer [Hin85], p. 178ff.

also applies to the case where X is $(\mathcal{G})$-measurable (see Lemma 2.6c)) which can be interpreted as X being known given $\mathcal{G}$. Following this argument, what is shown in Lemma 2.6d) is that we can take out of the conditional expectation everything that is known. Lemma 2.6e) states that the effect of the coarser sigma-algebra - the one representing less information - wipes out the effect of the finer. Setting $\mathcal{G}_0 = \{\emptyset, \Omega\}$ in e) gives us part f) in Lemma 2.6.

As we will see later on, it may be comfortable to change the point of view and use a different measure for calculating conditional expectations. The following Bayes formula gives us a relation between the conditional expectations with respect to different probability measures.

Theorem 2.7 (Bayes Formula) *Let Q and $\widetilde{Q}$ be two probability measures on the same measurable space $(\Omega, \mathcal{F})$ and let $f = d\widetilde{Q}/dQ$ be the Radon-Nikodým derivative of $\widetilde{Q}$ with respect to Q. Furthermore, let X be an integrable random variable on the probability space $\left(\Omega, \mathcal{F}, \widetilde{Q}\right)$ and $\mathcal{G} \subset \mathcal{F}$ a sub-sigma-algebra of $\mathcal{F}$. Then the following generalized version of the Bayes formula holds:*

$$E_Q\left[X \cdot f | \mathcal{G}\right] = E_{\widetilde{Q}}\left[X | \mathcal{G}\right] \cdot E_Q\left[f | \mathcal{G}\right].$$

Proof. Since $f \geq 0$ and X is $\widetilde{Q}$–integrable, we know that $X \cdot f$ is Q–integrable. By definition of the conditional expectation (see Definition 2.5) and since $f = d\widetilde{Q}/dQ$ we get for all $A \in \mathcal{G}$

$$
\begin{aligned}
\int_A X \cdot f \, dQ &= \int_A X \, d\widetilde{Q} = \int_A E_{\widetilde{Q}}\left[X | \mathcal{G}\right] d\widetilde{Q} \\
&= \int_A E_{\widetilde{Q}}\left[X | \mathcal{G}\right] \cdot f \, dQ \\
&= \int_A E_Q\left[E_{\widetilde{Q}}\left[X | \mathcal{G}\right] \cdot f | \mathcal{G}\right] dQ \\
&= \int_A E_{\widetilde{Q}}\left[X | \mathcal{G}\right] \cdot E_Q\left[f | \mathcal{G}\right] dQ \quad \text{using Lemma 2.6d).}
\end{aligned}
$$

Hence, by definition of the conditional expectation we conclude that

$$E_Q\left[X \cdot f | \mathcal{G}\right] = E_{\widetilde{Q}}\left[X | \mathcal{G}\right] \cdot E_Q\left[f | \mathcal{G}\right].$$

$\square$

Collecting information is not something which is done at one point in time only. Consequently, sigma-algebras such as $\mathcal{G}_0, \mathcal{G}$ or $\mathcal{F}$ could be thought of representing the state of information at different points in time. To add this idea to our framework, we equip our probability space $(\Omega, \mathcal{F}, Q)$ with a so-called filtration $\mathbb{F}$.

Definition 2.8 (Filtration) *A filtration $I\!F$ is a non-decreasing family of sub-sigma-algebras $(\mathcal{F}_t)_{t\geq0}$ with $\mathcal{F}_t \subset \mathcal{F}$ and $\mathcal{F}_s \subset \mathcal{F}_t$ for all $0 \leq s < t < \infty$. We call $(\Omega, \mathcal{F}, Q, I\!F)$ a filtered probability space, and require that*

a) *$\mathcal{F}_0$ contains all subsets of the $(Q-)$ null sets of $\mathcal{F}$,*

b) *$I\!F$ is right-continuous, i.e. $\mathcal{F}_t = \mathcal{F}_{t+} := \cap_{s>t}\mathcal{F}_s$.*

$\mathcal{F}_t$ represents the information available at time t, and $I\!F = (\mathcal{F}_t)_{t\geq0}$ describes the flow of information over time, where we suppose that we don't lose information as time passes by ($\mathcal{F}_s \subset \mathcal{F}_t$ for all $0 \leq s < t < \infty$). Again, each sigma-algebra $\mathcal{F}_t$, $0 \leq t < \infty$ not already complete can be completed using the set $\mathcal{N}_Q$ of all subsets of the $(Q-)$ null sets. This is then called a $(Q-)$ completion of the filtration $I\!F$. We say that $(\Omega, \mathcal{F}, Q, I\!F)$ is a *complete filtered probability space*, if $\mathcal{F}$ as well as each $\mathcal{F}_t$, $0 \leq t < \infty$, is complete. Note that for a $(Q-)$ completion of the filtration $I\!F$, by definition of the filtration, it is sufficient to complete the sigma-algebra $\mathcal{F}_0$. Hence, assumption a) of Definition 2.8 is dedicated to the fact that we will only work on complete filtered probability spaces. If, on the other hand, assumption b) of Definition 2.8 is not already satisfied, we may adjust the $(Q-$completed) filtration by setting $\widehat{\mathcal{F}}_t := \mathcal{F}_{t+}$ for all $0 \leq t < \infty$. The process of making a filtration complete and right-continuous is usually called $(Q-)$ augmentation of $I\!F$. Assumption b) of Definition 2.8 is to ensure that such a $(Q-)$ augmentation of $I\!F$ has already been done if we claim to work on a filtered probability space.

The price behaviour of financial products over time is usually described by a so-called stochastic process. It is therefore time to generally define this expression.

Definition 2.9 (Stochastic Process) *A stochastic process (vector process) is a family $X = (X_t)_{t\geq0} = (X(t))_{t\geq0}$ of random variables[2] (vectors) defined on the filtered probability space $(\Omega, \mathcal{F}, Q, I\!F)$. We say that*

a) *X is **adapted** (to the filtration $I\!F$) if $X_t = X(t)$ is $(\mathcal{F}_t -)$ measurable for all $t \geq 0$,*

b) *X is **measurable** if the mapping $X : [0, \infty) \times \Omega \to I\!R^k$, $k \in I\!N$, is $\big(\mathcal{B}([0,\infty)) \otimes \mathcal{F} - \mathcal{B}(I\!R^k)-\big)$ measurable with $\mathcal{B}([0,\infty)) \otimes \mathcal{F}$ denoting the product sigma-algebra created by $\mathcal{B}([0,\infty))$ and $\mathcal{F}$, i.e. the smallest sigma-algebra which contains all sets $A_1 \times A_2 \in \mathcal{B}([0,\infty)) \times \mathcal{F}$,*

c) *X is **progressively measurable** if the mapping $X : [0, t] \times \Omega \to I\!R^k$, $k \in I\!N$, is $\big(\mathcal{B}([0,t]) \otimes \mathcal{F}_t - \mathcal{B}(I\!R^k)-\big)$ measurable for each $t \geq 0$.*

---

[2] More generally, we may define a stochastic process on an index set $\mathcal{I} \subset [0, \infty)$ writing $(X(t))_{t\in\mathcal{I}}$. For the ease of exposition we concentrate our interest on the case $\mathcal{I} = [0, \infty)$ here but will come back to other subsets later on.

Note that we either write X_t or $X(t)$, whichever is more comfortable. Also note that a stochastic process is a function in t for each fixed or realized $\omega \in \Omega$. If the stochastic process X is measurable, the mapping $X(\cdot, \omega) : [0, \infty) \to I\!\!R^k$, $k \in I\!\!N$, is $(\mathcal{B}([0, \infty)) - \mathcal{B}(I\!\!R^k) -)$ measurable for each fixed $\omega \in \Omega$. For each fixed $\omega \in \Omega$ we call $X(\omega) = (X_t(\omega))_{t \geq 0} = (X(t, \omega))_{t \geq 0}$ a *path* or realization of the stochastic process X. If X is a progressively measurable and integrable stochastic process, then the stochastic process $Y = (Y_t)_{t \geq 0} = (Y(t))_{t \geq 0}$, defined by $Y(t, \omega) := \int_0^t X(s, \omega)\, ds$, $t \in [0, \infty)$, $\omega \in \Omega$, is progressively measurable. Note that each progressively measurable stochastic process (vector process) is also measurable. The proof of the following theorem may be found, e.g., in Korn and Korn [KK99], p. 36-37.

Theorem 2.10 *Let $(\Omega, \mathcal{F}, Q, I\!\!F)$ be a filtered probability space and X be a stochastic process (vector process) adapted to the filtration $I\!\!F$. If all paths of X are right-continuous then X is progressively measurable.*

Another very important class of stochastic processes is given in the following definition.

Definition 2.11 *Let $(\Omega, \mathcal{F}, Q, I\!\!F)$ be a filtered probability space and X be a stochastic process (vector process) adapted to the filtration $I\!\!F$. We call a stochastic process X a $\mathbf{L_2}[0, \mathbf{T}] -$process, if X is progressively measurable and*

$$\|X\|_{\mathbf{T}}^2 := E_Q\left[\int_0^T X^2(t)\, dt\right] < \infty.$$

Basically, $\|X\|_T$ is the L_2-norm on the measure space $([0, T] \times \Omega, \mathcal{B}[0, T] \otimes \mathcal{F}, \lambda \otimes Q)$ with $\lambda \otimes Q$ denoting the (unique) measure on the measure space $([0, T] \times \Omega, \mathcal{B}[0, T] \otimes \mathcal{F})$ determined by the Lebesgue measure λ and Q (see, e.g. Hinderer [Hin85], p. 128ff for more details). We say that two $L_2[0, T] -$processes X and Y are equivalent, if $X = Y$ $\lambda \otimes Q - a.s.$ If we identify equivalent $L_2[0, T] -$processes, the set of all $L_2[0, T] -$processes spans a linear vector space $\mathbf{V}_T$ with Norm $\|\|_{\mathbf{T}}$ and inner product

$$< X, Y >:= E_Q\left[\int_0^T X(t) \cdot Y(t)\, dt\right], \quad X, Y \in V_T.$$

Because of Schwarz' inequality we know that $|< X, Y >| \leq \|X\|_T \cdot \|Y\|_T < \infty$ for all $X, Y \in V_T$. Furthermore, $d_T(X, Y) := \|X - Y\|_T$ defines a metric on V_T. The space $(V_T, \|\|_T)$ or (V_T, d_T) is called $L_2[0, T]$-space.

Lemma 2.12 *Let $(\Omega, \mathcal{F}, Q, I\!\!F)$ be a filtered probability space and let X, Y be two measurable stochastic processes (vector processes) with*

$$Q(\{\omega \in \Omega : X_t(\omega) = Y_t(\omega) \text{ for all } t \in [0, T]\}) = 1.$$

Then X and Y are equivalent, i.e. $X = Y$ (in V_T).

Proof. Let $A := \{(t, \omega) \in [0, T] \times \Omega : X_t(\omega) \neq Y_t(\omega)\}$. Then

$$A_t := \{\omega \in \Omega : (t, \omega) \in A\} = \{\omega \in \Omega : X_t(\omega) \neq Y_t(\omega)\}, \ t \in [0, T].$$

Thus, for any $t \in [0, T]$ we have

$$
\begin{aligned}
Q(A_t) &= Q(\{\omega \in \Omega : X_t(\omega) \neq Y_t(\omega)\}) \\
&\leq Q(\{\omega \in \Omega : \text{there is a } s \in [0, T] \text{ with } X_s(\omega) \neq Y_s(\omega)\}) \\
&= 1 - Q(\{\omega \in \Omega : X_s(\omega) = Y_s(\omega) \text{ for all } s \in [0, T]\}) \\
&= 0
\end{aligned}
$$

Using Fubini's theorem (see Theorem 20.5 in Hinderer [Hin85]) we get

$$\lambda \otimes Q(A) = \int_0^T Q(A_t) \, dt = \int_0^T 0 \, dt = 0,$$

i.e. $X = Y \ \lambda \otimes Q - a.s.$ $\qquad\qquad\qquad\qquad\qquad\qquad\qquad\qquad\qquad\qquad\qquad$ □

A very important example of a filtration is given in the following definition.

Definition 2.13 (Natural Filtration) *Let $(\Omega, \mathcal{F}, Q, \mathbb{F})$ be a filtered probability space and X be a stochastic process (vector process) adapted to the filtration $\mathbb{F}$. The natural filtration $\mathbb{F}(X)$ is defined by the set of sigma-algebras*

$$\mathcal{F}_t := \mathcal{F}(X_s : 0 \leq s \leq t), \, 0 \leq t < \infty,$$

with $\mathcal{F}(X_s : 0 \leq s \leq t)$ being the smallest sigma-algebra which contains all sets $X_s^{-1}(B) = \{\omega \in \Omega : X_s(\omega) \in B\}, 0 \leq s \leq t$ where B runs through the Borel sigma-algebra $\mathcal{B}(\mathbb{R})$ $(\mathcal{B}(\mathbb{R}^k), k \in \mathbb{N})$. Again, we claim that $\mathbb{F}(X)$ has undergone a $(Q-)$ augmentation, if necessary, to ensure that conditions a) and b) of Definition 2.8 are satisfied.

Let us now look at a special stochastic process which may be considered as one of the atoms of modern finance. It was first used as a tool to describe the behaviour of stock prices by L. Bachelier [Bac00] in 1900. In honour of Norbert Wiener, who was the first to rigorously define and construct this process in 1923 (see [Wie23]), we call it the Wiener process and denote it by $W = (W_t)_{t \geq 0} = (W(t))_{t \geq 0}$. As this motion was already observed by the Botanist Robert Brown in 1829 watching pollen particles under a microscope the Wiener process is also very often called a Brownian motion.

Definition 2.14 (Wiener Process) *Let $(\Omega, \mathcal{F}, Q, \mathbb{F})$ be a filtered probability space. The stochastic process $W = (W_t)_{t \geq 0} = (W(t))_{t \geq 0}$ is called a $(Q-)$ Brownian motion or $(Q-)$ Wiener process if*

a) $W(0) = 0 \ Q - a.s.$,

b) W has independent increments, i.e. $W(t) - W(s)$ is independent of $W(t') - W(s')$ for all $0 \le s' \le t' \le s \le t < \infty$,

c) W has stationary increments, i.e. the distribution of $W(t+u) - W(t)$ only depends on u for $u \ge 0$,

d) Under Q, W has Gaussian increments, i.e. $W(t+u) - W(t)$ is normally distributed with mean 0 and variance u or $W(t+u) - W(t) \sim N(0, u)$,

e) W has continuous paths $Q - a.s.$, i.e. $t \longmapsto W(t, \omega)$ is a continuous function for $Q-$almost all $\omega \in \Omega$.

Note that if $I\!F = I\!F(W)$ Definition 2.14b) is equivalent to

b') $W(t) - W(s)$ is independent of $\mathcal{F}_s$ for $0 \le s < t < \infty$.

Also note that, under Q, $W(t)$ is normally distributed with mean 0 and variance t or $W(t) \sim N(0, t)$. Norbert Wiener has proved that Brownian motion really exists. Karatzas and Shreve [KS91], Section 2.7 show that the (Q)−completion of the natural filtration $I\!F(W)$, W being a $(Q-)$ Wiener process, is right- and left-continuous, i.e. $\mathcal{F}_t = \mathcal{F}_{t+} = \mathcal{F}_{t-}$ where $\mathcal{F}_{t-}$ is the smallest sigma-algebra containing $\cup_{s<t}\mathcal{F}_s$.

Definition 2.15 Let $(\Omega, \mathcal{F}, Q, I\!F)$ be a filtered probability space. We call $W = (W_1, ..., W_m) = (W_1(t), ..., W_m(t))_{t \ge 0}$ a **m-dimensional Wiener process**, if its components W_i, $i = 1, ..., m$, are independent Wiener processes.

One of the basic concepts we will need for modelling in finance is that of the so-called martingales. They are typically used to model situations where there is no tendency for a drift in one direction or another.

Definition 2.16 (Martingale) Let $(\Omega, \mathcal{F}, Q, I\!F)$ be a filtered probability space. A stochastic process $X = (X(t))_{t \ge 0}$ is called a

a) **martingale** relative to $(Q, I\!F)$ if X is adapted, $E_Q[|X(t)|] < \infty$ for all $t \ge 0$, and

$$E_Q[X(t) \mid \mathcal{F}_s] = X(s) \quad Q - a.s. \text{ for all } 0 \le s \le t < \infty,$$

b) **supermartingale** relative to $(Q, I\!F)$ if X is adapted, $E_Q[|X(t)|] < \infty$ for all $t \ge 0$, and

$$E_Q[X(t) \mid \mathcal{F}_s] \le X(s) \quad Q - a.s. \text{ for all } 0 \le s \le t < \infty,$$

c) **submartingale** relative to $(Q, I\!F)$ if X is adapted, $E_Q[|X(t)|] < \infty$ for all $t \ge 0$, and

$$E_Q[X(t) \mid \mathcal{F}_s] \ge X(s) \quad Q - a.s. \text{ for all } 0 \le s \le t < \infty.$$

Very useful examples of a martingale are stated in the following lemma.

Lemma 2.17 *Let* $(\Omega, \mathcal{F}, Q, I\!F)$ *be a filtered probability space.*

a) *Let* Y *be an integrable random variable[3]. Then the stochastic process* $X = (X(t))_{t\geq 0}$ *with* $X(t) = E_Q[Y | \mathcal{F}_t]$, $t \geq 0$, *is a martingale relative to* $(Q, I\!F)$.

b) *Let* $W = (W(t))_{t\geq 0}$ *be a* $(Q-)$ *Wiener process and* $I\!F = I\!F(W)$. *Then* W *is a martingale relative to* $(Q, I\!F)$.

c) *Let* $W = (W(t))_{t\geq 0}$ *be a* $(Q-)$ *Wiener process and* $I\!F = I\!F(W)$. *Then the stochastic process* $X = (X(t))_{t\geq 0}$ *with* $X(t) = W^2(t) - t$ *is a martingale relative to* $(Q, I\!F)$.

d) *Let* $I\!F = I\!F(W)$ *and* $W = (W(t))_{t\geq 0}$ *be a* $(Q-)$ *Wiener process. The stochastic process* $X = (X(t))_{t\geq 0}$ *with* $X(t) = \mu \cdot t + \sigma \cdot W(t)$ *with real numbers* μ *and* σ *is a martingale relative to* $(Q, I\!F)$ *if* $\mu = 0$, *a supermartingale relative to* $(Q, I\!F)$ *if* $\mu \leq 0$, *and a submartingale relative to* $(Q, I\!F)$ *if* $\mu \geq 0$.

Proof.

a) By definition we know that $X(t)$ is $(\mathcal{F}_t-)$ measurable for all $t \geq 0$. Thus, X is adapted. Using Lemma 2.6 we also know that for $0 \leq s \leq t < \infty$, as $\mathcal{F}_s \subset \mathcal{F}_t$, we have $Q - a.s.$

$$E_Q[X(t) | \mathcal{F}_s] = E_Q[E_Q[Y | \mathcal{F}_t] | \mathcal{F}_s] = E_Q[Y | \mathcal{F}_s] = X(s).$$

b) Because of $I\!F = I\!F(W)$, $W(t) - W(s)$ is independent of $\mathcal{F}_s$ for $0 \leq s < t < \infty$. Therefore, using Lemma 2.6, $Q - a.s.$

$$
\begin{aligned}
E_Q[W(t) | \mathcal{F}_s] &= E_Q[W(t) - W(s) + W(s) | \mathcal{F}_s] \\
&= E_Q[W(t) - W(s) | \mathcal{F}_s] + E_Q[W(s) | \mathcal{F}_s] \\
&= E_Q[W(t) - W(s)] + W(s) \\
&= W(s).
\end{aligned}
$$

---

[3] It should be mentioned that, as Lemma 2.6, this statement can be extended to $Q-$ quasi integrable random variables Y using Lemma 20.4 in Hinderer [Hin85].

c) As in b) we get for all $0 \leq s < t < \infty$, $Q - a.s.$

$$
\begin{aligned}
E_Q\left[X\left(t\right)\mid\mathcal{F}_s\right] &= E_Q\left[W^2\left(t\right)\mid\mathcal{F}_s\right] - t \\
&= E_Q\left[\left(W\left(t\right) - W\left(s\right) + W\left(s\right)\right)^2\mid\mathcal{F}_s\right] - t \\
&= E_Q\left[\left(W\left(t\right) - W\left(s\right)\right)^2\mid\mathcal{F}_s\right] + E_Q\left[W^2\left(s\right)\mid\mathcal{F}_s\right] \\
&\quad -t + 2\cdot E_Q\left[W\left(s\right)\cdot\left(W\left(t\right) - W\left(s\right)\right)\mid\mathcal{F}_s\right] \\
&= t - s + W^2\left(s\right) - t \\
&\quad + 2\cdot W\left(s\right)\cdot E_Q\left[\left(W\left(t\right) - W\left(s\right)\right)\mid\mathcal{F}_s\right] \\
&= W^2\left(s\right) - s = X\left(s\right).
\end{aligned}
$$

d) Using b) and Lemma 2.6h), we get for all $0 \leq s < t < \infty$, $Q - a.s.$

$$
\begin{aligned}
E_Q\left[X\left(t\right)\mid\mathcal{F}_s\right] &= \mu\cdot t + \sigma\cdot E_Q\left[W\left(t\right)\mid\mathcal{F}_s\right] = \mu\cdot t + \sigma\cdot W\left(s\right) \\
&= \mu\cdot\left(t - s\right) + X\left(s\right).
\end{aligned}
$$

Hence, $E_Q\left[X\left(t\right)\mid\mathcal{F}_s\right] \geq X\left(s\right)$ if $\mu \geq 0$ and $E_Q\left[X\left(t\right)\mid\mathcal{F}_s\right] \leq X\left(s\right)$ if $\mu \leq 0$ which completes the proof. □

2.2   Stopped Stochastic Processes

Another important building block in stochastic analysis is the stopping time which is, roughly speaking, the time when a stochastic process is stopped. The price process of a financial derivative may, for instance, be stopped because the owner of this derivative exercises an included option to cash the product. The stopping time is also used for defining the so-called local martingales. Following is a summary of some important definitions and results on this topic.

Definition 2.18 (Stopping Time) *Let* $(\Omega, \mathcal{F}, Q, \mathbb{F})$ *be a filtered proba-bility space. A stopping time with respect to the filtration* $\mathbb{F} = (\mathcal{F}_t)_{t \geq 0}$ *is a* $(\mathcal{F} - \mathcal{B}\left([0, \infty]\right) -)$ *measurable function (random variable)* $\tau : \Omega \rightarrow [0, \infty]$ *with*

$$\{\tau \leq t\} := \{\omega \in \Omega : \tau\left(\omega\right) \leq t\} \in \mathcal{F}_t \text{ for all } t \in [0, \infty).$$

Intuitively the measurability assumption of the previous definition means that the decision whether to stop a stochastic process at time $t \in [0, \infty)$ depends on the information up to time t only. Using this definition the following lemma holds.

Lemma 2.19 *Let τ_1 and τ_2 be two stopping times. Then $\tau_1 \wedge \tau_2 :=$ $\min\{\tau_1, \tau_2\}$ is a stopping time. Especially $\tau_1 \wedge t$ is a stopping time for all $t \in [0, \infty)$.*

Proof. For all $t \in [0, \infty)$ we have

$$\{\omega \in \Omega : \tau_1 \wedge \tau_2(\omega) \leq t\} = \{\omega \in \Omega : \tau_1(\omega) \leq t\} \cup \{\omega \in \Omega : \tau_2(\omega) \leq t\} \in \mathcal{F}_t.$$

□

Definition 2.20 (Stopped Process) *Let $(\Omega, \mathcal{F}, Q, \mathbb{F})$ be a filtered probability space and τ be a stopping time.*

a) The stochastic process $(X_{t \wedge \tau})_{t \geq 0} = (X(t \wedge \tau))_{t \geq 0}$ defined by

$$X_{t \wedge \tau}(\omega) := \begin{cases} X_t(\omega), & \text{if } t \leq \tau(\omega) \\ X_\tau(\omega), & \text{if } t > \tau(\omega) \end{cases}$$

is called a stopped process.

b) Let the sigma-algebra of the events up to time τ be defined by

$$\mathcal{F}_\tau := \{A \in \mathcal{F} : A \cap \{\tau \leq t\} \in \mathcal{F}_t \text{ for all } t \in [0, \infty)\}.$$

*The filtration $(\mathcal{F}_{t \wedge \tau})_{t \geq 0}$ is called a **stopped filtration**.*

Note that τ is $\mathcal{F}_\tau$-measurable and $\mathcal{F}_{t \wedge \tau} \subset \mathcal{F}_t$ for all $t \geq 0$. The following theorem shows what influence the stopping will have on martingales or submartingales. A proof can be found in Karatzas and Shreve [KS91], Theorem 1.3.22.

Theorem 2.21 (Optional Sampling) *Let $(\Omega, \mathcal{F}, Q, \mathbb{F})$ be a filtered probability space and $X = (X_t)_{t \geq 0}$ be a right-continuous martingale (submartingale) which means that all paths of X_t are right-continuous. Furthermore, let τ_1 and τ_2 be two stopping times with $\tau_1 \leq \tau_2$. Then $Q - a.s.$ for all $t \geq 0$ we have*

$$E_Q[X_{t \wedge \tau_2} | \mathcal{F}_{t \wedge \tau_1}] = X_{t \wedge \tau_1} \quad (E_Q[X_{t \wedge \tau_2} | \mathcal{F}_{t \wedge \tau_1}] \geq X_{t \wedge \tau_1}).$$

Especially, for each stopping time τ and for $0 \leq s \leq t < \infty$ we have $Q - a.s.$ for all $t \geq 0$

$$E_Q[X_{t \wedge \tau} | \mathcal{F}_s] = X_{s \wedge \tau} \quad (E_Q[X_{t \wedge \tau} | \mathcal{F}_s] \geq X_{s \wedge \tau}).$$

The following theorem gives an interesting characterization of martingales. For a proof see Korn and Korn [KK99], p. 23.

Theorem 2.22 *Let $(\Omega, \mathcal{F}, Q, \mathbb{F})$ be a filtered probability space and $X = (X_t)_{t \geq 0}$ be a right-continuous stochastic process. Then X is a martingale if and only if*

$$E_Q[X_\tau] = E_Q[X_0] \text{ for all bounded stopping times } \tau.$$

We are now able to generalize the definition of a local martingale.

Definition 2.23 (Local Martingale) *Let $(\Omega, \mathcal{F}, Q, \mathbb{F})$ be a filtered probability space and $X = (X_t)_{t \geq 0}$ be a stochastic process with $X_0 = 0$. If there is a sequence $(\tau_n)_{n \in \mathbb{N}}$ of non-decreasing stopping times with*

$$Q\left(\lim_{n \to \infty} \tau_n = \infty \right) = 1$$

such that

$$X^n = (X_t^n)_{t \geq 0} := (X_{t \wedge \tau_n})_{t \geq 0}$$

is a martingale relative to $(Q, \mathbb{F})$ for all $n \in \mathbb{N}$, then we call X a local martingale. The sequence $(\tau_n)_{n \in \mathbb{N}}$ is called localizing sequence. If X is a local martingale with continuous paths, we call X a continuous local martingale.

Note that each martingale is also a local martingale. Also note that the expectation $E_Q [X_t]$, $t \geq 0$, of a local martingale may not exist, hence the need for a localization concept. A proof of the following theorem may be found in Korn and Korn [KK99], p. 24-25 and p. 37-38.

Theorem 2.24 *Let $(\Omega, \mathcal{F}, Q, \mathbb{F})$ be a filtered probability space and τ be a stopping time.*

a) *If the stochastic process $X = (X_t)_{t \geq 0}$ is progressively measurable, then the stopped process $(X_{t \wedge \tau})_{t \geq 0}$ is also progressively measurable. Especially, $X_{t \wedge \tau}$ is $\mathcal{F}_t-$ and $\mathcal{F}_{t \wedge \tau}-$measurable.*

b) *If the stochastic process $X = (X_t)_{t \geq 0}$ is a non-negative local martingale, then X is a supermartingale.*

c) *If the stochastic process $X = (X_t)_{t \geq 0}$ is a non-negative right-continuous supermartingale, then for each $\lambda > 0$ we have*

$$Q\left(\left\{ \omega \in \Omega : \sup_{0 \leq t \leq T} X_t(\omega) \geq \lambda \right\} \right) \leq \frac{E_Q [X_0]}{\lambda}.$$

d) *If the stochastic process $X = (X_t)_{t \geq 0}$ is a right-continuous martingale with $E_Q [X_T^2] < \infty$ for all $T \geq 0$, then Doob's inequality holds:*

$$E_Q \left[\left(\sup_{0 \leq t \leq T} |X_t| \right)^2 \right] \leq 4 \cdot E_Q [X_T^2].$$

2.3   Stochastic Integrals

In 1944 K. Itô [Itô44] introduced the concept of stochastic integration. In a suitable environment he defined what should be the meaning of the expression $\int_0^t X(t,\omega)\, dY(t,\omega)$ with $X = (X_t)_{t \geq 0} = (X(t))_{t \geq 0}$ and $Y = (Y_t)_{t \geq 0} = (Y(t))_{t \geq 0}$ two stochastic processes. Here, we will concentrate on the special case where $Y = W$ is a Wiener process. To do this, we assume that we work on a filtered probability space $(\Omega, \mathcal{F}, Q, I\!\!F)$ where $(\Omega, \mathcal{F}, Q)$ is complete, $W = (W_t)_{t \geq 0} = (W(t))_{t \geq 0}$ is a $(Q-)$ Wiener process and $I\!\!F = I\!\!F(W)$. It can be shown that for Q-almost all paths of a Wiener process there is no $t \geq 0$ where these paths are differentiable. It can also be shown that the paths of a Wiener process have $Q - a.s.$ a variation of ∞ on each interval, i.e. $\sum_{i=1}^{2^n} \left| W_{\frac{i}{2^n}} - W_{\frac{i-1}{2^n}} \right| \to \infty$ Q-a.s. as $n \to \infty$. Thus, we are not able to define an integral such as $\int_0^t X(t,\omega)\, dW(t,\omega)$ by a path-by-path procedure. We need a new way of defining such an integral, and start doing this with a so-called simple process specified in the following definition.

Definition 2.25 (Simple Process) *A stochastic process* $(X(t))_{t \in [0,T]}$ *is called a simple process, if there are a partition* $\mathcal{Z}_n := \{t_0, \ldots, t_n : 0 = t_0 < t_1 < \cdots < t_n = T\}$, $n \in I\!\!N$, *and bounded random variables* $\Upsilon_i : \Omega \to I\!\!R$, $i = 0, \ldots, n$, *such that*

a) Υ_0 *is* $\mathcal{F}_0$*-measurable,* Υ_i *is* $\mathcal{F}_{t_{i-1}}$*-measurable,* $i = 1, \ldots, n$, *and*

b) $X_t(\omega) = X(t,\omega) = \Upsilon_0(\omega) \cdot 1_{\{0\}}(t) + \sum_{i=1}^n \Upsilon_i(\omega) \cdot 1_{(t_{i-1},t_i]}(t)$ *for all* $\omega \in \Omega$ *and* $t \in [0,T]$, *where* $1_M(t) = 1$ *if* $t \in M$ *and* 0 *else for any set* $M \subset I\!\!R$.

Note that for a simple process $X = (X_t)_{t \geq 0} = (X(t))_{t \geq 0}$ the random variable $X(t)$ is always $\mathcal{F}_{t_{i-1}}$-measurable for $t \in (t_{i-1}, t_i]$. Using the notation $t \wedge t' := \min\{t, t'\}$, $t, t' \in [0, \infty)$ we now define the stochastic integral for simple processes.

Definition 2.26 (Stochastic Integral for Simple Processes) *For a simple process* $X = (X(t))_{t \in [0,T]}$, *the stochastic integral* $I_t(X)$, $t \in (t_k, t_{k+1}] \subset [0, T]$, *is defined by*

$$I_t(X) \quad := \quad \int_0^t X(s)\, dW(s)$$

$$:= \quad \Upsilon_{k+1} \cdot (W(t) - W(t_k)) + \sum_{i=1}^k \Upsilon_i \cdot (W(t_i) - W(t_{i-1}))$$

or, more general, for $t \in [0, T]$:

$$I_t(X) := \int_0^t X(s)\, dW(s) := \sum_{i=1}^n \Upsilon_i \cdot (W(t_i \wedge t) - W(t_{i-1} \wedge t)).$$

Furthermore, we define

$$\int_t^T X(s)\, dW(s) := \int_0^T X(s)\, dW(s) - \int_0^t X(s)\, dW(s) = I_T(X) - I_t(X).$$

The following theorem shows some interesting properties of the stochastic integral defined so far. For a proof see, e.g., Korn and Korn [KK99], p. 32-35 or Bingham and Kiesel [BK98], p. 145ff.

Theorem 2.27 *Let* $X = (X(t))_{t\in[0,T]}$ *be a simple process and* $I_t(X)$ *be the stochastic integral as defined above.*

a) *The stochastic process* $I(X) = (I_t(X))_{t\in[0,T]}$ *is a continuous martingale, i.e.* $E_Q[I_t(X)|\mathcal{F}_s] = I_s(X)$ $Q-a.s.$ *for all* $0 \le s \le t \le T < \infty.$

b) $E_Q[I_t^2(X)] = E_Q\left[\left(\int_0^t X(s)\, dW(s)\right)^2\right] = E_Q\left[\int_0^t X^2(s)\, ds\right]$ *for all* $t \in [0, T].$ *This is called the* **Itô isometry.**

c) $E_Q\left[(I_t(X) - I_s(X))^2\right] = E_Q\left[\int_s^t X^2(u)\, du\right]$ *for all* $0 \le s \le t \le T < \infty.$

d) *For each* $A \in \mathcal{F}_t,$ $1_A : \Omega \to \mathbb{R},$ *and for all* $t \in [0, T],$ $\omega \in \Omega,$ *we have*

$$\int_0^T 1_A \cdot X(s) \cdot 1_{[t,T]}(s)\, dW(s) = 1_A \cdot \int_t^T X(s)\, dW(s).$$

e) *Let* $X = (X(t))_{t\in[0,T]}$ *and* $Y = (Y(t))_{t\in[0,T]}$ *be two simple processes and* a, b *be two real numbers. Then*

$$I_t(a \cdot X + b \cdot Y) = a \cdot I_t(X) + b \cdot I_t(Y) \quad \text{for all } t \in [0, T],$$

i.e. the stochastic integral is linear.

For the proof of part c) note that for all $0 \le s \le t \le T < \infty$ we have

$$E_Q\left[(I_t(X) - I_s(X))^2\right] = E_Q\left[E_Q\left[(I_t(X) - I_s(X))^2 |\mathcal{F}_s\right]\right]$$
$$= E_Q\left[E_Q\left[I_t^2(X) - 2 \cdot I_t(X) \cdot I_s(X) + I_s^2(X)|\mathcal{F}_s\right]\right]$$
$$= E_Q\left[E_Q\left[I_t^2(X)|\mathcal{F}_s\right] - 2 \cdot I_s(X) \cdot E_Q\left[I_t(X)|\mathcal{F}_s\right] + E_Q\left[I_s^2(X)|\mathcal{F}_s\right]\right]$$
$$= E_Q\left[E_Q\left[I_t^2(X)|\mathcal{F}_s\right]\right] - 2 \cdot E_Q\left[I_s^2(X)\right] + E_Q\left[E_Q\left[I_s^2(X)|\mathcal{F}_s\right]\right]$$
$$= E_Q\left[I_t^2(X)\right] - E_Q\left[I_s^2(X)\right]$$
$$= E_Q\left[\int_0^t X^2(s)\, ds\right] - E_Q\left[\int_0^s X^2(s)\, ds\right]$$
$$= E_Q\left[\int_s^t X^2(s)\, ds\right].$$

Also note that, with $X(t) \equiv 1$ for all $t \in [0, T]$, we get

$$\int_0^t 1 dW(s) = W(t) - W(0) = W(t) \quad Q - a.s. \text{ for all } t \in [0, T].$$

Furthermore, we see that $E_Q \left[\int_0^t 1 dW(s) \right]^2 = E_Q \left[\int_0^t 1 ds \right] = t$ for all $t \in [0, T]$. If we look at Theorem 2.27b) in a little more detail, we see that for simple processes the map $X \to I(X)$ induces a norm for stochastic integrals by

$$\| I(X) \|_{I_T}^2 := E_Q \left[\int_0^T X(t) dW(t) \right]^2 = E_Q \left[\int_0^T X^2(t) dt \right] = \| X \|_T^2,$$

which tells us that this map is linear and norm preserving, i.e. an isometry. The following theorem shows that each element of the $L_2([0, T])$–space can be approximated by a sequence of simple processes.

Theorem 2.28 *Let* $X = (X(t))_{t \geq 0}$ *be an arbitrary* $L_2([0, T])$*–process. Then there is a sequence of simple processes* $(X^n)_{n \in \mathbb{N}}$ *such that*

$$\lim_{n \to \infty} \| X - X^n \|_T^2 = \lim_{n \to \infty} E_Q \left[\int_0^T (X(t) - X^n(t))^2 dt \right] = 0.$$

For a proof see, e.g., Korn and Korn [KK99], p. 40-41. They also give a proof of the following theorem (p. 42-46).

Theorem 2.29 *There is a linear mapping* J *from the* $L_2([0, T])$*–space to the space of all continuous martingales on* $[0, T]$ *such that:*

a) *if* X *is a simple process, then*

$$Q(\{\omega \in \Omega : J_t(X, \omega) = I_t(X, \omega) \text{ for all } t \in [0, T]\}) = 1.$$

b) $E_Q[J_t(X)]^2 = E_Q \left[\int_0^t X^2(s) ds \right]$ *for all* $t \in [0, T]$ (**Itô isometry**).

c) *The following special case of Doob's inequality holds:*

$$E_Q \left[\left(\sup_{0 \leq t \leq T} |J_t(X)| \right)^2 \right] \leq 4 \cdot \| X \|_T^2.$$

d) *For* J *and* J' *two linear mappings from the* $L_2([0, T])$*–space to the space of all continuous martingales on* $[0, T]$ *such that a) and b) hold, then*

$$Q(\{\omega \in \Omega : J_t(X, \omega) = J'_t(X, \omega) \text{ for all } t \in [0, T]\}) = 1.$$

In this sense, the linear mapping J *is unique.*

Using Lemma 2.12, we know that each pair $J(X)$ and $J'(X)$ of continuous martingales on $[0,T]$ with

$$Q(\{\omega \in \Omega : J_t(X,\omega) = J'_t(X,\omega) \text{ for all } t \in [0,T]\}) = 1$$

is equivalent. Thus, we can make the following definition of the stochastic integral.

Definition 2.30 (Stochastic Integral) *Let $X = (X_t)_{t \geq 0} = (X(t))_{t \geq 0}$ be an arbitrary $L_2([0,T])$ $-$process and J as in Theorem 2.29. Then the stochastic integral or Itô integral of X with respect to the Wiener process W is defined by*

$$\int_0^t X(s)\, dW(s) := J_t(X) \text{ for all } 0 \leq t \leq T.$$

This definition of the stochastic integral for $L_2([0,T])$ $-$processes can be extended to the linear vector space of all $L_2^Q[0,T]$ $-$processes, which are defined to be all progressively measurable stochastic processes X with

$$\left(\|X\|_{\mathbf{T}}^Q\right)^2 := \int_0^T X^2(t)\, dt < \infty \quad Q - a.s.$$

In this case the stochastic integral is still a linear mapping and a continuous local martingale, but the Itô isometry no longer holds as the integral $E_Q\left[\int_0^t X^2(s)\, ds\right]$ may not exist for all $t \in [0,T]$. Note that $\|X\|_T^2 = E_Q\left[\int_0^T X^2(s)\, ds\right] < \infty$ implies $\left(\|X\|_{\mathbf{T}}^Q\right)^2 = \int_0^T X^2(s)\, ds < \infty$ $Q - a.s.$, and thus

$$\int_0^t X^2(s)\, ds \leq \int_0^T X^2(s)\, ds < \infty \quad Q - a.s. \text{ for all } t \in [0,T].$$

Another generalization of the stochastic integral to higher dimensions is given in the following definition.

Definition 2.31 *Let $W = (W_1(t), ..., W_m(t))_{t \geq 0}$ be a $m-$dimensional Wiener process, $m \in I\!N$, and let for $n \in I\!N$*

$$X = (X_{ij})_{\substack{i=1,...,n, \\ j=1,...,m}} = \left((X_{ij}(t))_{\substack{i=1,...,n, \\ j=1,...,m}}\right)_{t \geq 0}$$

be an $n \times m-$dimensional progressively measurable stochastic process with components X_{ij} which are $L_2([0,T])$ $-$processes. Then the stochastic integral or Itô integral of X with respect to the Wiener process W is defined *for all $0 \leq t \leq T$ by*

$$\int_0^t X(s)\,dW(s) := \begin{pmatrix} \sum_{j=1}^m \int_0^t X_{1j}(s)\,dW_j(s) \\ \vdots \\ \sum_{j=1}^m \int_0^t X_{nj}(s)\,dW_j(s) \end{pmatrix}.$$

2.4   Itô Calculus

We now move on to one of the most important tools for working with stochastic integrals, the Itô formula. It can be applied to processes which we will call Itô processes and will be used extensively later on to derive the price behaviour of financial derivatives.

Definition 2.32 (Itô Process) *Let $W = (W_1(t), ..., W_m(t))_{t \geq 0}$, $m \in I\!N$ be a $m-$dimensional Wiener process. A stochastic process $X = (X(t))_{t \geq 0}$ is called an Itô process if for all $t \geq 0$ we have*

$$\begin{aligned} X(t) &= X(0) + \int_0^t \mu(s)\,ds + \int_0^t \sigma(s)\,dW(s) & (2.1) \\ &= X(0) + \int_0^t \mu(s)\,ds + \sum_{j=1}^m \int_0^t \sigma_j(s)\,dW_j(s), \end{aligned}$$

where $X(0)$ is $(\mathcal{F}_0-)$ measurable and $\mu = (\mu(t))_{t \geq 0}$ and $\sigma = (\sigma(t))_{t \geq 0}$ with $\sigma(t) = (\sigma_1(t), ..., \sigma_m(t))_{t \geq 0}$ are $(m-$dimensional) progressively measurable stochastic processes with

$$\int_0^t |\mu(s)|\,ds < \infty \ and \ \int_0^t \sigma_j^2(s)\,ds < \infty \ Q-a.s. \ for \ all \ t \geq 0, \ j = 1, ..., m.$$

A $n-$dimensional Itô process is given by a vector $X = (X_1, ..., X_n)$, $n \in I\!N$, with each X_i being an Itô process, $i = 1, ..., n$.

For convenience we write symbolically instead of (2.1)

$$dX(t) = \mu(t)\,dt + \sigma(t)\,dW(t) = \mu(t)\,dt + \sum_{j=1}^m \sigma_j(t)\,dW_j(t),$$

and call this a *stochastic differential equation (SDE)*. It is interesting to note that there is only one pair of stochastic processes μ and σ which satisfies (2.1), in the sense that if μ^* and σ^* is another pair of stochastic processes satisfying

$$dX(t) = \mu^*(t)\,dt + \sigma^*(t)\,dW(t),$$

then μ is equivalent to μ^* and σ is equivalent to σ^*.

Definition 2.33 (Quadratic Covariance Process) *Let $m \in I\!N$ and $W = (W_1(t),...,W_m(t))_{t\geq 0}$ be a $m-$dimensional Wiener process. Furthermore, let $X_1 = (X_1(t))_{t\geq 0}$ and $X_2 = (X_2(t))_{t\geq 0}$ be two Itô processes with*

$$dX_i(t) = \mu_i(t)\,dt + \sigma_i(t)\,dW(t) = \mu_i(t)\,dt + \sum_{j=1}^{m}\sigma_{ij}(t)\,dW_j(t),\ i=1,2.$$

Then we call the stochastic process $< X_1, X_2 >= (< X_1, X_2 >_t)_{t\geq 0}$ defined by

$$< X_1, X_2 >_t := \sum_{j=1}^{m} \int_0^t \sigma_{1j}(s) \cdot \sigma_{2j}(s)\,ds$$

*the quadratic covariance (process) of X_1 and X_2. If $X_1 = X_2 =: X$ we call the stochastic process $< X >:=< X, X >$ the **quadratic variation** (process) of X, i.e.*

$$< X, X >_t := \sum_{j=1}^{m} \int_0^t \sigma_j^2(s)\,ds = \int_0^t \|\sigma(s)\|^2\,ds,$$

where $\|\sigma(t)\| := \sqrt{\sum_{j=1}^m \sigma_j^2(t)}$, $t \in [0,\infty)$ denotes the Euclidean norm in $I\!R^m$ and $\sigma := \sigma_1$.

Again for convenience, if X is an Itô process with notation as in (2.1) and Y is a progressively measurable process, we write

$$\int_0^t Y(s)\,dX(s) := \int_0^t Y(s) \cdot \mu(s)\,ds + \int_0^t Y(s) \cdot \sigma(s)\,dW(s),$$

if the integrals on the right hand side of the equation exist, or briefly

$$Y(s)\,dX(s) := Y(s) \cdot \mu(s)\,ds + Y(s) \cdot \sigma(s)\,dW(s).$$

Theorem 2.34 (Itô´s Lemma)   *Let $W = (W_1(t),...,W_m(t))_{t\geq 0}$ be a $m-$dimensional Wiener process, $m \in I\!N$, and $X = (X(t))_{t\geq 0}$ be an Itô process with*

$$dX(t) = \mu(t)\,dt + \sigma(t)\,dW(t) = \mu(t)\,dt + \sum_{j=1}^{m}\sigma_j(t)\,dW_j(t).$$

Furthermore, let $G : \mathbb{R} \times [0, \infty) \to \mathbb{R}$ be twice continuously differentiable[4] in the first variable, with derivatives denoted by G_x and G_{xx}, and once continuously differentiable in the second, with derivative denoted by G_t. Then we have for all $t \in [0, \infty)$

$$G\left(X\left(t\right), t\right) = G\left(X\left(0\right), 0\right) + \int_0^t G_t\left(X\left(s\right), s\right) ds$$

$$+ \int_0^t G_x\left(X\left(s\right), s\right) dX\left(s\right)$$

$$+ \frac{1}{2} \int_0^t G_{xx}\left(X\left(s\right), s\right) d < X > \left(s\right)$$

or briefly

$$dG\left(X\left(t\right), t\right) = \left(G_t\left(X\left(t\right), t\right) + G_x\left(X\left(t\right), t\right) \cdot \mu\left(t\right)\right.$$

$$+ \frac{1}{2} \cdot G_{xx}\left(X\left(t\right), t\right) \cdot \|\sigma\left(t\right)\|^2\right) dt$$

$$+ G_x\left(X\left(t\right), t\right) \cdot \sigma\left(t\right) dW\left(t\right).$$

Note that the components W_j, $j = 1, ..., m$, are assumed to be independent. For a proof of this and the following theorem see, e.g., Korn and Korn [KK99], p. 48ff.

Theorem 2.35 (Itô's Lemma in Higher Dimensions) *Let $m \in \mathbb{N}$ and $W = (W_1, ..., W_m) = (W_1(t), ..., W_m(t))_{t \geq 0}$ be a $m-$dimensional Wiener process. Furthermore, let $X = (X_1, ..., X_n) = (X_1(t), ..., X_n(t))_{t \geq 0}$ be an $n-$dimensional Itô process, $n \in \mathbb{N}$, with*

$$dX_i\left(t\right) = \mu_i\left(t\right) dt + \sigma_i\left(t\right) dW\left(t\right) = \mu_i\left(t\right) dt + \sum_{j=1}^m \sigma_{ij}\left(t\right) dW_j\left(t\right), \; i = 1, ..., n.$$

Also, let $G : \mathbb{R}^n \times [0, \infty) \to \mathbb{R}$ be twice continuously differentiable in the first n variables, with derivatives denoted by G_{x_i} and $G_{x_i x_j}$, $i, j = 1, ..., n$, and once continuously differentiable in the last variable, with derivative denoted by G_t. Then we have for all $t \geq 0$

$$dG\left(X\left(t\right), t\right) = G_t\left(X_1\left(t\right), ..., X_n\left(t\right), t\right) dt$$

$$+ \sum_{i=1}^n G_{x_i}\left(X_1\left(t\right), ..., X_n\left(t\right), t\right) dX_i\left(t\right)$$

$$+ \frac{1}{2} \sum_{i=1}^n \sum_{j=1}^n G_{x_i x_j}\left(X_1\left(t\right), ..., X_n\left(t\right), t\right) d < X_i, X_j > \left(t\right).$$

---

[4] Using this formulation it is sufficient to suppose that the partial derivatives have continuous extensions at $t = 0$, which we will assume if necessary.

An easy application of Itô's lemma in higher dimensions gives us the following product rule.

Lemma 2.36 (Product Rule) *Let* $X = (X_1(t), ..., X_n(t))_{t \geq 0}$ *be a* $n-$*dimensional Itô process,* $n \in I\!N$, *with*

$$dX_i(t) = \mu_i(t)\, dt + \sigma_i(t)\, dW(t) = \mu_i(t)\, dt + \sum_{j=1}^m \sigma_{ij}(t)\, dW_j(t), \ i = 1, ..., n.$$

Furthermore, let $G : I\!R^n \times [0, \infty) \to I\!R$ *be given by* $G(x, t) = x_1 \cdot x_2$. *Then*

$$
\begin{aligned}
dG(X(t), t) &= X_1(t)\, dX_2(t) + X_2(t)\, dX_1(t) + d < X_1, X_2 > (t) \\
&= (X_2(t) \cdot \mu_1(t) + X_1(t) \cdot \mu_2(t) + \sum_{k=1}^m \sigma_{1k}(t) \cdot \sigma_{2k}(t))dt \\
&\quad + (X_2(t) \cdot \sigma_1(t) + X_1(t) \cdot \sigma_2(t))\, dW(t).
\end{aligned}
$$

Proof. For all $(x, t) \in I\!R^n \times [0, \infty)$, $i, j = 1, ..., n$ we know that

$$G_t(x, t) = 0, \quad G_{x_i}(x, t) = \begin{cases} x_j, & \text{if } i \in \{1, 2\} \\ 0, & \text{else} \end{cases}$$

and

$$G_{x_i x_j}(x, t) = \begin{cases} 1, & \text{if } i \neq j, \ i, j \in \{1, 2\} \\ 0, & \text{else.} \end{cases}$$

Furthermore,

$$< X_i, X_j >_t := \sum_{k=1}^m \int_0^t \sigma_{ik}(s) \cdot \sigma_{jk}(s)\, ds.$$

Therefore, we have, using Theorem 2.35, for all $t \geq 0$

$$
\begin{aligned}
dG(X(t), t) &= X_2(t)\, dX_1(t) + X_1(t)\, dX_2(t) + d < X_1, X_2 > (t) \\
&= (X_2(t) \cdot \mu_1(t) + X_1(t) \cdot \mu_2(t))\, dt \\
&\quad + \sum_{k=1}^m (X_2(t) \cdot \sigma_{1k}(t) + X_1(t) \cdot \sigma_{2k}(t))\, dW_k(t) \\
&\quad + \sum_{k=1}^m \sigma_{1k}(t) \cdot \sigma_{2k}(t)\, dt \\
&= \left(X_2(t)\, \mu_1(t) + X_1(t)\, \mu_2(t) + \sum_{k=1}^m \sigma_{1k}(t)\, \sigma_{2k}(t) \right) dt \\
&\quad + (X_2(t) \cdot \sigma_1(t) + X_1(t) \cdot \sigma_2(t))\, dW(t).
\end{aligned}
$$

$\square$

Applying Theorem 2.34 we get the following special form of Itô's lemma.

Lemma 2.37 (Itô's Lemma, Special Case) *Let* $W = (W(t))_{t \geq 0}$ *be a (one-dimensional) Wiener process and let* $X = (X(t))_{t \geq 0}$ *be an Itô process with*

$$dX(t) = \mu(X(t), t) dt + \sigma(X(t), t) dW(t).$$

Furthermore, let $G : \mathbb{R} \times [0, \infty) \to \mathbb{R}$ *be twice continuously differentiable in the first variable, with derivatives denoted by* G_x *and* G_{xx}, *and once continuously differentiable in the second, with derivative denoted by* G_t. *Then we have for all* $t \in [0, T]$

$$
\begin{aligned}
dG(X(t), t) \;=\; & (G_x(X(t), t) \cdot \mu(X(t), t) + G_t(X(t), t) \\
& + \tfrac{1}{2} \cdot G_{xx}(X(t), t) \cdot \sigma^2(X(t), t)) dt \\
& + G_x(X(t), t) \cdot \sigma(X(t), t) dW(t).
\end{aligned}
$$

2.5   Martingale Representation

The following theorem shows that each continuous local martingale relative to $(Q, \mathbb{F}(W))$, i.e. relative to Q and the natural filtration $\mathbb{F}(W)$, can be written as an Itô process. Whenever we will work on the filtered probability space $(\Omega, \mathcal{F}, Q, \mathbb{F}(W))$ we call these martingales briefly $(Q-)$ martingales. A proof of this theorem can be found in Korn and Korn [KK99], p. 81ff.

Theorem 2.38 (Martingale representation I)   *Let* $m \in \mathbb{N}$, $W = (W_1(t), ..., W_m(t))_{t \geq 0}$ *be a* $m-$*dimensional Wiener process, and* $M - (M(t))_{t \in [0,T]}$ *be a continuous local* $(Q-)$ *martingale. Then there is a progressively measurable process* $\phi = (\phi(t))_{t \in [0,T]}$, $\phi : [0,T] \times \Omega \to \mathbb{R}^m$ *such that*

a) $\displaystyle \int_0^T \|\phi(t)\|^2 dt < \infty \;\; Q - a.s.,$

b) $\displaystyle M(t) = M(0) + \int_0^t \phi(s)' dW(s) \;$ *or briefly* $\; dM(t) = \phi(t)' dW(t)$
   $Q - a.s.$ *for all* $t \in [0, T]$ *where* $\phi(s)'$ *is the transposed of* $\phi(s)$.

If $M = (M(t))_{t \in [0,T]}$ *is a continuous* $(Q-)$ *martingale with* $E_Q[M^2(t)] < \infty$ *for all* $t \in [0, T]$, *then a) is strengthened to*

a') $\displaystyle E_Q \left[\int_0^T \|\phi(t)\|^2 dt \right] < \infty$

while b) still holds.

Using Theorem 2.38 we can show the following relation between two continuous local martingales.

Theorem 2.39 (Martingale Representation II) *Let $W = (W(t))_{t \geq 0}$ be a (one-dimensional) Wiener process and $M_i = (M_i(t))_{t \in [0,T]}$, $i = 1, 2$ be two continuous local $(Q-)$ martingales. Furthermore, let $\phi \neq 0$ where we assume that ϕ is given by $dM_1(t) = \phi(t)dW(t)$, $t \in [0, T]$ as stated in Theorem 2.38. Then there is a progressively measurable process $\varphi = (\varphi(t))_{t \in [0,T]}$, $\varphi : [0, t] \times \Omega \to \mathbb{R}$ such that*

a) $\displaystyle\int_0^T \varphi^2(t) \cdot \phi^2(t)dt < \infty \quad Q - a.s.,$

b) $M_2(t) = M_2(0) + \displaystyle\int_0^t \varphi(s)dM_1(s) \quad Q - a.s.$ *for all $t \in [0, T]$.*

Proof. Using Theorem 2.38 there are two progressively measurable stochastic processes $\phi = (\phi(t))_{t \in [0,T]}$, where we assume that $\phi \neq 0$, and $\psi = (\psi(t))_{t \in [0,T]}$, $\phi, \psi : [0, t] \times \Omega \to \mathbb{R}$ such that

$$\int_0^T \phi^2(t)dt < \infty \quad \text{and} \quad \int_0^T \psi^2(t)dt < \infty \quad Q - a.s.$$

as well as

$$M_1(t) = M_1(0) + \int_0^t \phi(s)dW(s) \text{ and } M_2(t) = M_2(0) + \int_0^t \psi(s)dW(s),$$

or briefly
$$dM_1(t) = \phi(t)dW(t) \text{ and } dM_2(t) = \psi(t)dW(t)$$
$Q - a.s.$ for all $t \in [0, T]$. Let $\varphi : [0, t] \times \Omega \to \mathbb{R}$ be defined by

$$\varphi(t) = \psi(t) \cdot \phi^{-1}(t) \text{ for all } [0, T].$$

Then $\varphi = (\varphi(t))_{t \geq 0}$ is a progressively measurable processes on $[0, T]$ with

$$\varphi(t)\, dM_1(t) = \left(\psi(t) \cdot \phi^{-1}(t)\right) \cdot \phi(t)dW(t) = \psi(t)dW(t)$$

$Q - a.s.$ for all $t \in [0, T]$. So we have $Q - a.s.$ for all $t \in [0, T]$

$$
\begin{aligned}
M_2(t) &= M_2(0) + \int_0^t \psi(s)dW(s) \\
&= M_2(0) + \int_0^t \varphi(s)dM_1(s).
\end{aligned}
$$

Furthermore,

$$\int_0^T \varphi^2(t) \cdot \phi^2(t)dt = \int_0^T \psi^2(t)dt < \infty \quad Q - a.s. \quad \square$$

Now let $\gamma = (\gamma(t))_{t\geq 0}$ be a m–dimensional progressively measurable stochastic process, $m \in I\!N$, with

$$\int_0^t \gamma_j^2(s)\,ds < \infty \; Q - a.s. \text{ for all } t \geq 0, j = 1, ..., m.$$

Let the stochastic process $L(\gamma) = (L(\gamma,t))_{t\geq 0} = (L(\gamma(t),t))_{t\geq 0}$ for all $t \geq 0$ be defined by

$$L(\gamma,t) = e^{-\int_0^t \gamma(s)'dW(s) - \frac{1}{2}\int_0^t \|\gamma(s)\|^2 ds}.$$

Note that the stochastic process $X(\gamma) = (X(\gamma,t))_{t\geq 0} = (X(\gamma(t),t))_{t\geq 0}$ with

$$X(\gamma,t) := \int_0^t \gamma(s)'\,dW(s) + \frac{1}{2}\int_0^t \|\gamma(s)\|^2\,ds$$

or

$$dX(\gamma,t) := \frac{1}{2} \cdot \|\gamma(t)\|^2\,dt + \gamma(t)'\,dW(t)$$

for all $t \geq 0$ is an Itô process with $\mu(\gamma(t),t) = \frac{1}{2}\cdot\|\gamma(t)\|^2 = \frac{1}{2}\cdot\sum_{j=1}^m \gamma_j^2(t)$ and $\sigma(\gamma(t),t) = \gamma(t)'$ for all $t \geq 0$. Thus, using the transformation $G : I\!R \times [0,\infty) \to I\!R$ with $G(x,t) = e^{-x}$ we know that $G_t(x,t) = 0$, $G_x(x,t) = -e^{-x}$, $G_{xx}(x,t) = e^{-x}$ for all $t \geq 0$. Using Itô's lemma (Theorem 2.34) with $G(X(\gamma,t),t) = e^{-X(\gamma,t)} = L(\gamma,t)$ we get for all $t \geq 0$

$$\begin{aligned}
dL(\gamma,t) &= dG(X(\gamma,t),t) \\
&= \left(-L(\gamma,t)\cdot\frac{1}{2}\cdot\|\gamma(s)\|^2 + \frac{1}{2}\cdot L(\gamma,t)\cdot\|\gamma(s)\|^2\right)dt \\
&\quad -L(\gamma,t)\cdot\gamma(t)'\,dW(t) \\
&= -L(\gamma,t)\cdot\gamma(t)'\,dW(t).
\end{aligned}$$

Hence,

$$L(\gamma,t) = L(\gamma,0) + \int_0^t dL(\gamma,s) = 1 - \int_0^t L(\gamma,s)\cdot\gamma(s)'\,dW(s)$$

for all $t \geq 0$. The following lemma gives a condition under which $L(\gamma)$ is a continuous $(Q-)$ martingale.

Lemma 2.40 (Novikov Condition) *Let γ and $L(\gamma)$ be as defined above. Then $L(\gamma) = (L(\gamma,t))_{t\in[0,T]}$ is a continuous $(Q-)$ martingale if*

$$E_Q\left[e^{\frac{1}{2}\int_0^T \|\gamma(s)\|^2 ds}\right] < \infty.$$

For a proof see Karatzas and Shreve [KS91], Corollary 3.5.13. It can be easily seen that the Novikov condition holds if there is a constant real number $K > 0$ with

$$\int_0^T \|\gamma(s)\|^2 \, ds < K.$$

Note that under Novikov's condition

$$\frac{1}{\sqrt{2}} \cdot E_Q \left[\int_0^T \|\gamma(s)\|^2 \, ds \right] = E_Q \left[\left(\left(\frac{1}{2} \cdot \left(\int_0^T \|\gamma(s)\|^2 \, ds \right)^2 \right) \right)^{\frac{1}{2}} \right]$$

$$\leq E_Q \left[\left(\left(1 + \int_0^T \|\gamma(s)\|^2 \, ds + \frac{1}{2} \cdot \left(\int_0^T \|\gamma(s)\|^2 \, ds \right)^2 \right) \right)^{\frac{1}{2}} \right]$$

$$\leq E_Q \left[e^{\frac{1}{2} \int_0^T \|\gamma(s)\|^2 \, ds} \right] < \infty.$$

Especially, under Novikov's condition,

$$\int_0^T \|\gamma(s)\|^2 \, ds < \infty \quad Q - a.s.$$

and so,

$$\int_0^t \|\gamma(s)\|^2 \, ds < \infty \quad Q - a.s. \text{ for all } t \in [0,T].$$

For each $T \geq 0$ we define the measure $\widetilde{Q} = Q_{L(\gamma,T)}$ on the measure space $(\Omega, \mathcal{F}_T)$ by

$$\widetilde{Q}(A) := E_Q \left[1_A \cdot L(\gamma,T) \right] = \int_A L(\gamma,T) \, dQ \text{ for all } A \in \mathcal{F}_T,$$

which is a probability measure if $L(\gamma,T)$ is a $(Q-)$ martingale. In this case, $L(\gamma,T)$ is the Q–density of $\widetilde{Q}$, i.e. $L(\gamma,T) = d\widetilde{Q}/dQ$ on $(\Omega, \mathcal{F}_T)$. The following Girsanov theorem shows how we can construct a $\left(\widetilde{Q}- \right)$ Wiener process $\widetilde{W} = \left(\widetilde{W}(t) \right)_{t \in [0,T]}$ starting with a $(Q-)$ Wiener process $W = (W(t))_{t \geq 0}$.

Theorem 2.41 (Girsanov Theorem) *Let* $W = (W_1(t), ..., W_m(t))_{t \geq 0}$ *be a* m–*dimensional* $(Q-)$ *Wiener process,* $m \in \mathbb{N}$, γ, $L(\gamma)$, $\widetilde{Q}$, *and* $T \in [0, \infty)$ *be as defined above, and the* m–*dimensional stochastic process* $\widetilde{W} = \left(\widetilde{W}_1, ..., \widetilde{W}_m \right) = \left(\widetilde{W}_1(t), ..., \widetilde{W}_m(t) \right)_{t \in [0,T]}$ *be defined by*

$$\widetilde{W}_j(t) := W_j(t) + \int_0^t \gamma_j(s) \, ds, \, t \in [0,T], \, j = 1, ..., m,$$

i.e.

$$d\widetilde{W}(t) := \gamma(t)\,dt + dW(t)\,,\ t \in [0,T]\,.$$

If the stochastic process $L(\gamma) = (L(\gamma,t))_{t \in [0,T]}$ *is a* $(Q-)$ *martingale, then the stochastic process* $\widetilde{W}$ *is a* $m-$ *dimensional* $\left(\widetilde{Q}-\right)$ *Wiener process on the measure space* $(\Omega, \mathcal{F}_T)$.

For a proof see Korn and Korn [KK99], p. 108ff. Using Theorem 2.38 and Lemma 2.40 we can easily prove the following theorem (see, e.g., Baxter and Rennie [BR96], p. 79).

Theorem 2.42 (Martingale Characterization) *Let* $Z = (Z(t))_{t \geq 0}$ *be an Itô process on* $[0,T]$ *with* $dZ(t) = \mu(t)\,dt + \sigma_Z(t)dW(t)$. *Then:*

a) If Z *is a martingale, then* $\mu \equiv 0$ $Q-a.s.$

b) If $\mu \equiv 0$ *and if* $E_Q\left[\int_0^T \|\sigma_Z(s)\|^2\,ds\right] < \infty$, *then* Z *is a martingale.*

c) If $dZ(t) = Z(t) \cdot \sigma(t)\,dW(t)$, *i.e.* $\sigma_Z := Z \cdot \sigma$ *on* $[0,T]$, *for a progressively measurable process* σ *and if Novikov's condition holds for* σ, *i.e.*

$$E_Q\left[e^{\frac{1}{2}\int_0^T \|\sigma(s)\|^2 ds}\right] < \infty,$$

then Z *is a martingale. In this case, the solution of the stochastic differential equation is*

$$Z(t) = Z(0) \cdot e^{\int_0^t \sigma(s)dW(s) - \frac{1}{2}\int_0^t \|\sigma(s)\|^2 ds}.$$

There is also an interesting theorem telling us something about the distribution of a special stochastic integral (see, e.g. Karatzas and Shreve [KS91], p. 354-355 and Lamberton and Lapeyre [LL97], p. 57-58).

Theorem 2.43 (Distribution) *Let* $W = (W_1, ..., W_m)$ *be a* $m-$ *dimensional* $(Q-)$ *Wiener process and* $\sigma = (\sigma_1, ..., \sigma_m)$ *a* $m-$ *dimensional deterministic, real, Borel-measurable function with* $\int_0^\infty \|\sigma(s)\|^2\,ds < \infty$. *Then* $I(\sigma) = \int_0^\infty \sigma(s)dW(s)$ *is a normally distributed random variable with expectation* $E_Q[I(\sigma)] = 0$ *and variance* $E_Q[I^2(\sigma)] = \int_0^\infty \|\sigma(s)\|^2\,ds$.

2.6 The Feynman-Kac Representation

In the previous sections we studied $n-$dimensional stochastic processes $X = (X(t))_{t \geq 0}$ on the filtered probability space $(\Omega, \mathcal{F}, Q, \mathbb{F})$ of the form

$$X(t) = X(0) + \int_0^t \mu(s)\,ds + \int_0^t \sigma(s)\,dW(s),$$

where $W = (W_1, ..., W_m) = (W_1(t), ..., W_m(t))_{t \geq 0}$ is a m-dimensional Wiener process, $m \in I\!\!N$, $I\!\!F = I\!\!F(W)$, $X(0)$ is $(\mathcal{F}_0-)$ measurable and $\mu = (\mu(t))_{t \geq 0}$, $\sigma = (\sigma(t))_{t \geq 0}$ are progressively measurable stochastic processes with

$$\int_0^t |\mu_i(s)| \, ds < \infty \text{ and } \int_0^t \sigma_{ij}^2(s) \, ds < \infty \quad Q - a.s. \qquad (2.2)$$

for all $t \geq 0$, $i = 1, ..., n$, and $j = 1, ..., m$. For convenience we symbolically wrote

$$dX(t) = \mu(t) \, dt + \sigma(t) \, dW(t) = \mu(t) \, dt + \sum_{j=1}^m \sigma_j(t) \, dW_j(t),$$

and called this a stochastic differential equation. We now move on to a special form of stochastic differential equation. To do this let $\mu : I\!\!R^n \times [0, \infty) \to I\!\!R^n$ and $\sigma : I\!\!R^n \times [0, \infty) \to I\!\!R^{n \times m}$ be measurable functions (with respect to the corresponding Borel sigma-algebras).

Definition 2.44 (Strong Solution) *If there exists a n-dimensional stochastic process $X = (X(t))_{t \geq 0}$ on the filtered probability space $(\Omega, \mathcal{F}, Q, I\!\!F)$ satisfying (2.2), i.e. an Itô process, such that for all $t \geq 0$*

$$\begin{aligned} X(t) &= x + \int_0^t \mu(X(s), s) \, ds + \int_0^t \sigma(X(s), s) \, dW(s) \quad Q - a.s., \\ X(0) &= x \in I\!\!R^n, \text{ fixed,} \end{aligned}$$

we call X a strong solution of the stochastic differential equation

$$\begin{aligned} dX(t) &= \mu(X(t), t) \, dt + \sigma(X(t), t) \, dW(t) \quad \text{for all } t \geq 0 \text{ (SDE)} \\ X(0) &= x. \end{aligned}$$

The following theorem gives conditions for the existence and uniqueness of a strong solution for (SDE). It is the analogue of the deterministic Picard-Lindelöf theorem (see, e.g., Bingham and Kiesel [BK98], Theorem 5.7.1 or Lamberton and Lapeyre [LL97], p. 49ff).

Theorem 2.45 (Existence and Uniqueness) *Let μ and σ of the stochastic differential equation (SDE) be continuous functions such that for all $t \geq 0$, $x, y \in I\!\!R^n$ and for some constant $K > 0$ the following conditions hold:*

(i) $\|\mu(x, t) - \mu(y, t)\| + \|\sigma(x, t) - \sigma(y, t)\| \leq K \cdot \|x - y\|$

$\qquad\qquad\qquad\qquad\qquad\qquad\qquad\qquad$ *(Lipschitz condition),*

(ii) $\|\mu(x, t)\|^2 + \|\sigma(x, t)\|^2 \leq K^2 \cdot \left(1 + \|x\|^2\right)$ $\qquad$ *(growth condition).*

Then there exists a unique, continuous strong solution $X = (X(t))_{t \geq 0}$ of (SDE) and a constant C, depending only on K and $T > 0$, such that

$$E_Q \left[\|X(t)\|^2 \right] \leq C \cdot \left(1 + \|x\|^2 \right) \cdot e^{C \cdot t} \quad \text{for all } t \in [0, T]. \qquad (2.3)$$

Moreover,

$$E_Q \left[\sup_{0 \leq t \leq T} \|X(t)\|^2 \right] < \infty.$$

For a detailed proof see, e.g., Korn and Korn [KK99], p. 127-133. Note that the constant K in conditions (i) and (ii) applies for all t simultaneously. Theorem 2.45 may be extended to the case where $X(0) = \xi$ with ξ denoting a random vector, independent of W, with $E_Q \left[\|\xi\|^2 \right] < \infty$. In this case we have to replace $\|x\|^2$ by $E_Q \left[\|\xi\|^2 \right]$ in (2.3). Lipschitz condition (i) can be weakened to what is called a local Lipschitz condition (see, e.g. Duffie [Duf92], p. 239ff) still supplying us with the existence and uniqueness of a strong solution $X = (X(t))_{t \geq 0}$ of (SDE):

(i') For each constant $K^* > 0$ there is a constant $K > 0$ such that the Lipschitz condition (i) is satisfied for all $t \geq 0$ and for all $x, y \in \mathbb{R}^n$ with $\|x\| \leq K^*$ and $\|y\| \leq K^*$.

For the special case $n = m = 1$, Yamada and Watanabe [YW71] showed that it is enough to claim that

(i") $|\mu(t, x) - \mu(t, y)| \leq K \cdot |x - y|$ and $|\sigma(t, x) - \sigma(t, y)| \leq \rho(|x - y|)$ for all $t \geq 0$ and for all $x, y \in \mathbb{R}$, where $\rho : [0, \infty) \to [0, \infty)$ is a strictly increasing function with $\rho(0) = 0$ such that, for any $\varepsilon > 0$,

$$\int_0^\varepsilon \rho^{-2}(x) \, dx = \infty.$$

One possible function ρ in assumption (ii') is $\rho(x) = \sqrt{x}$. It should be mentioned that the only information used to evaluate μ and σ at time t is the actual value $X(t)$ of the stochastic process. It can be shown (see, e.g., Lamberton and Lapeyre [LL97], p. 54ff) that the strong solution X of (SDE) is a Markov process, i.e. for all measurable (with respect to the corresponding Borel sigma-algebra), bounded functions D and fixed $0 \leq t \leq T < \infty$ we have

$$E_Q \left[D(X(T)) | \mathcal{F}_t \right] = E_Q \left[D(X(T)) | X(t) \right] = g(X(t)),$$

with

$$g(x) := E_Q^{x,t} \left[D(X(T)) \right] := E_Q \left[D \left(X^{x,t}(T) \right) \right]$$

and $X^{x,t}(T)$ denoting the strong solution of the stochastic differential equation (SDE) with initial condition $X(t) = x$. For a proof and more details see, e.g. Rogers and Williams [RW87]. Under the assumptions of Theorem 2.45 a strong solution $X^{x,t} = (X^{x,t}(T))_{T \geq t}$ of the stochastic differential equation (SDE) with initial condition $X(t) = x$ exists as a stochastic process which is almost surely continuous in t, x and T. To move on to the so-called Feynman-Kac representation the following definition is very helpful.

Definition 2.46 (Characteristic Operator) *Let $X = (X(t))_{t \geq 0}$ be the unique strong solution of the stochastic differential equation $(S D \bar{E})$ under the conditions (i) and (ii) of Theorem 2.45. Then the operator $\mathcal{D}$ defined by*

$$(\mathcal{D}v)(x,t) := v_t(x,t) + \sum_{i=1}^{n} \mu_i(x,t) \cdot v_{x_i}(x,t) + \frac{1}{2} \cdot \sum_{i=1}^{n} \sum_{j=1}^{n} a_{ij}(x,t) \cdot v_{x_i x_j}(x,t),$$

with $v : \mathbb{R}^n \times [0,\infty) \to \mathbb{R}$ twice continuously differentiable in x, once continuously differentiable in t, and

$$a_{ij}(x,t) := \sum_{k=1}^{m} \sigma_{ik}(x,t) \cdot \sigma_{jk}(x,t) = d < X_i, X_j >_t,$$

is called the characteristic operator for $X(t)$.

This operator is used to define the so-called Cauchy problem.

Definition 2.47 (Cauchy Problem) *Let $D : \mathbb{R}^n \to \mathbb{R}$, $r : \mathbb{R}^n \times [0,T] \to \mathbb{R}$ be continuous and $T > 0$ be arbitrary but fixed. Then the Cauchy problem is stated as follows: Find a function $v : \mathbb{R}^n \times [0,T] \to \mathbb{R}$ which is continuously differentiable in t and twice continuously differentiable in x and solves the partial differential equation (sometimes called backward Kolmogorov equation)*

$$\begin{aligned} \mathcal{D}v(x,t) &= r(x,t) \cdot v(x,t) \quad \text{for all } (x,t) \in \mathbb{R}^n \times [0,T], \\ v(x,T) &= D(x) \quad \text{for all } x \in \mathbb{R}^n. \end{aligned}$$

We now use the Cauchy problem to state the so-called Feynman-Kac representation. We do this under sufficient regularity conditions on μ, σ, r, v, and D, ensuring that there is a unique solution of the stochastic differential equation (SDE) (see Theorem 2.45) and that all functions are well-behaved. For more details and possible sets of regularity conditions see, e.g. Duffie [Duf92], p. 242ff. Regularity conditions are also given in Karatzas and Shreve [KS91], Korn and Korn [KK99], A. Friedman [Fri75] or Krylov [Kry80]. Let, for $0 \leq t \leq s \leq T$,

$$P_0(t,s) := e^{\int_t^s r(X(u),u)du}.$$

Then we know that

$$d_s P_0(t,s) = r(X(s),s) \cdot P_0(t,s) \, ds$$

and

$$d_s P_0^{-1}(t,s) = -r(X(s),s) \cdot P_0^{-1}(t,s) \, ds.$$

Let the function $\widetilde{v} : \mathbb{R}^n \times [t,T] \to \mathbb{R}$ be defined by

$$\widetilde{v}(x,s) := P_0^{-1}(t,s) \cdot v(x,s) \quad \text{for all } s \in [t,T].$$

Then, using the product rule in Lemma 2.36, we get

$$
\begin{aligned}
d_s \widetilde{v}(X(s),s) &= P_0^{-1}(t,s) \, d_s v(X(s),s) + v(X(s),s) \, d_s P_0^{-1}(t,s) \\
&= P_0^{-1}(t,s) \cdot \left[\begin{array}{c} dv(X(s),s) \\ -r(X(s),s) \cdot v(X(s),s) \, ds \end{array} \right].
\end{aligned}
$$
(2.4)

Using Itô's lemma (Theorem 2.34), we get

$$
\begin{aligned}
dv(X(s),s) &= v_s(X(s),s) \, ds + \sum_{i=1}^{n} v_{x_i}(X(s),s) \, dX_i(s) \\
&\quad + \frac{1}{2} \sum_{i=1}^{n} \sum_{j=1}^{n} v_{x_i x_j}(X(s),s) \, d < X_i(s), X_j(s) > (s) \\
&= (\mathcal{D}v)(X(s),s) \, ds \\
&\quad + \sum_{i=1}^{n} v_{x_i}(X(s),s) \cdot \sigma_i(X(s),s) \, dW(s).
\end{aligned}
$$
(2.5)

Combining equations (2.4) and (2.5), we have

$$
\begin{aligned}
d\widetilde{v}(X(s),s) &= P_0^{-1}(t,s) \cdot [dv(X(s),s) - r(X(s),s) \cdot v(X(s),s) \, ds] \\
&= P_0^{-1}(t,s) \cdot (\mathcal{D}v)(X(s),s) \, ds \\
&\quad - P_0^{-1}(t,s) \cdot r(X(s),s) \cdot v(X(s),s) \, ds \\
&\quad + P_0^{-1}(t,s) \cdot \sum_{i=1}^{n} v_{x_i}(X(s),s) \cdot \sigma_i(X(s),s) \, dW(s).
\end{aligned}
$$
(2.6)

Let us now suppose that v is a solution of the Cauchy problem. Then equation (2.6) simplifies to

$$d\widetilde{v}(X(s),s) = P_0^{-1}(t,s) \cdot \sum_{i=1}^{n} v_{x_i}(X(s),s) \cdot \sigma_i(X(s),s) \, dW(s),$$

i.e.

$$\tilde{v}\left(X\left(T\right),T\right) \;=\; \tilde{v}\left(X\left(t\right),t\right)$$
$$+ \sum_{i=1}^{n} \int_{t}^{T} P_{0}^{-1}\left(t,s\right) \cdot v_{x_i}\left(X\left(s\right),s\right) \cdot \sigma_i\left(X\left(s\right),s\right) dW\left(s\right).$$

Now let us assume that for each $i \in \{1,...,n\}$ and for all $\tau \in [t,T]$

$$E_Q\left[\int_{t}^{\tau}\left(P_{0}^{-1}\left(t,s\right) \cdot v_{x_i}\left(X\left(s\right),s\right) \cdot \sigma_i\left(X\left(s\right),s\right)\right)^2 ds\right] < \infty. \qquad (2.7)$$

Then the stochastic process $M = \left(M\left(\tau\right)\right)_{\tau \geq t}$, defined by

$$M\left(\tau\right) \;:=\; M\left(X\left(\tau\right),\tau\right)$$
$$:=\; \int_{t}^{\tau} P_{0}^{-1}\left(t,s\right) \cdot v_{x_i}\left(X\left(s\right),s\right) \cdot \sigma_i\left(X\left(s\right),s\right) dW\left(s\right)$$

is a continuous $(Q-)$ martingale, and we know that

$$E_Q^{t,x}\left[M\left(T\right)\right] = E_Q\left[M\left(T\right) | M\left(t\right)\right] = M\left(t\right) = M\left(X\left(t\right),t\right) = 0 \;\; Q-a.s.$$

Hence $E_Q^{t,x}\left[\tilde{v}\left(X\left(T\right),T\right)\right] = \tilde{v}\left(x,t\right) = v\left(x,t\right)$, i.e.

$$v\left(x,t\right) \;=\; E_Q^{x,t}\left[P_{0}^{-1}\left(t,T\right) \cdot D\left(X\left(T\right)\right)\right] \qquad (2.8)$$
$$=\; E_Q^{x,t}\left[e^{-\int_{t}^{T} r(X(u),u)du} \cdot D\left(X\left(T\right)\right)\right]$$

on $\mathbb{R}^n \times [0,T]$. Thus, under sufficient regularity conditions, the solution v of the Cauchy problem is given by the *Feynman-Kac representation* (2.8). Under these regularity conditions: If we can show that there is a (unique) solution of the Cauchy problem, this solution is given by the expected value (2.8) as a function of the initial parameters (x,t) of the stochastic differential equation (SDE). The opposite direction is not always true. Nevertheless, if we can derive the expected value (2.8) and show that it solves the Cauchy problem, it is the unique solution to the Cauchy problem. And indeed, under technical conditions on μ, σ, and r, there is a function $G : \mathbb{R}^n \times [0,T] \times \mathbb{R}^n \times [0,T] \rightarrow \mathbb{R}$, the so-called fundamental solution of the Cauchy problem, sometimes also called *Green's function*, which has the following properties:

a) For any $(x',t') \in \mathbb{R}^n \times [0,T]$ the function $G_{x',t'}\left(x,t\right) := G\left(x,t,x',t'\right)$ is continuously differentiable in t, twice continuously differentiable in x, and solves the partial differential equation (*backward Kolmogorov equation*)

$$\mathcal{D}G_{x',t'}\left(x,t\right) = r\left(x,t\right) \cdot G_{x',t'}\left(x,t\right) \;\; \text{for all} \;\; (x,t) \in \mathbb{R}^n \times [0,T].$$

b) For any $(x,t) \in \mathbb{R}^n \times [0,T]$ the function $G_{x,t}(x',t') := G(x,t,x',t')$ is continuously differentiable in t', twice continuously differentiable in x', and solves the partial differential equation (sometimes called *Fokker-Planck* or *forward Kolmogorov equation*)

$$\mathcal{D}^* G_{x,t}(x',t') = r(x',t') \cdot G_{x,t}(x',t') \quad \text{for all } (x',t') \in \mathbb{R}^n \times [t,T] \tag{2.9}$$

with

$$
\begin{aligned}
\mathcal{D}^* G_{x,t}(x',t') \;=\; & -\frac{\partial}{\partial t'} G_{x,t}(x',t') - \sum_{i=1}^{n} \mu_i(x',t') \cdot \frac{\partial}{\partial x_i'} G_{x,t}(x',t') \\
& + \frac{1}{2} \cdot \sum_{i=1}^{n} \sum_{j=1}^{n} a_{ij}(x',t') \cdot \frac{\partial^2}{\partial x_i' \partial x_j'} G_{x,t}(x',t')
\end{aligned}
$$

and a_{ij}, $i,j = 1, ..., n$, as in Definition 2.46.

c) Under technical conditions on D the solution of the Cauchy problem is given by

$$v(x,t) = \int_{\mathbb{R}^n} G(x,t,x',T) \cdot D(x')\, dx'. \tag{2.10}$$

A sufficient set of technical and terminal conditions may be found in Friedman [Fri64] and [Fri75]. Essentially, the relevant boundary condition says that, at a fixed time T, the function $x' \longmapsto G_{x,t}(x',T)$ is the density of a measure on $\mathbb{R}^n$ that converges to a probability measure ν with $\nu(\{x\}) = 1$ as $t \to T$, i.e. $\lim_{t \to T} v(x,t) = D(x)$ for all $x \in \mathbb{R}^n$. Thus, property c) is the key to turn the fundamental solution G into the solution with given boundary condition $v(x,T) = D(x)$ for all $x \in \mathbb{R}^n$.

3

Financial Markets

This chapter introduces the basic building blocks and assumptions used to set up a consistent framework for describing financial markets. We do this by starting with a general model of the basic financial instruments, called the primary traded assets, in Section 3.1. Their prices are described by stochastic processes or, to be more precise, by a corresponding stochastic differential equation. Trading with these financial instruments requires some basic trading principles defining the possibilities of how we can put the assets together (over time) and build a so-called portfolio (process). Given a specific financial product, we may be interested in replicating the cash flow paid by this product over time using our primary traded assets. In a heavily traded market we would hope that there will be no riskless profit which could be earned by selling the financial product and replicating it with the primary traded assets over time. If so, the market is called arbitrage-free. In Section 3.2 we show under which conditions there are no arbitrage opportunities in the financial market. We will then change the price scale of the primary traded assets and rather observe the prices relative to a specific unit price or numéraire. If the resulting normalized or discounted market prices can be described by martingales with respect to a so-called martingale measure we will show that the financial market is free of arbitrage opportunities. Furthermore, we will show that today's price of a (discounted) primary traded asset is equal to the expected value of any corresponding and adequately discounted future market price where the expectation is taken with respect to the martingale measure. Another important characteristic of a financial market is its completeness, i.e. the possibility of replicating every financial product traded in the market. As we will learn in Section 3.3, this important feature is closely related to the uniqueness of the martingale measure. In Section 3.4 we will prove that financial derivatives can be uniquely priced if the financial market is complete. As an application, one of the most famous market models, the Black-Scholes model, is discussed in Section 3.5. It is used to derive the prices for (European) options on contingent claims by taking expectations with respect to the corresponding martingale measure. However, depending on the specific financial product, it may be convenient to change the numéraire or unit price and corresponding measure to significantly simplify the calculation of the expected value or price. This important technique is described and applied in Sections 3.6 and 3.7.

If you are already familiar with the basic modelling of financial markets and rather interested in the specific modelling of interest-rate markets, you may immediately switch to Chapter 4. If you do feel, however, that you would like to read more modelling details than those presented here, you may also refer to Bingham and Kiesel [BK98], Lamberton and Lapeyre [LL97], or Musiela and Rutkowski [MR97]. Again, to simplify the search for more information, at least one reference including the exact location of the respective statement or proof is listed wherever it seemed to be appropriate and helpful.

3.1 The Financial Market Model

We suppose that we are dealing with a frictionless security market where investors are allowed to trade continuously up to some fixed finite planning horizon T. Uncertainty in our financial market is modelled by a complete probability space $(\Omega, \mathcal{F}, Q)$ where all prices are driven by a $m-$dimensional Wiener process $W = (W(t))_{t \in [0,T]}$. We assume that the probability space is filtered by the natural filtration $I\!F = I\!F(W)$ with $\mathcal{F} = \mathcal{F}_T = \mathcal{F}_T(W)$. There are $n+1, n \in I\!N$, primary traded assets in this market with prices that can be described by non-negative Itô processes $P_0, ..., P_n$ on $[0,T]$. Thus, for all $i = 0, ..., n$ and for all $t \in [0,T]$ we have

$$dP_i(t) = \mu_i(t)\, dt + \sigma_i(t)\, dW(t) = \mu_i(t)\, dt + \sum_{j=1}^{m} \sigma_{ij}(t)\, dW_j(t),$$

with progressively measurable stochastic processes μ_i and σ_{ij} such that

$$\int_0^T |\mu_i(s)|\, ds < \infty \quad Q - a.s. \tag{M1}$$

and

$$\int_0^T \sigma_{ij}^2(s)\, ds < \infty \quad Q - a.s. \text{ for all } j = 1, ..., m. \tag{M2'}$$

For technical convenience with respect to the trading strategies which will be considered here, we furthermore assume that $\sigma_{ij} \in L_2[0,T]$ for all $j = 1, ..., m$, i.e.

$$E_Q\left[\int_0^T \sigma_{ij}^2(s)\,ds\right] < \infty \quad \text{for all } j = 1, ..., m. \tag{M2}$$

We will refer to this model under conditions $(M1)$ and $(M2)$ as the (primary) *financial market* $\mathcal{M} = \mathcal{M}(Q)$. Note that in $\mathcal{M}$ the stochastic integrals $J_t(\sigma_{ij}) = \int_0^t \sigma_{ij}(s)\,dW_j(s)$ do exist for all $t \in [0,T]$ and the processes $J(\sigma_{ij})$ are continuous Q–martingales with

$$E_Q[J_t(\sigma_{ij})]^2 = E_Q\left[\int_0^t \sigma_{ij}^2(s)\,ds\right] \quad \text{for all } t \in [0,T]$$

and for all $i \in \{0, ..., n\}$, $j \in \{1, ..., m\}$.

Remark. *Using Theorem 2.24d), we know that under condition $(M2)$*

$$E_Q\left[\left(\sup_{0\leq t\leq T}\left|\int_0^t \sigma_{ij}(s)\,dW(s)\right|\right)^2\right] \leq 4 \cdot E_Q\left[\int_0^T \sigma_{ij}^2(s)\,ds\right] < \infty.$$

Hence, for all $i \in \{0, ..., n\}$, $j \in \{1, ..., m\}$, we have

$$\sup_{0\leq t\leq T}\left|\int_0^t \sigma_{ij}(s)\,dW(s)\right| < \infty \quad Q-a.s. \tag{3.1}$$

Combining $(M1)$ and (3.1), we get

$$\begin{aligned}
\sup_{0\leq t\leq T} |P_i(t)| &= \sup_{0\leq t\leq T}\left|P_i(0) + \int_0^t \mu_i(s)\,ds + \int_0^t \sigma_{ij}(s)\,dW(s)\right| \\
&\leq |P_i(0)| + \int_0^T |\mu_i(s)|\,ds + \sup_{0\leq t\leq T}\left|\int_0^t \sigma_{ij}(s)\,dW(s)\right| \\
&< \infty \quad Q-a.s. \tag{3.2}
\end{aligned}$$

$\square$

As we will see later on, it can make sense to change the price scale in our financial market. In other words, we may want to observe the prices of our primary traded financial assets relative to another unit price or numéraire. The mathematical definition of this instrument follows.

Definition 3.1 (Numéraire) *A numéraire in $\mathcal{M} = \mathcal{M}(Q)$ is a price process $X = (X(t))_{t\in[0,T]}$ with*

$$X(t) > 0 \quad \text{for each } t \in [0,T].$$

We will choose the price process P_0 to be our numéraire, and claim that

$$P_0(0) = 1, \quad \mu_0(t) = r(t) \cdot P_0(t), \tag{N1}$$

with a progressively measurable stochastic process r such that $\int_0^T |r(s)|\, ds < \infty$ $Q - a.s.$, and

$$\sigma_{0j}(t) = 0 \text{ for all } j = 1, ..., m \text{ and for all } t \in [0, T]. \tag{N2}$$

Hence, our numéraire P_0 is a special Itô process with

$$dP_0(t) = r(t) \cdot P_0(t)\, dt, \text{ i.e. } P_0(t) = e^{\int_0^t r(s)ds}$$

for all $t \in [0, T]$. Furthermore,

$$dP_0^{-1}(t) = -r(t) \cdot P_0^{-1}(t)\, dt, \text{ i.e. } P_0^{-1}(t) = e^{-\int_0^t r(s)ds}$$

for all $t \in [0, T]$.

Remark. *Using (N1), we get for each $z \in \mathbb{R}$*

$$0 < \sup_{0 \le t \le T} P_0^z(t) \le e^{|z| \cdot \int_0^T |r(s)| ds} < \infty \quad Q - a.s. \tag{3.3}$$

□

Because of the absence of a diffusion term (although r may be stochastic) the numéraire P_0 is usually called *safe* (investment) and may be thought of as a banking or cash account. On the other hand, due to the presence of their diffusion terms, the assets $1, ..., n$ are called *risky* (investments). The price processes $\widetilde{P}_0, ..., \widetilde{P}_n$ on $[0, T]$ with

$$\widetilde{P}_i(t) := P_0^{-1}(t) \cdot P_i(t), \ t \in [0, T], \ i = 0, ..., n$$

are called the *discounted* or *normalized prices* of the primary traded assets. Normalization corresponds to the idea that we consider the price $P_0(t)$ of the safe investment to be the unit price in our financial market. The price $\widetilde{P}_i(t)$ then is the price of asset $i \in \{0, ..., n\}$ in terms of this unit price with $\widetilde{P}_0 \equiv 1$. Using the product rule (Lemma 2.36), we get the following behaviour of the discounted price process:

$$
\begin{aligned}
d\widetilde{P}_i(t) &= d\left(P_0^{-1}(t) \cdot P_i(t)\right) = P_i(t)dP_0^{-1}(t) + P_0^{-1}(t)\, dP_i(t) \\
&= P_0^{-1}(t) \cdot (dP_i(t) - r(t) \cdot P_i(t)dt) \\
&= P_0^{-1}(t) \cdot ((\mu_i(t) - r(t) \cdot P_i(t))\, dt + \sigma_i(t)\, dW(t)) \\
&=: \widetilde{\mu}_i(t)\, dt + \widetilde{\sigma}_i(t)\, dW(t).
\end{aligned} \tag{3.4}
$$

with $\widetilde{\mu}_i(t) = (\mu_i(t) - r(t) \cdot P_i(t)) \cdot P_0^{-1}(t)$ and $\widetilde{\sigma}_{ij}(t) = \sigma_{ij}(t) \cdot P_0^{-1}(t)$ for all $i = 1, ..., n$, $j = 1, ..., m$, $t \in [0, T]$. Correspondingly, we set $\widetilde{\mu}_0 \equiv 0$ and $\widetilde{\sigma}_{0j} \equiv 0$, $j = 1, ..., m$. Using $(M1)$, $(N1)$, (3.3), and (3.2) we get[1] for all $i = 1, ..., n$

$$
\int_0^T |\widetilde{\mu}_i(s)|\, ds \;\leq\; \left(\int_0^T |\mu_i(s)|\, ds + \int_0^T |r(s)| \cdot |P_i(s)|\, ds \right)
$$
$$
\cdot \; \sup_{0 \leq t \leq T} P_0^{-1}(t)
$$
$$
\leq\; \left(\int_0^T |\mu_i(s)|\, ds + \sup_{0 \leq t \leq T} |P_i(t)| \cdot \int_0^T |r(s)|\, ds \right)
$$
$$
\cdot \; \sup_{0 \leq t \leq T} P_0^{-1}(t)
$$
$$
<\; \infty \quad Q - a.s., \tag{$\widetilde{M1}$}
$$

and, using condition $(M2)$ and (3.3),

$$
\int_0^T \widetilde{\sigma}_{ij}^2(s)\, ds \;=\; \int_0^T \sigma_{ij}^2(s) \cdot P_0^{-2}(s)\, ds \leq \sup_{0 \leq t \leq T} P_0^{-2}(t) \cdot \int_0^T \sigma_{ij}^2(s)\, ds
$$
$$
\leq\; e^{2 \cdot \int_0^T |r(s)| ds} \cdot \int_0^T \sigma_{ij}^2(s)\, ds < \infty \quad Q - a.s.
$$

for all $j = 1, ..., m$. We therefore know that the stochastic integrals $J_t(\widetilde{\sigma}_{ij}) = \int_0^t \widetilde{\sigma}_{ij}(s)\, dW_j(s)$ exist for all $t \in [0, T]$ and the processes $J(\widetilde{\sigma}_{ij})$ are continuous local $Q-$martingales for all $i \in \{1, ..., n\}$, $j \in \{1, ..., m\}$. They are continuous $Q-$martingales with

$$
E_Q[J_t(\widetilde{\sigma}_{ij})]^2 = E_Q\left[\int_0^t \widetilde{\sigma}_{ij}^2(s)\, ds \right] \quad \text{for all } t \in [0, T],
$$

if we assume that $\widetilde{\sigma}_{ij} \in L_2[0, T]$ for all $i = 1, ..., n$, $j = 1, ..., m$, i.e.

$$
E_Q\left[\int_0^T \widetilde{\sigma}_{ij}^2(s)\, ds \right] < \infty, \quad i = 1, ..., n, j = 1, ..., m. \tag{$\widetilde{M2}$}
$$

We will refer to this model for the discounted prices of the primary traded assets under conditions $\left(\widetilde{M1}\right)$ and $\left(\widetilde{M2}\right)$ as the *discounted* or *normalized* (primary) *financial market* $\widetilde{\mathcal{M}} = \widetilde{\mathcal{M}}(Q)$.

---

[1] Especially, we get

$$
\int_0^t |r(s) \cdot P_i(s)|\, ds < \infty \quad Q - a.s. \text{ for all } i \in \{1, ..., n\}.
$$

Remark. *As in $M(Q)$ we also know in $\widetilde{M}(Q)$ that*

$$\sup_{0 \leq t \leq T} \left| \widetilde{P}_i(t) \right| < \infty \quad Q-a.s. \tag{3.5}$$

□

Having defined the primary traded assets, we now consider the holdings $\varphi_i(t)$ in each asset $i \in \{0, ..., n\}$ at any time $t \in [0, T]$. These holdings may be altered over time to reach a specific gain or cash flow.

Definition 3.2 (Trading Strategy) *A trading strategy or portfolio process $\varphi = (\varphi_0, ..., \varphi_n) = (\varphi_0(t), ..., \varphi_n(t))_{t \in [0,T]}$ (in $M(Q)$ and $\widetilde{M}(Q)$) is a $(n+1)-$dimensional progressively measurable stochastic process such that for all $t \in [0, T]$, $i \in \{0, ..., n\}$:*

(T1) $\displaystyle \int_0^T |\varphi_i(t) \cdot \mu_i^*(t)| \, dt < \infty \; Q-a.s., \; \mu_i^* \in \{\mu_i, r\}, i = 1, ..., n, \mu_0^* = r,$

(T2) $\displaystyle \sum_{j=1}^m \int_0^T (\varphi_i(t) \cdot \sigma_{ij}(t))^2 \, dt < \infty \quad Q-a.s. \text{ for all } j = 1, ..., m.$

The vector $\varphi(t)$ is called the **portfolio** at time $t \in [0, T]$, and the stochastic process $V(\varphi) = (V(\varphi, t))_{t \in [0,T]}$ defined by

$$V(\varphi, t) := \sum_{i=0}^n \varphi_i(t) \cdot P_i(t) = \varphi(t)' P(t), \; t \in [0, T]$$

is called the **price process** of the trading strategy or portfolio process φ (in $M(Q)$). Consequently, we call $V(\varphi, 0)$ the **initial price** of the portfolio at time $t = 0$. The **gains process** of the trading strategy or portfolio process φ (in $M(Q)$), $G(\varphi) = (G(\varphi, t))_{t \in [0,T]}$ is defined by

$$G(\varphi, t) := \int_0^t \varphi(s)' \, dP(s) = \sum_{i=0}^n \int_0^t \varphi_i(s) \, dP_i(s). \tag{3.6}$$

Note that, under conditions $(T1)$, (3.2), and (3.3), it follows that

$$\int_0^T |\varphi_i(t) \cdot \tilde{\mu}_i(t)| \, dt \ \leq \ \left(\int_0^T |\varphi_i(t) \cdot \mu_i(t)| \, dt \right.$$

$$\left. + \int_0^T |\varphi_i(t) \cdot r(t) \cdot P_i(t)| \, dt \right) \cdot \sup_{0 \leq t \leq T} P_0^{-1}(t)$$

$$\leq \ \sup_{0 \leq t \leq T} P_0^{-1}(t) \cdot \left(\int_0^T |\varphi_i(t) \cdot \mu_i(t)| \, dt \right.$$

$$\left. + \sup_{0 \leq t \leq T} |P_i(t)| \cdot \int_0^T |\varphi_i(t) \cdot r(t)| \, dt \right)$$

$$< \ \infty \quad Q - a.s.$$

and

$$\int_0^T |\varphi_0(t) \cdot \mu_0(t)| \, dt \ = \ \int_0^T |\varphi_0(t) \cdot r(t) \cdot P_0(t)| \, dt$$

$$\leq \ \sup_{0 \leq t \leq T} P_0(t) \cdot \int_0^T |\varphi_0(t) \cdot r(t)| \, dt$$

$$< \ \infty \quad Q - a.s.$$

Furthermore, under conditions $(T2)$ and (3.3), we get

$$\sum_{j=1}^m \int_0^T (\varphi_i(t) \cdot \tilde{\sigma}_{ij}(t))^2 \, dt \ \leq \ \sup_{0 \leq t \leq T} P_0^{-2}(t) \cdot \sum_{j-1}^m \int_0^T (\varphi_i(t) \cdot \sigma_{ij}(t))^2 \, dt$$

$$< \ \infty \quad Q - a.s.$$

Hence, the assumptions $(T1)$ and $(T2)$ ensure that the stochastic integrals $\int_0^t \varphi_i(s) \, dP_i(s)$ and $\int_0^t \varphi_i(s) \, d\tilde{P}_i(s)$ exist as local Q-martingales for all $i = 0, ..., n$ and for all $t \in [0, T]$. If φ is bounded[2] $\lambda \otimes Q - a.s.$, $(T1)$ and $(T2)$ are always satisfied because of the conditions $(M1)$ and $(M2)$. On the other hand, if μ_i and r are bounded $\lambda \otimes Q - a.s.$, then $\int_0^T |\varphi_i(t)| \, dt < \infty$ $Q - a.s.$ is sufficient for $(T1)$, $i \in \{1, ..., n\}$. If σ_{ij} is bounded $\lambda \otimes Q - a.s.$ for all $j \in \{1, ..., m\}$, then $\int_0^T \varphi_i^2(t) \, dt < \infty$ $Q - a.s.$ is sufficient for $(T2)$, $i \in \{1, ..., n\}$.

---

[2] It should be mentioned that conditions $(T1)$ and $(T2)$ can be relaxed with $\int_0^t \varphi_i(s) \, dP_i(s)$ and $\int_0^t \varphi_i(s) \, d\tilde{P}_i(s)$ becoming local Q-martingales if we concentrate on locally bounded trading strategies φ, i.e. there is a localizing sequence $(\tau_n)_{n \in IN}$ of stopping times such that $\varphi^n = (\varphi_t^n)_{t \geq 0} := (\varphi_{t \wedge \tau_n})_{t \geq 0}$ is bounded for all $n \in IN$. For more details see, e.g. Musiela and Rutkowski [MR97], p. 231ff., Protter [Pro92], p. 34ff., or Bingham and Kiesel [BK98], p. 159ff.

Remark. *Let $i \in \{1, ..., n\}$. For the special case that $\mu_i(t) = \widehat{\mu}_i(t) \cdot P_i(t)$ and $\sigma_{ij}(t) = \widehat{\sigma}_{ij}(t) \cdot P_i(t)$ for all $t \in [0, T]$ and for all $j \in \{1, ..., m\}$, i.e.*

$$
\begin{aligned}
dP_i(t) &= P_i(t) \cdot \left[\widehat{\mu}_i(t)\, dt + \sum_{j=1}^{m} \widehat{\sigma}_{ij}(t)\, dW_j(t) \right] \qquad (3.7) \\
&= P_i(t) \cdot \left[\widehat{\mu}_i(t)\, dt + \widehat{\sigma}_i(t)\, dW(t) \right]
\end{aligned}
$$

with $\widehat{\sigma}_i := (\widehat{\sigma}_{i1}, ..., \widehat{\sigma}_{im})$, we get by applying Itô's lemma (Theorem 2.34) with $X(t) := P_i(t)$ and $G(x, t) = \ln x$:

$$
\begin{aligned}
d\ln P_i(t) &= dG(X(t), t) \\
&= \left(\widehat{\mu}_i(t) - \frac{1}{2} \cdot \|\widehat{\sigma}_i(t)\|^2 \right) dt + \widehat{\sigma}_i(t)\, dW(t).
\end{aligned}
$$

Hence, if we assume that

$$
\widehat{\sigma}_i \in L_2[0, T], \qquad\qquad (\widehat{M2})
$$

we know by Theorem 2.24d) that

$$
E_Q\left[\left(\sup_{0 \le t \le T} \left| \int_0^t \widehat{\sigma}_{ij}(s)\, dW(s) \right| \right)^2 \right] \le 4 \cdot E_Q\left[\int_0^T \widehat{\sigma}_{ij}^2(s)\, ds \right] < \infty
$$

and therefore,

$$
\sup_{0 \le t \le T} \left| \int_0^t \widehat{\sigma}_{ij}(s)\, dW(s) \right| < \infty \quad Q - a.s.
$$

for all $j \in \{1, ..., m\}$. If we furthermore assume that

$$
\int_0^T |\widehat{\mu}_i(s)|\, ds < \infty \quad Q - a.s. \qquad\qquad (\widehat{M1})
$$

we get

$$
\begin{aligned}
\sup_{0 \le t \le T} |\ln P_i(t)| &\le \sup_{0 \le t \le T} \left| \ln P_i(0) + \int_0^t \left(\widehat{\mu}_i(s) - \frac{1}{2} \cdot \|\widehat{\sigma}_i(s)\|^2 \right) dt \right| \\
&\quad + \sup_{0 \le t \le T} \left| \int_0^t \widehat{\sigma}_i(s)\, dW(s) \right| \\
&\le |\ln P_i(0)| + \int_0^T |\widehat{\mu}_i(s)|\, dt + \frac{1}{2} \cdot \sum_{j=1}^{m} \int_0^T \widehat{\sigma}_{ij}^2(s)\, dt \\
&\quad + \sup_{0 \le t \le T} \left| \int_0^t \widehat{\sigma}_i(s)\, dW(s) \right| \\
&< \infty \quad Q - a.s.,
\end{aligned}
$$

and thus

$$\sup_{0 \le t \le T} |P_i(t)| = \sup_{0 \le t \le T} \left| e^{\ln P_i(t)} \right| \le e^{\sup_{0 \le t \le T} |\ln P_i(t)|} < \infty \quad Q - a.s.$$

Correspondingly, we learn from equation (3.4) that

$$
\begin{aligned}
d\widetilde{P}_i(t) &= \widetilde{\mu}_i(t) \, dt + \widetilde{\sigma}_i(t) \, dW(t) \\
&= \widetilde{P}_i(t) \cdot [(\widehat{\mu}_i(t) - r(t)) \, dt + \widehat{\sigma}_i(t) \, dW(t)]
\end{aligned}
$$

or

$$d\ln \widetilde{P}_i(t) = \left(\widehat{\mu}_i(t) - r(t) - \frac{1}{2} \cdot \|\widehat{\sigma}_i(t)\|^2 \right) dt + \widehat{\sigma}_i(t) \, dW(t).$$

Hence,

$$\sup_{0 \le t \le T} \left| \widetilde{P}_i(t) \right| < \infty \quad Q - a.s.$$

under the assumptions $\left(\widehat{M1} \right)$, *(N1), (N2), and* $\left(\widehat{M2} \right)$. *With respect to (T1) and (T2) we can easily see that*

$$\int_0^T |\varphi_i(t) \cdot \mu_i(t)| \, dt \le \sup_{0 \le t \le T} |P_i(t)| \cdot \int_0^T |\varphi_i(t) \cdot \widehat{\mu}_i(t)| \, dt < \infty$$

under the assumption

$$\int_0^T |\varphi_i(t) \cdot \widehat{\mu}_i(t)| \, dt < \infty \quad Q - a.s. \tag{$\widehat{T1}$}$$

and

$$
\sum_{j=1}^m \int_0^T (\varphi_i(t)\sigma_{ij}(t))^2 \, dt \le \left(\sup_{0 \le t \le T} |P_i(t)| \right)^2 \cdot \sum_{j=1}^m \int_0^T (\varphi_i(t)\widehat{\sigma}_{ij}(t))^2 \, dt
$$
$$< \infty$$

$Q - a.s.$ *under the assumption*

$$\sum_{j=1}^m \int_0^T (\varphi_i(t) \cdot \widehat{\sigma}_{ij}(t))^2 \, dt < \infty \quad Q - a.s. \text{ for all } j = 1, ..., m. \tag{$\widehat{T2}$}$$

To conclude, if the price process of a primary traded asset is given by a stochastic differential equation such as in (3.7), it is sufficient to claim that $\left(\widehat{M1} \right)$, $\left(\widehat{M2} \right)$, $\left(\widehat{T1} \right)$, *and* $\left(\widehat{T2} \right)$ *hold for this asset instead of (M1), (M2),* $\left(\widetilde{M2} \right)$, *(T1), and (T2).* $\square$

Definition 3.3 (Self-Financing Trading Strategy) *A trading strategy or portfolio process* $\varphi = (\varphi_0, ..., \varphi_n)$ *is called self-financing (in* $\mathcal{M}(Q)$*), if*

$$V(\varphi, t) = V(\varphi, 0) + G(\varphi, t) \text{ for all } t \in [0, T]$$

or

$$dV(\varphi, t) = dG(\varphi, t) = \varphi(t)' dP(t) = \sum_{i=0}^{n} \varphi_i(t) dP_i(t). \qquad (3.8)$$

If, for a moment of explanation, we look at a discrete version of (3.8), i.e.

$$\sum_{i=0}^{n} \varphi_i(t) \cdot (P_i(t + \Delta t) - P_i(t)) = V(\varphi, t + \Delta t) - V(\varphi, t)$$

$$= \sum_{i=0}^{n} \varphi_i(t + \Delta t) \cdot P_i(t + \Delta t) - \sum_{i=0}^{n} \varphi_i(t) \cdot P_i(t)$$

which is equivalent to

$$\sum_{i=0}^{n} \varphi_i(t + \Delta t) \cdot P_i(t + \Delta t) = \sum_{i=0}^{n} \varphi_i(t) \cdot P_i(t + \Delta t),$$

we see that a trading strategy is self-financing (in $\mathcal{M}(Q)$) if and only if the change in its value only depends on changes of the asset prices. For a self-financing trading strategy we can hold our portfolio for an instant time step and then, after the asset prices may have changed, restructure the portfolio consistent with our trading strategy without additional inflow of money. Note that for a self-financing trading strategy the price process $V(\varphi)$ is a (local) Q–martingale if and only if the gains process $G(\varphi)$ is a (local) Q–martingale.

Lemma 3.4 *Let* $\widetilde{V}(\varphi) = \left(\widetilde{V}(\varphi, t)\right)_{t \in [0,T]}$ *with*

$$\widetilde{V}(\varphi, t) := \sum_{i=0}^{n} \varphi_i(t) \cdot \widetilde{P}_i(t) = \varphi(t)' \widetilde{P}(t), \ t \in [0, T],$$

*be the **discounted price process** of the trading strategy* φ*. Then the trading strategy* φ *is self-financing in* $\mathcal{M}(Q)$ *if and only if* φ *is self-financing in the discounted (primary) financial market* $\widetilde{\mathcal{M}}(Q)$*, i.e.*

$$d\widetilde{V}(\varphi, t) = \varphi(t)' d\widetilde{P}(t) = \sum_{i=0}^{n} \varphi_i(t) d\widetilde{P}_i(t)$$

$$= \sum_{i=1}^{n} \varphi_i(t) d\widetilde{P}_i(t) \text{ for all } t \in [0, T].$$

In other words, a trading strategy remains self-financing after the change of numéraire. Especially, we have for all $t \in [0, T]$:

$$V(\varphi, t) = V(\varphi, 0) + \int_0^t \varphi(s)' dP(s)$$

if and only if

$$\widetilde{V}(\varphi, t) = V(\varphi, 0) + \int_0^t \varphi(s)' d\widetilde{P}(s)$$

and

$$V(\varphi, t) \geq 0 \quad \text{if and only if} \quad \widetilde{V}(\varphi, t) \geq 0.$$

Proof. By definition we have

$$\widetilde{V}(\varphi, t) = \varphi(t)' \widetilde{P}(t) = P_0^{-1}(t) \cdot \varphi(t)' P(t) = P_0^{-1}(t) \cdot V(\varphi, t)$$

for all $t \in [0, T]$. Thus, since $P_0^{-1} > 0$, $\widetilde{V}(\varphi) \geq 0$ if and only if $V(\varphi) \geq 0$. Now let $\varphi = (\varphi_0, ..., \varphi_n)$ be a self-financing trading strategy in $\mathcal{M}(Q)$. Then, using the product rule (Lemma 2.36), we get

$$
\begin{aligned}
d\widetilde{V}(\varphi, t) &= d\left(P_0^{-1}(t) \cdot V(\varphi, t)\right) = V(\varphi, t) dP_0^{-1}(t) + P_0^{-1}(t) \, dV(\varphi, t) \\
&= -r(t) \cdot P_0^{-1}(t) \cdot V(\varphi, t) dt + P_0^{-1}(t) \cdot \varphi(t)' dP(t) \text{ using (3.8)} \\
&= \varphi(t)' \left(P_0^{-1}(t) \, dP(t) - P(t) \cdot r(t) \cdot P_0^{-1}(t) \, dt\right) \\
&= \varphi(t)' d\widetilde{P}(t) \text{ using (3.4)},
\end{aligned}
$$

i.e. $\varphi = (\varphi_0, ..., \varphi_n)$ is a self-financing trading strategy in the discounted (primary) financial market $\widetilde{\mathcal{M}}(Q)$. The opposite direction is proved by similar arguments. $\qquad\square$

The following lemma shows that, if $(\varphi_1, ..., \varphi_n)$ is chosen according to the conditions $(T1)$ and $(T2)$, we can always extend $(\varphi_1, ..., \varphi_n)$ to a self-financing trading strategy $(\varphi_0, \varphi_1, ..., \varphi_n)$.

Lemma 3.5 *Let $(\varphi_1, ..., \varphi_n)$ be a $n-$dimensional $\lambda \otimes Q - a.s.$ bounded progressively measurable stochastic process on $[0, T]$. Then for any $v \in I\!R$, the stochastic process $\varphi = (\varphi_0, ..., \varphi_n)$ with*

$$\varphi_0(t) := v + \sum_{i=1}^{n} \int_0^t \varphi_i(s) \, d\widetilde{P}_i(s) - \sum_{i=1}^{n} \varphi_i(t) \cdot \widetilde{P}_i(t) \quad \text{for all } t \in [0, T]$$

is a self-financing trading strategy with an initial price of $V(\varphi, 0) = \widetilde{V}(\varphi, 0) = v$.

Proof. Since $\widetilde{P}_0(t) \equiv 1$, we have

$$
\begin{aligned}
\widetilde{V}(\varphi, t) &= \sum_{i=0}^{n} \varphi_i(t) \cdot \widetilde{P}_i(t) = \varphi_0(t) + \sum_{i=1}^{n} \varphi_i(t) \cdot \widetilde{P}_i(t) \\
&= v + \sum_{i=1}^{n} \int_0^t \varphi_i(s) \, d\widetilde{P}_i(s) \\
&= \widetilde{V}(\varphi, 0) + \sum_{i=1}^{n} \int_0^t \varphi_i(s) \, d\widetilde{P}_i(s)
\end{aligned}
$$

with $\widetilde{V}(\varphi, 0) = v$. Furthermore, $(T1)$ and $(T2)$ are satisfied for $i = 1, ..., n$, $j = 1, ..., m$. Hence, using Lemma 3.4, the trading strategy $\varphi = (\varphi_0, ..., \varphi_n)$ is self-financing with $V(\varphi, 0) = \widetilde{V}(\varphi, 0) = v$ if we can assure that condition $(T1)$ also holds for[3] $i = 0$. To show this, let $(\varphi_1, ..., \varphi_n)$ be $\lambda \otimes Q - a.s.$ bounded with $|\varphi_i(t)| \leq K_i < \infty \; \lambda \otimes Q - a.s., \; i = 1, ..., n$. Then, using $\left(\widetilde{M2}\right)$,

$$
E_Q\left[\int_0^T (\varphi_i(s) \cdot \widetilde{\sigma}_{ij}(s))^2 \, ds\right] \leq K_i^2 \cdot E_Q\left[\int_0^T \widetilde{\sigma}_{ij}^2(s) \, ds\right] < \infty. \qquad (3.9)
$$

Hence, using $\left(\widetilde{M1}\right)$, (3.9) and Theorem 2.24d) we get

$$
\begin{aligned}
\sup_{0 \leq t \leq T} \left| \int_0^t \varphi_i(s) \, d\widetilde{P}_i(s) \right| &= \sup_{0 \leq t \leq T} \left| \int_0^t \varphi_i(s) \cdot \widetilde{\mu}_i(s) \, ds \right. \\
&\quad \left. + \int_0^t \varphi_i(s) \cdot \widetilde{\sigma}_{ij}(s) \, dW(s) \right| \\
&\leq K_i \cdot \int_0^T |\widetilde{\mu}_i(s)| \, ds \\
&\quad + \sup_{0 \leq t \leq T} \left| \int_0^t \varphi_i(s) \cdot \widetilde{\sigma}_{ij}(s) \, dW(s) \right| \\
&< \infty \quad Q - a.s.
\end{aligned}
$$

---

[3] Note, that this is automatically true if $r \equiv 0$, i.e. $P_0 \equiv 1$ in $\mathcal{M}(Q)$. In this case, we can relax assumptions $(M2)$ and $\left(\widetilde{M2}\right)$ to assumption $(M2')$.

Together with $(M1)$ and (3.5) we conclude that

$$
\int_0^T |\varphi_0(t) \cdot r(t)| \, dt \quad \leq \quad \int_0^T |v \cdot r(t)| \, dt
$$

$$
+ \sum_{i=1}^n \int_0^T \left| r(t) \cdot \int_0^t \varphi_i(s) \, d\widetilde{P}_i(s) \right| dt
$$

$$
+ \sum_{i=1}^n \int_0^T \left| \varphi_i(t) \cdot \widetilde{P}_i(t) \cdot r(t) \right| dt
$$

$$
\leq \quad |v| \cdot \int_0^T |r(t)| \, dt
$$

$$
+ \sum_{i=1}^n \left(\sup_{0 \leq t \leq T} \left| \int_0^t \varphi_i(s) \, d\widetilde{P}_i(s) \right| \right) \cdot \int_0^T |r(t)| \, dt
$$

$$
+ \sum_{i=1}^n \left(K_i \cdot \sup_{0 \leq t \leq T} \left| \widetilde{P}_i(t) \right| \right) \cdot \int_0^T |r(t)| \, dt
$$

$$
< \quad \infty \quad Q - a.s.
$$

$\square$

Remark. *Note that for* $i \in \{1, ..., n\}$ *and* $j \in \{1, ..., m\}$ *we directly get from Theorem 2.24d) that*

$$
\sup_{0 \leq t \leq T} \left| \int_0^t \varphi_i(s) \cdot \widetilde{\sigma}_{ij}(s) \, dW(s) \right| < \infty \quad Q - a.s.
$$

if $\widetilde{\sigma}_{ij}$ *is bounded* $\lambda \otimes Q - a.s.$ *and if we assume that* $\varphi_i \in L_2[0, T]$. $\quad\square$

3.2   Absence of Arbitrage

An important step to the evaluation of a financial asset is the possibility of replicating it by a specific trading strategy. If we do this, we expect the financial market not to deliver us with a riskless profit or "free lunch". This assumption is closely related to the absence of arbitrage, which we will discuss in this section. However, there are natural restrictions to the set of all (self-financing) trading strategies which have to be considered in real life. Usually there is a limit to how much debt a bank or creditor will tolerate. In other words there must be a lower bound for the (discounted) price process of a portfolio or trading strategy.

Definition 3.6 (Admissible Trading Strategy) *A self-financing trading strategy* $\varphi = (\varphi_0, ..., \varphi_n)$ *is called admissible (in* $\mathcal{M}(Q)$ *and* $\widetilde{\mathcal{M}}(Q)$*) if* *the corresponding price processes* $V(\varphi) = (V(\varphi, t))_{t \in [0,T]}$ *and* $\widetilde{V}(\varphi) = \left(\widetilde{V}(\varphi, t) \right)_{t \in [0,T]}$ *(in* $\mathcal{M}(Q)$ *and* $\widetilde{\mathcal{M}}(Q)$*) are* $\lambda \otimes Q - a.s.$ *bounded below, i.e. there are finite real numbers* $K = K(\varphi)$ *and* $\widetilde{K} = \widetilde{K}(\varphi)$ *such that*

$$V(\varphi) \geq K \text{ and } \widetilde{V}(\varphi) \geq \widetilde{K} \quad \lambda \otimes Q - a.s.$$

The set of all admissible trading strategies (in $\mathcal{M}(Q)$ *and* $\widetilde{\mathcal{M}}(Q)$*) is denoted by* $\Phi(Q)$*. Furthermore, the set of all* $\varphi \in \Phi(Q)$ *which are* $\lambda \otimes Q - a.s.$ *bounded is denoted by* $\Phi^*(Q)$*.*

Using Lemma 3.4 we know that we may either look at the original or the discounted price process if we work with admissible trading strategies. Note that if we choose $K \equiv \widetilde{K} \equiv 0$, or if r is $\lambda \otimes Q - a.s.$ bounded, each trading strategy φ with $V(\varphi) \geq K \in \mathbb{R}$ $\lambda \otimes Q - a.s.$ also satisfies the inequality $\widetilde{V}(\varphi) \geq \widetilde{K}$ $\lambda \otimes Q - a.s.$ for a suitable $\widetilde{K} \in \mathbb{R}$ and vice versa.

Definition 3.7 (Arbitrage Opportunity) *An admissible trading strategy* $\varphi = (\varphi_0, ..., \varphi_n)$ *is called an arbitrage opportunity (in* $\mathcal{M}(Q)$*) if the price process* $V(\varphi)$ *of the trading strategy satisfies the following conditions:*

(i) $V(\varphi, 0) = 0$ $Q - a.s.$,

(ii) $V(\varphi, T) \geq 0$ $Q - a.s.$ *and* $Q(\{\omega \in \Omega : V(\varphi, T, \omega) > 0\}) > 0$.

The (primary) financial market $\mathcal{M}(Q)$ *is called* **arbitrage-free** *if there is no arbitrage opportunity in* $\Phi(Q)$*.*

Since $P_0^{-1}(t) > 0$ for all $t \in [0, T]$ we know that $V(\varphi, T) > 0$ if and only if $\widetilde{V}(\varphi, T) > 0$ and thus, using Lemma 3.4, the trading strategy φ is an arbitrage opportunity in $\mathcal{M}(Q)$ if and only if the conditions (i) and (ii) of Definition 3.7 are satisfied for the discounted price process $\widetilde{V}(\varphi)$, i.e. φ is an arbitrage opportunity in $\widetilde{\mathcal{M}}(Q)$. Furthermore, it can be shown that φ is an arbitrage opportunity in $\widetilde{\mathcal{M}}(Q)$ if and only if

$$\widetilde{V}(\varphi, T) \geq \widetilde{V}(\varphi, 0) \quad Q-a.s. \text{ and } Q\left(\left\{ \omega \in \Omega : \widetilde{V}(\varphi, T, \omega) > \widetilde{V}(\varphi, 0, \omega) \right\} \right) > 0$$

(see Øksendal [Øks98], p. 284 for more details).

Definition 3.8 (Equivalent Martingale Measure) *A probability measure* $\widetilde{Q}$ *on* $(\Omega, \mathcal{F})$ *is called an equivalent martingale measure to* Q *if:*

(i) $\widetilde{Q}$ *is equivalent to* Q*.*

(ii) The discounted price process $\widetilde{P} = \left(\widetilde{P}_1(t), ..., \widetilde{P}_n(t) \right)_{t \in [0,T]}$ *is an* n*-dimensional* $\widetilde{Q}$*-martingale.*

The set of equivalent martingale measures to Q is denoted by $I\!M = I\!M(Q)$.

Using the martingale representation Theorem 2.38 we can obtain the following lemma (see, e.g. Bingham and Kiesel [BK98], p. 159, Korn and Korn [KK99], Proposition 31, or Musiela and Rutkowski [MR97], Proposition B.2.1) which shows that all equivalent martingale measures are of the same structure.

Lemma 3.9 (Equivalent Martingale Measure Characterization)
Let $\widetilde{Q} \in I\!M$. *Then there is a* $m-$*dimensional progressively measurable stochastic process* γ *with* $\int_0^T \|\gamma(s)\|^2 \, ds < \infty$ $Q-a.s.$ *such that*

$$\widetilde{Q}(A) = Q_{L(\gamma,T)}(A) = E_Q[1_A \cdot L(\gamma,T)] \quad \text{for all} \ \ A \in \mathcal{F}_T$$

with

$$L(\gamma,T) := e^{-\int_0^T \gamma(s)'dW(s) - \frac{1}{2}\int_0^T \|\gamma(s)\|^2 ds}.$$

Another interesting result showing the relation between the discounted price process and the corresponding equivalent martingale measure is summarized in the following lemma.

Lemma 3.10

a) *Let* $\widetilde{Q} \in I\!M$ *and let the martingale condition*[4]

$$E_{\widetilde{Q}}\left[\widetilde{P}_i^2(t)\right] < \infty \ \ \text{for all} \ t \in [0,T] , \ i = 1,...,n \qquad (\widetilde{M3})$$

be satisfied. Then the discounted price process $\widetilde{V}(\varphi)$ *is a* $\widetilde{Q}-$*martingale for all* $\varphi \in \Phi^*(Q)$.

b) *Let the discounted price process* $\widetilde{V}(\varphi)$ *be a* $\widetilde{Q}-$*martingale for all* $\varphi \in \Phi^*(Q)$ *and for a* $\widetilde{Q}$ *which is equivalent to* Q. *Then* $\widetilde{Q} \in I\!M$.

Proof. Using Lemma 3.4 we know that each $\varphi \in \Phi(Q)$ is also admissible in the discounted market and vice versa. Especially,

$$\widetilde{V}(\varphi,t) = \widetilde{V}(\varphi,0) + \int_0^t \varphi(s)' \, d\widetilde{P}(s) = V(\varphi,0) + \sum_{i=1}^n \int_0^t \varphi_i(s) d\widetilde{P}_i(s)$$

for all $t \in [0,T]$.

---

[4]It should be mentioned that the martingale condition $\left(\widetilde{M3}\right)$ can be relaxed with $\widetilde{V}(\varphi)$ becoming a local $\widetilde{Q}-$martingale if we allow for locally bounded trading strategies $\varphi \in \Phi(Q)$. See, e.g. Musiela and Rutkowski [MR97], p. 231ff., Protter [Pro92], p. 34ff., or Bingham and Kiesel [BK98], p. 159ff., for more details.

a) Let $\widetilde{Q} \in I\!M$, $\left(\widetilde{M3}\right)$ be satisfied, and $\varphi \in \Phi^*(Q)$ with

$$|\varphi_i(t)| \leq K_i < \infty \quad \lambda \otimes Q - a.s. \text{ for all } i = 1, ..., n.$$

Then, since $\widetilde{P}_i$ is a $\widetilde{Q}$−martingale for all $i = 1, ..., n$ we know by Theorem 2.38 that there are progressively measurable stochastic processes $\phi_i = (\phi_i(t))_{t \in [0,T]}$, $\phi_i : [0,T] \times \Omega \to I\!R^m$, $i = 1, ..., n$, such that $E_{\widetilde{Q}} \left[\int_0^T \|\phi_i(t)\|^2 \, dt \right] < \infty$ and

$$d\widetilde{P}_i(t) = \phi_i(t)' \, d\widetilde{W}(t) \quad \widetilde{Q} - a.s. \text{ for all } t \in [0,T], \, i = 1, ..., n,$$

where $\widetilde{W}$ is a Wiener process under $\widetilde{Q}$. Thus,

$$\widetilde{V}(\varphi, t) = V(\varphi, 0) + \sum_{i=1}^n \int_0^t \varphi_i(s) \cdot \phi_i(s)' \, d\widetilde{W}(s) \text{ for all } t \in [0,T]$$

with

$$E_{\widetilde{Q}} \left[\int_0^T \|\varphi_i(t) \cdot \phi_i(t)\|^2 \, dt \right] = E_{\widetilde{Q}} \left[\int_0^T |\varphi_i(t)|^2 \cdot \|\phi_i(t)\|^2 \, dt \right]$$

$$\leq K_i^2 \cdot E_{\widetilde{Q}} \left[\int_0^T \|\phi_i(t)\|^2 \, dt \right] < \infty$$

since $\widetilde{Q} \sim Q$. Hence, $\widetilde{V}(\varphi, t)$ is a $\widetilde{Q}$−martingale for all $\varphi \in \Phi^*(Q)$.

b) On the other hand, let $\widetilde{V}(\varphi)$ be a $\widetilde{Q}$−martingale for all $\varphi \in \Phi^*(Q)$. Obviously, for each $i \in \{1, ..., n\}$, the trading strategy $\varphi^i = (\varphi_0^i, ..., \varphi_n^i)$ with

$$\varphi_j^i(t) = 1_{A_{ij}}, \quad A_{ij} = \Omega \text{ for } j = i \text{ and } A_{ij} = \emptyset \text{ for } j \neq i, t \in [0,T],$$

$j = 0, ..., n$, is progressively measurable and bounded. Furthermore,

$$\widetilde{V}(\varphi^i, t) = \sum_{j=0}^n \varphi_j^i(t) \cdot \widetilde{P}_j(t) = \widetilde{P}_i(t) \geq 0, \ t \in [0,T],$$

and therefore $\varphi^i \in \Phi^*(Q)$, $i = 1, ..., n$. Since $\widetilde{V}(\varphi^i)$ is a $\widetilde{Q}$−martingale, so is $\widetilde{P}_i(t) = \widetilde{V}(\varphi^i, t)$ for all $i = 1, ..., n$, i.e. $\widetilde{Q} \in I\!M$. □

We now show that the existence of an equivalent martingale measure is sufficient for a financial market to be free of arbitrage. A little more precisely, we get the following theorem.

Theorem 3.11 (No-Arbitrage) *Let* $\widetilde{Q} \in \mathbb{M} \neq \emptyset$ *and* $\left(\widetilde{M3}\right)$ *be satisfied. Then the financial market model* $\mathcal{M}(Q)$ *contains no arbitrage opportunities in* $\Phi^*(Q)$.

Proof. Let $\widetilde{Q} \in \mathbb{M}$ be an equivalent martingale measure and $\varphi \in \Phi^*(Q)$. Then using Lemma 3.10a), $\widetilde{V}(\varphi)$ is a $\widetilde{Q}$−martingale because of $\left(\widetilde{M3}\right)$, i.e.

$$E_{\widetilde{Q}}\left[\widetilde{V}(\varphi,t)|\mathcal{F}_s\right] = \widetilde{V}(\varphi,s) \quad \widetilde{Q} - a.s. \text{ for all } 0 \leq s \leq t \leq T. \qquad (3.10)$$

Now suppose that $\widetilde{V}(\varphi,0) = 0$. Using (3.10) with $s := 0$, we get, $\widetilde{Q} - a.s.$,

$$E_{\widetilde{Q}}\left[\widetilde{V}(\varphi,t)\right] = \widetilde{V}(\varphi,0) = 0 \quad \text{for all } 0 \leq t \leq T. \qquad (3.11)$$

Furthermore, we claim that $\widetilde{V}(\varphi,T) \geq 0$. Combining this with (3.11), we see that

$$\widetilde{V}(\varphi,T) = 0 \quad \widetilde{Q} - a.s.,$$

and so, since $\widetilde{Q}$ and Q are equivalent,

$$\widetilde{V}(\varphi,T) = 0 \quad Q - a.s.$$

Consequently, there is no arbitrage opportunity in $\Phi^*(Q)$. □

Remark. *Since* $\widetilde{V}(\varphi)$ *is a local* $\widetilde{Q}$−*martingale for all* $\varphi \in \Phi(Q)$, *we know by Theorem 2.24b) that* $\widetilde{V}(\varphi)$ *is a* $\widetilde{Q}$−*supermartingale if* $\widetilde{V}(\varphi)$ *is non-negative. If we therefore choose* $K \equiv \widetilde{K} \equiv 0$ *in Definition 3.6, we can show that there are no arbitrage opportunities in* $\Phi(Q)$. *To see this, we simply substitute equation (3.10) in the proof of Theorem 3.11 by*

$$E_{\widetilde{Q}}\left[\widetilde{V}(\varphi,t)|\mathcal{F}_s\right] \leq \widetilde{V}(\varphi,s) \quad \widetilde{Q} - a.s. \text{ for all } 0 \leq s \leq t \leq T.$$

□

The following lemma gives a characterization of the discounted (primary) financial market.

Lemma 3.12 (Discounted Market Characterization) *Suppose that there exists a* $m-$*dimensional progressively measurable stochastic process* γ *such that the no-arbitrage condition*

$$\mu_i(t) - \sigma_i(t) \cdot \gamma(t) = r(t) \cdot P_i(t) \quad \lambda \otimes Q - a.s. \text{ on } [0,T], \qquad \text{(NA)}$$

$i = 1, ..., n$, *and the Novikov condition*

$$E_Q\left[e^{\frac{1}{2}\int_0^T \|\gamma(s)\|^2 \, ds}\right] < \infty \qquad \text{(NV)}$$

hold. Furthermore, let the probability measure $\widetilde{Q}$ on $(\Omega, \mathcal{F}_T)$ be defined as in Lemma 3.9. Then the stochastic process $\widetilde{W} = \left(\widetilde{W}(t)\right)_{t \in [0,T]}$, defined by

$$d\widetilde{W}(t) := \gamma(t)\, dt + dW(t) \quad on \; [0,T]\,,$$

is a $\widetilde{Q}-$ Wiener process and the discounted price processes $\widetilde{P}_i$, $i = 0, ..., n$, have the following representation in terms of $\widetilde{W}$:

$$d\widetilde{P}_0(t) = 0 \;\; and \;\; d\widetilde{P}_i(t) = \widetilde{\sigma}_i(t)\, d\widetilde{W}(t)\,, \; i = 1, ..., n.$$

Furthermore,

$$dP_i(t) = r(t) \cdot P_i(t)\, dt + \sigma_i(t)\, d\widetilde{W}(t)\,.$$

If the martingale condition[5]

$$E_{\widetilde{Q}}\left[\int_0^T \widetilde{\sigma}_{ij}^2(t)\, dt\right] < \infty \quad for \; all \; i = 1, ..., n, \; j = 1, ..., m \qquad (\widetilde{M3}')$$

is satisfied, then $\widetilde{Q}$ is an equivalent martingale measure, i.e. $\widetilde{Q} \in M$.

Proof. The first statement follows by the Girsanov Theorem 2.41. Using (3.4) we get for all $t \in [0,T]$, $i = 1, ..., n$:

$$
\begin{aligned}
d\widetilde{P}_i(t) &= P_0^{-1}(t) \cdot [(\mu_i(t) - r(t) \cdot P_i(t))\, dt + \sigma_i(t) dW(t)] \\
&= P_0^{-1}(t) \cdot [(\mu_i(t) - r(t) \cdot P_i(t))\, dt \\
&\qquad + \sigma_i(t) \cdot (-\gamma(t)\, dt + d\widetilde{W}(t))] \\
&= P_0^{-1}(t) \cdot [(\mu_i(t) - \sigma_i(t) \cdot \gamma(t) - r(t) \cdot P_i(t))\, dt + \sigma_i(t) d\widetilde{W}(t)] \\
&= P_0^{-1}(t) \cdot \sigma_i(t) d\widetilde{W}(t) \;\; \text{using (NA)} \\
&= \widetilde{\sigma}_i(t)\, d\widetilde{W}(t)\,.
\end{aligned}
$$

---

[5] If $d\widetilde{P}_i(t) = \widetilde{\sigma}_i(t)\, d\widetilde{W}(t)$, $i \in \{1, ..., n\}$, with $E\left[\int_0^T \widetilde{\sigma}_{ij}^2(s)\, ds\right] < \infty \;\; \widetilde{Q} - a.s.$ for all $j = 1, ..., m$, we know that with $I_t(\widetilde{\sigma}_{ij}) := \int_0^t \widetilde{\sigma}_{ij}(s)\, d\widetilde{W}_j(s)$, $t \in [0,T]$, we have

$$
\begin{aligned}
E_{\widetilde{Q}}\left[\widetilde{P}_i^2(t)\right] &= E_{\widetilde{Q}}\left[\sum_{j=1}^m \int_0^t \widetilde{\sigma}_{ij}(s)\, d\widetilde{W}_j(s)\right]^2 \\
&= E_{\widetilde{Q}}\left[\sum_{j=1}^m I_t(\widetilde{\sigma}_{ij})\right]^2 = \left\|\sum_{j=1}^m I(\widetilde{\sigma}_{ij})\right\|_{L_t}^2 \\
&\leq \sum_{j=1}^m \|I(\widetilde{\sigma}_{ij})\|_{L_t}^2 = \sum_{j=1}^m E_{\widetilde{Q}}\left[\int_0^t \widetilde{\sigma}_{ij}^2(s)\, ds\right] < \infty
\end{aligned}
$$

because of the Itô isometry (see Theorem 2.29b). Therefore, under (NA) and (NV), condition $\left(\widetilde{M3}'\right)$ implies condition $\left(\widetilde{M3}\right)$.

Hence, because of $\left(\widetilde{M3}'\right)$, the stochastic processes $\widetilde{P}_i = \left(\widetilde{P}_i\left(t\right)\right)_{t \in [0,T]}$ are continuous $\widetilde{Q}$-martingales, $i = 1, ..., n$, i.e. $\widetilde{Q} \in I\!M$. Furthemore, using the product rule (Lemma 2.36), we get

$$
\begin{aligned}
dP_i\left(t\right) &= d\left(P_0\left(t\right) \cdot \widetilde{P}_i\left(t\right)\right) = P_0\left(t\right) d\widetilde{P}_i\left(t\right) + \widetilde{P}_i\left(t\right) dP_0\left(t\right) \\
&= P_0\left(t\right) \cdot \widetilde{\sigma}_i\left(t\right) d\widetilde{W}\left(t\right) + \widetilde{P}_i\left(t\right) \cdot r\left(t\right) \cdot P_0\left(t\right) dt \\
&= r\left(t\right) \cdot P_i\left(t\right) dt + \sigma_i\left(t\right) d\widetilde{W}\left(t\right).
\end{aligned}
$$

$\square$

A very useful result which is sufficient for our applications is stated in the following corollary.

Corollary 3.13 *Suppose that there exists a m-dimensional progressively measurable stochastic process γ such that (NA) and (NV) hold. Furthermore, let $\left(\widetilde{M3}'\right)$ be satisfied. Then the financial market $\mathcal{M}\left(Q\right)$ contains no arbitrage opportunities in $\Phi^*\left(Q\right)$.*

Proof. The existence of an equivalent martingale measure $\widetilde{Q} \in I\!M$ is an immediate consequence of Lemma 3.12. Using Theorem 3.11 we conclude that the financial market $\mathcal{M}\left(Q\right)$ contains no arbitrage opportunities in $\Phi^*\left(Q\right)$. $\square$

The inverse conclusion to Theorem 3.11, i.e. if the financial market $\mathcal{M}\left(Q\right)$ contains no arbitrage opportunities then $I\!M \neq \emptyset$, is only true under additional assumptions and is beyond the scope of this book. The interested reader may refer to Delbaen and Schachermayer [DS94a]. Leaving aside some technical assumptions we quote one of their results to give an impression of the developments in this area (see also Bingham and Kiesel [BK98], p. 176-177). A trading strategy φ is called *simple* if it can be represented by a finite linear combination of stochastic processes of the form $\psi \cdot 1_{(\tau_1, \tau_2]}$ where τ_1 and τ_2 are stopping times and ψ is a $\mathcal{F}_{\tau_1}$-measurable random variable. A simple trading strategy φ is said to be δ-*admissible* if the corresponding price process satisfies $V\left(\varphi, t\right) \geq -\delta$ for all $t \in [0, T]$. Then the following theorem can be shown.

Theorem 3.14 (Fundamental Theorem of Asset Pricing) $I\!M \neq \emptyset$ *if and only if for any sequence $\left(\varphi^n : n \in I\!N\right)$ of simple trading strategies such that φ^n is δ_n-admissible and δ_n tends to zero we have*

$$
\lim_{n \to \infty} Q\left(\{\omega \in \Omega : |V\left(\varphi^n, T, \omega\right)| > \varepsilon\}\right) = 0 \text{ for each } \varepsilon > 0,
$$

i.e. $V\left(\varphi^n, T\right) \to 0$ in probability as $n \to \infty$. This condition is known as the no free lunch with vanishing risk (NFLVR) condition.

3.3   Market Completeness

Roughly speaking, the absence of arbitrage ensures that we can not find two self-financing trading strategies which start at the same initial portfolio value and end up with different values at time T. Another important question for the pricing of financial assets is whether there even exists a self-financing trading strategy paying the same cash flows as the financial asset. To answer this question we focus on those assets which are characterized by the following defintion.

Definition 3.15 (Contingent Claim) *We call a random variable $D = D(T)$ on $(\Omega, \mathcal{F}_T)$ a (European) contingent claim (with maturity T) if $P_0(t) \cdot \widetilde{D}(T)$ is lower bounded for all $t \in [0, T]$.*

Note that $P_0(t) \cdot \widetilde{D}(T)$ is lower bounded for any lower bounded random variable $D = D(T)$ on $(\Omega, \mathcal{F}_T)$ if r is $\lambda \otimes Q - a.s.$ bounded. It should also be noted that the conditions for (lower) boundedness in Definitions 3.15 and 3.17 are made for technical convenience and could be relaxed by other related definitions. To give an example, if we had defined admissible trading strategies for $K \equiv \widetilde{K} \equiv 0$, we could define a contingent claim simply to be a non-negative random variable on $(\Omega, \mathcal{F}_T)$. For more details and other definitions see, e.g., Bingham and Kiesel [BK98], p. 177. We can now define what is considered to be a hedging strategy.

Definition 3.16 (Replicating Strategy) *A replicating or hedging strategy for a contingent claim $D = D(T)$ with maturity T in $\mathcal{M}(Q)$ is an admissible trading strategy φ with*

$$V(\varphi, T) = D(T).$$

*A contingent claim $D = D(T)$ with maturity T is called $(\Phi^* -)$ **attainable** in $\mathcal{M}(Q)$ if there is a replicating strategy φ $(\in \Phi^*(Q))$ for D in $\mathcal{M}(Q)$.*

Using Lemma 3.4, we know that a trading strategy φ is a replicating strategy for a contingent claim $D = D(T)$ with maturity T in $\mathcal{M}(Q)$ if and only if φ is a replicating strategy for the discounted contingent claim $\widetilde{D}(T)$ in $\widetilde{\mathcal{M}}(Q)$. Furthermore, $D(T)$ is attainable in $\mathcal{M}(Q)$ if and only if $\widetilde{D}(T)$ is attainable in $\widetilde{\mathcal{M}}(Q)$.

Definition 3.17 (Completeness) *The financial market $\mathcal{M}(Q)$ $\left[\widetilde{\mathcal{M}}(Q)\right]$ is said to be $(\Phi^* -)$ **complete** if any contingent claim $D(T)$ $\left[\widetilde{D}(T)\right]$ with maturity T such that $\widetilde{D}(T)$ is bounded, is $(\Phi^* -)$ attainable in $\mathcal{M}(Q)$ $\left[\widetilde{\mathcal{M}}(Q)\right]$.*

Hence, $\mathcal{M}(Q)$ is complete if and only if $\widetilde{\mathcal{M}}(Q)$ is complete. As an example for an attainable claim notice that the stochastic process

$\varphi^i = \left(\varphi_0^i, ..., \varphi_n^i\right)$, $i \in \{0, ..., n\}$, with

$$\varphi_j^i(t) = 1_{A_{ij}}, \ A_{ij} = \Omega \text{ for } j = i \text{ and } A_{ij} = \emptyset \text{ for } j \neq i, t \in [0, T],$$

$j = 0, ..., n$, is progressively measurable and bounded, i.e. φ^i is a bounded trading strategy in $\mathcal{M}(Q)$. Furthermore, φ^i, $i = 0, ..., n$, satisfies

$$V(\varphi^i, t) = \sum_{j=1}^{n} \varphi_j^i(t) \cdot P_j(t) = P_i(t) \geq 0 \text{ for each } t \in [0, T].$$

So $\varphi^i \in \Phi^*(Q)$ is a replicating strategy for P_i in $\mathcal{M}(Q)$, $i = 0, ..., n$, which shows that each primary traded asset is (Φ^*-) attainable in $\mathcal{M}(Q)$.

Lemma 3.18 *Let $\widetilde{Q} \in I\!M$ and $D = D(T)$ be a contingent claim with maturity T. Suppose there exists an $n-$dimensional $\lambda \otimes Q - a.s.$ bounded progressively measurable stochastic process $(\varphi_1, ..., \varphi_n)$ on $[0, T]$ and $v \in I\!R$ such that for all $t \in [0, T]$*

$$E_{\widetilde{Q}}\left[\widetilde{D}(T)|\mathcal{F}_t\right] = v + \sum_{i=1}^{n} \int_0^t \varphi_i(s) \, d\widetilde{P}_i(s).$$

Then the contingent claim D is attainable in $\mathcal{M}(Q)$, and the discounted price process $\widetilde{V}(\varphi)$ of the corresponding trading strategy $\varphi = (\varphi_0, ..., \varphi_n)$ is a $\widetilde{Q}-$martingale.

Proof. Let $(\varphi_1, ..., \varphi_n)$ be given as in the assumption. Then, using Lemma 3.5, $(\varphi_1, ..., \varphi_n)$ can be extended to a self-financing trading strategy $\varphi = (\varphi_0, \varphi_1, ..., \varphi_n)$ such that

$$\widetilde{V}(\varphi, t) = v + \sum_{i=1}^{n} \int_0^t \varphi_i(s) \, d\widetilde{P}_i(s) = E_{\widetilde{Q}}\left[\widetilde{D}(T)|\mathcal{F}_t\right]$$

for all $t \in [0, T]$. Thus, since

$$V(\varphi, t) = P_0(t) \cdot \widetilde{V}(\varphi, t) = E_{\widetilde{Q}}\left[P_0(t) \cdot \widetilde{D}(T)|\mathcal{F}_t\right]$$

and $P_0(t) \cdot \widetilde{D}(T)$ is lower bounded for all $t \in [0, T]$, φ is an admissible trading strategy in $\mathcal{M}(Q)$, i.e. $\varphi \in \Phi(Q)$. Furthermore,

$$V(\varphi, T) = P_0(T) \cdot \widetilde{V}(\varphi, T) = P_0(T) \cdot E_{\widetilde{Q}}\left[\widetilde{D}(T)|\mathcal{F}_T\right] = D(T).$$

Hence, φ is a replicating strategy for D in $\mathcal{M}(Q)$, i.e. D is attainable in $\mathcal{M}(Q)$. By Lemma 2.17, $\left(E_{\widetilde{Q}}\left[\widetilde{D}(T)|\mathcal{F}_t\right]\right)_{t \in [0, T]}$ is a $\widetilde{Q}-$martingale and so is $\widetilde{V}(\varphi)$. □

The following theorem is an important step to the characterization of market completeness. More precisely, it shows the relation between market completeness and the invertibility of the matrix σ.

Theorem 3.19 *Suppose that there exists a m-dimensional progressively measurable stochastic process γ such that the conditions (NA) and (NV) hold. Furthermore, let $\tilde{Q}$ be defined as in Theorem 3.12.*

a) *Let $\tilde{\sigma}$ have a bounded left inverse $\tilde{\sigma}^{-1}$ $\lambda \otimes Q - a.s.$, i.e. there exists an adapted matrix-valued stochastic process σ^{-1} on $[0,T]$ with values in $\mathbb{R}^{m \times n}$ such that*

$$\sigma^{-1}(t,\omega) \cdot \sigma(t,\omega) = I_m \quad \lambda \otimes Q - a.s.$$

and each matrix element of $\tilde{\sigma}^{-1} = \sigma^{-1} \cdot P_0$ is bounded $\lambda \otimes Q - a.s.$ Then the financial market $\mathcal{M}(Q)$ is complete.

b) *Let $\mathcal{M}(Q)$ be complete. Then σ has a left inverse $\lambda \otimes Q - a.s.$ In particular,*

$$rank\,(\sigma(t,\omega)) = m \quad \lambda \otimes Q - a.s.,$$

i.e. $n \geq m$. Moreover, the m-dimensional progressively measurable stochastic process γ is unique ($\lambda \otimes Q - a.s.$). If $n = m$, in addition to the completeness of $\mathcal{M}(Q)$, then $\sigma(t,\omega)$ is invertible $\lambda \otimes Q - a.s.$

Proof. Let $D(T)$ be a contingent claim with maturity T such that $\tilde{D}(T)$ is bounded, i.e. $\left|\tilde{D}(T,\omega)\right| \leq K$ for all $\omega \in \Omega$. Since $\left(E_{\tilde{Q}}\left[\tilde{D}(T)|\mathcal{F}_t\right]\right)_{t \in [0,T]}$ is a bounded $\tilde{Q}$-martingale, we know by Theorem 2.38 that there is a progressively measurable stochastic process $\phi = (\phi(t))_{t \in [0,T]}$, $\phi : [0,T] \times \Omega \to \mathbb{R}^m$, such that $E_{\tilde{Q}}\left[\int_0^T \|\phi(t)\|^2\,dt\right] < \infty$ and

$$E_{\tilde{Q}}\left[\tilde{D}(T)|\mathcal{F}_t\right] = E_{\tilde{Q}}\left[\tilde{D}(T)\right] + \int_0^t \phi(s)'\,d\widetilde{W}(s) \quad \tilde{Q} - a.s. \tag{3.12}$$

for all $t \in [0,T]$, where $\widetilde{W}$ is a Wiener process under $\tilde{Q}$. Especially,

$$\tilde{D}(T) = E_{\tilde{Q}}\left[\tilde{D}(T)|\mathcal{F}_T\right] = E_{\tilde{Q}}\left[\tilde{D}(T)\right] + \int_0^T \phi(s)'\,d\widetilde{W}(s) \quad \tilde{Q} - a.s.$$

Using the Itô isometry we get, because $\tilde{D}(T)$ is bounded,

$$E_{\tilde{Q}}\left[\int_0^T \|\phi(t)\|^2\,dt\right] = E_{\tilde{Q}}\left[\tilde{D}^2(T)\right] - \left(E_{\tilde{Q}}\left[\tilde{D}(T)\right]\right)^2 \leq K^2.$$

Since $Q \sim \tilde{Q}$, ϕ is bounded $\lambda \otimes Q - a.s.$ Now let

$$(\varphi_1(t), ..., \varphi_n(t)) := \phi(t)'\tilde{\sigma}^{-1}(t) \text{ for all } t \in [0,T].$$

Because ϕ and $\tilde{\sigma}^{-1}$ are bounded $\lambda \otimes Q - a.s.$, $(\varphi_1, ..., \varphi_n)$ is also $\lambda \otimes Q - a.s.$ bounded with, using Lemma 3.12,

$$\sum_{i=1}^{n} \int_0^t \varphi_i(s)\, d\tilde{P}_i(s) = \sum_{i=1}^{n} \int_0^t \varphi(s)'\, d\tilde{P}(s) = \sum_{i=1}^{n} \int_0^t \varphi(s)'\, \tilde{\sigma}(t)\, d\tilde{W}(s)$$

$$= \int_0^t \phi(s)'\, d\tilde{W}(s) \quad \tilde{Q} - a.s.$$

In combination with equation (3.12), we thus get

$$E_{\tilde{Q}}\left[\tilde{D}(T)\,|\mathcal{F}_t\right] = E_{\tilde{Q}}\left[\tilde{D}(T)\right] + \sum_{i=1}^{n} \int_0^t \varphi_i(s)\, d\tilde{P}_i(s)$$

for all $t \in [0, T]$. So using Lemma 3.18, the contingent claim $D(T)$ is attainable in $\mathcal{M}(Q)$, i.e. the financial market $\mathcal{M}(Q)$ is complete. For the proof of part b) see, e.g., Øksendal [Øks98], p. 263-264.   ☐

From the proof of Theorem 3.19 we can directly extract the following corollary.

Corollary 3.20 *Let the assumptions of Theorem 3.19 be satisfied and suppose that σ has a left inverse $\lambda \otimes Q - a.s.$ Then for any random variable $\tilde{D}(T)$ on $(\Omega, \mathcal{F}_T)$ with $E_{\tilde{Q}}\left[\tilde{D}^2(T)\right] < \infty$, there exists an $(n+1)-$ dimensional progressively measurable stochastic process $\varphi = (\varphi_0, ..., \varphi_n)$ $= (\varphi_0(t), ..., \varphi_n(t))_{t\in[0,T]}$ such that for all $t \in [0, T]$*

$$E_{\tilde{Q}}\left[\int_0^T \|\varphi(t)'\tilde{\sigma}(t)\|^2\, dt\right] < \infty^6, \tag{3.12}$$

---

[6] In the case that $\|\varphi\| = \sqrt{\sum_{i=1}^n \varphi_i^2(t)} \le K_\varphi$ $Q-a.s.$ for a fixed $K_\varphi \in [0, \infty)$, we get

$$E_{\tilde{Q}}\left[\int_0^T \|\varphi(t)'\tilde{\sigma}(t)\|^2\, dt\right] = \sum_{j=1}^{m} E_{\tilde{Q}}\left[\int_0^T \left(\varphi(t)'\tilde{\sigma}(t)\right)_j^2\, dt\right]$$

$$= \sum_{j=1}^{m} E_{\tilde{Q}}\left[\int_0^T \left(\sum_{i=1}^{n} \varphi_i(t) \cdot \tilde{\sigma}_{ij}(t)\right)^2\, dt\right]$$

$$\le \sum_{j=1}^{m} E_{\tilde{Q}}\left[\int_0^T \left(\sum_{i=1}^{n} \varphi_i^2(t)\right) \cdot \left(\sum_{i=1}^{n} \tilde{\sigma}_{ij}^2(t)\right)\, dt\right]$$

$$\le K_\varphi^2 \cdot \sum_{j=1}^{m}\sum_{i=1}^{n} E_{\tilde{Q}}\left[\int_0^T \tilde{\sigma}_{ij}^2(t)\, dt\right]$$

$$< \infty \text{ if condition } \left(\widetilde{M3}'\right) \text{ is satisfied.}$$

and

$$\tilde{V}(\varphi,t) = E_{\tilde{Q}}\left[\tilde{D}(T)\right] + \int_0^t \varphi(s)' \, d\tilde{P}(s)$$

$$= E_{\tilde{Q}}\left[\tilde{D}(T)|\mathcal{F}_t\right] \quad \text{for all } t \in [0,T].$$

Especially, $\left(\tilde{V}(\varphi,t)\right)_{t\in[0,T]}$ *is a* $\tilde{Q}$*-martingale.*

Proof. Let $\tilde{D}(T)$ be a random variable on $(\Omega,\mathcal{F}_T)$ with $E_{\tilde{Q}}\left[\tilde{D}^2(T)\right] < \infty$. Furthermore, let

$$(M(t))_{t\in[0,T]} := \left(E_{\tilde{Q}}\left[\tilde{D}(T)|\mathcal{F}_t\right]\right)_{t\in[0,T]}.$$

Then, since[7]

$$M^2(t) = \left(E_{\tilde{Q}}\left[\tilde{D}(T)|\mathcal{F}_t\right]\right)^2 \leq E_{\tilde{Q}}\left[\tilde{D}^2(T)|\mathcal{F}_t\right] \quad \tilde{Q} - a.s.$$

for all $t \in [0,T]$, we know that

$$E_{\tilde{Q}}\left[M^2(t)\right] \leq E_{\tilde{Q}}\left[E_{\tilde{Q}}\left[\tilde{D}^2(T)|\mathcal{F}_t\right]\right] = E_{\tilde{Q}}\left[\tilde{D}^2(T)\right] < \infty$$

for all $t \in [0,T]$. Using the Cauchy-Schwarz inequality, we conclude that

$$E_{\tilde{Q}}\left[|M(t)|\right] \leq \sqrt{E_{\tilde{Q}}\left[M^2(t)\right]} < \infty,$$

i.e. $(M(t))_{t\in[0,T]}$ is a $\tilde{Q}$-martingale. So following Theorem 2.38, there is a progressively measurable stochastic process $\phi = (\phi(t))_{t\in[0,T]}$, $\phi : [0,T] \times \Omega \to \mathbb{R}^m$, such that $E_{\tilde{Q}}\left[\int_0^T \|\phi(t)\|^2 \, dt\right] < \infty$ and

$$M(t) = E_{\tilde{Q}}\left[\tilde{D}(T)|\mathcal{F}_t\right] = E_{\tilde{Q}}\left[\tilde{D}(T)\right] + \int_0^t \phi(s)' \, d\widetilde{W}(s) \quad \tilde{Q} - a.s.$$

---

[7] To be precise, we can easily see, using Lemma 2.6d),h), and i), that $\tilde{Q} - a.s.$

$$0 \leq E_{\tilde{Q}}\left[\left(\tilde{D}(T) - E_{\tilde{Q}}\left[\tilde{D}(T)|\mathcal{F}_t\right]\right)^2 |\mathcal{F}_t\right]$$

$$= E_{\tilde{Q}}\left[\tilde{D}^2(T) - 2 \cdot \tilde{D}(T) \cdot E_{\tilde{Q}}\left[\tilde{D}(T)|\mathcal{F}_t\right] + \left(E_{\tilde{Q}}\left[\tilde{D}(T)|\mathcal{F}_t\right]\right)^2 |\mathcal{F}_t\right]$$

$$= E_{\tilde{Q}}\left[\tilde{D}^2(T)|\mathcal{F}_t\right] - \left(E_{\tilde{Q}}\left[\tilde{D}(T)|\mathcal{F}_t\right]\right)^2.$$

for all $t \in [0, T]$, where $\widetilde{W}$ is a Wiener process under $\widetilde{Q}$. Now let $\varphi = (\varphi_0, ..., \varphi_n)$ with

$$(\varphi_1(t), ..., \varphi_n(t)) := \phi(t)' \widetilde{\sigma}^{-1}(t) \text{ for all } t \in [0, T].$$

and

$$\varphi_0(t) := E_{\widetilde{Q}}\left[\widetilde{D}(T)\right] + \sum_{i=1}^{n} \int_0^t \varphi_i(s) \, d\widetilde{P}_i(s) - \sum_{i=1}^{n} \varphi_i(t) \cdot \widetilde{P}_i(t)$$

for all $t \in [0, T]$. Then, using Lemma 3.12, we know that

$$
\begin{aligned}
\widetilde{V}(\varphi, t) &= E_{\widetilde{Q}}\left[\widetilde{D}(T)\right] + \int_0^t \varphi(s)' \, d\widetilde{P}(s) \\
&= E_{\widetilde{Q}}\left[\widetilde{D}(T)\right] + \int_0^t \varphi(s)' \widetilde{\sigma}(s) \, d\widetilde{W}(s) \\
&= E_{\widetilde{Q}}\left[\widetilde{D}(T)\right] + \int_0^t \phi(s)' \, d\widetilde{W}(s) \\
&= E_{\widetilde{Q}}\left[\widetilde{D}(T) | \mathcal{F}_t\right] \text{ for all } t \in [0, T].
\end{aligned}
$$

Espcecially, $\left(\widetilde{V}(\varphi, t)\right)_{t \in [0, T]}$ is a $\widetilde{Q}$–martingale. $\qquad\square$

Remark. *If the function* $x \longmapsto x' \widetilde{\sigma}(t) \widetilde{\sigma}(t)' x$ *is strongly positive for all* $x \in \mathbb{R}^n$ $\lambda \otimes Q - a.s.$, *i.e. there is a constant* $K_\sigma > 0$ *such that* $\lambda \otimes Q - a.s.$

$$x' \widetilde{\sigma}(t) \widetilde{\sigma}(t)' x \geq K_\sigma \cdot x'x \text{ for all } x \in \mathbb{R}^n, \tag{SP}$$

then we get

$$
\begin{aligned}
E_{\widetilde{Q}}\left[\int_0^T \|\varphi(t)' \widetilde{\sigma}(t)\|^2 \, dt\right] &= E_{\widetilde{Q}}\left[\int_0^T \sum_{j=1}^{m} (\varphi(t)' \widetilde{\sigma}(t))_j^2 \, dt\right] \\
&= E_{\widetilde{Q}}\left[\int_0^T \varphi(t)' \widetilde{\sigma}(t) \widetilde{\sigma}(t)' \varphi(t) \, dt\right] \\
&\geq K_\sigma \cdot E_{\widetilde{Q}}\left[\int_0^T \varphi(t)' \varphi(t) \, dt\right] \\
&= K_\sigma \cdot \sum_{i=1}^{n} E_{\widetilde{Q}}\left[\int_0^T \varphi_i^2(t) \, dt\right].
\end{aligned}
$$

Hence, under the assumptions of Corollary 3.20 and if condition (SP) is satisfied, the stochastic process $\varphi = (\varphi_0, ..., \varphi_n)$ *defined in Corollary 3.20 satisfies* $\varphi_i \in L_2[0, T]$ *for all* $i \in \{1, ..., n\}$. $\qquad\square$

Note that for the important *special case* of

$$\sigma_i(t) = P_i(t) \cdot \hat{\sigma}_i(t) \text{ with } P_i(t) > 0 \text{ for all } t \in [0,T], \, i = 1, ..., n$$

we know that σ has a left inverse σ^{-1} $\lambda \otimes Q - a.s.$ if and only if $\hat{\sigma}$ has a left inverse $\hat{\sigma}^{-1}$ $\lambda \otimes Q - a.s.$ with

$$\sigma_i^{-1}(t) = \hat{\sigma}_i^{-1}(t) \cdot P_i^{-1}(t) \text{ for all } t \in [0,T], \, i = 1, ..., n.$$

Also note that $\tilde{\sigma}$ with $\tilde{\sigma}(t) = P_0^{-1}(t) \cdot \sigma(t)$ for all $t \in [0,T]$ has a left inverse $\tilde{\sigma}^{-1}$ $\lambda \otimes Q - a.s.$ if and only if σ has a left inverse σ^{-1} with

$$\tilde{\sigma}^{-1}(t) = \sigma_i^{-1}(t) \cdot P_0(t) \text{ for all } t \in [0,T], \, i = 1, ..., n.$$

In the following theorem we show that we can only have one unique equivalent martingale measure if the financial market $\mathcal{M}(Q)$ is complete.

Theorem 3.21 (Unique Martingale Measure) *Let the assumptions of Theorem 3.19 be satisfied and suppose that σ has a left inverse $\lambda \otimes Q - a.s.$ Then $|I\!M| = 1$. Especially, if the financial market $\mathcal{M}(Q)$ is complete, then $|I\!M| = 1$.*

Proof. a) Let $\tilde{Q}_1, \tilde{Q}_2 \in I\!M$ be arbitrary but fixed. Furthermore, let $\tilde{D}(T)$ be a non-negative, bounded (discounted) contingent claim. Using Corollary 3.20, we know that there exists an $(n+1)$−dimensional progressively measurable stochastic process $\varphi^{\tilde{Q}_i} = (\varphi_0^{\tilde{Q}_i}, ..., \varphi_n^{\tilde{Q}_i}) = (\varphi_0^{\tilde{Q}_i}(t), ..., \varphi_n^{\tilde{Q}_i}(t))_{t\in[0,T]}$ such that for all $t \in [0,T]$

$$E_{\tilde{Q}_i}\left[\int_0^T \left\|\varphi^{\tilde{Q}_i}(t)'\tilde{\sigma}(t)\right\|^2 dt\right] < \infty,$$

and

$$\begin{aligned}\tilde{V}(\varphi^{\tilde{Q}_i}, t) &= E_{\tilde{Q}_i}\left[\tilde{D}(T)\right] + \int_0^t \varphi^{\tilde{Q}_i}(s)' \tilde{\sigma}(s)d\tilde{W}(s) \qquad (3.13)\\ &= E_{\tilde{Q}_i}\left[\tilde{D}(T)|\mathcal{F}_t\right] \geq 0\end{aligned}$$

for all $t \in [0,T]$ and $i \in \{1,2\}$. Especially, we learn that for all $\tilde{Q} \in I\!M$ and $i \in \{1,2\}$

$$\int_0^T \left\|\varphi^{\tilde{Q}_i}(t)'\tilde{\sigma}(t)\right\|^2 dt < \infty \quad \tilde{Q} - a.s.,$$

i.e. $\left(\tilde{V}(\varphi^{\tilde{Q}_i}, t)\right)_{t\in[0,T]}$ is a non-negative, continuous local $\tilde{Q}$−martingale for all $\tilde{Q} \in I\!M$. Hence, by Theorem 2.24b), $\left(\tilde{V}(\varphi^{\tilde{Q}_i}, t)\right)_{t\in[0,T]}$ is a

$\widetilde{Q}$ – supermartingale for all $\widetilde{Q} \in I\!M$ and $i \in \{1,2\}$. So we know that for all $\widetilde{Q} \in I\!M$ and for all $0 \le s \le t \le T$

$$E_{\widetilde{Q}}\left[\widetilde{V}(\varphi^{\widetilde{Q}_i},t)\,|\,\mathcal{F}_s\right] \le \widetilde{V}(\varphi^{\widetilde{Q}_i},s) \quad \widetilde{Q} - a.s., \; i \in \{1,2\}. \qquad (3.14)$$

Combining equations (3.13) and (3.14), we get for $t := T$, $s := t$, and $i \in \{1,2\}$

$$E_{\widetilde{Q}}\left[\widetilde{D}(T)\,|\,\mathcal{F}_t\right] = E_{\widetilde{Q}}\left[\widetilde{V}(\varphi^{\widetilde{Q}_i},T)\,|\,\mathcal{F}_t\right] \le \widetilde{V}(\varphi^{\widetilde{Q}_i},t) = E_{\widetilde{Q}_i}\left[\widetilde{D}(T)\,|\,\mathcal{F}_t\right].$$

Hence, for all $t \in [0,T]$,

$$E_{\widetilde{Q}}\left[\widetilde{D}(T)\,|\,\mathcal{F}_t\right] \le E_{\widetilde{Q}_i}\left[\widetilde{D}(T)\,|\,\mathcal{F}_t\right], \; i \in \{1,2\}. \qquad (3.15)$$

Applying equation (3.15) for

$$\left(\widetilde{Q},\widetilde{Q}_i\right) := \left(\widetilde{Q}_1,\widetilde{Q}_2\right) \quad \text{and} \quad \left(\widetilde{Q},\widetilde{Q}_i\right) := \left(\widetilde{Q}_2,\widetilde{Q}_1\right),$$

we get

$$E_{\widetilde{Q}_1}\left[\widetilde{D}(T)\,|\,\mathcal{F}_t\right] \le E_{\widetilde{Q}_2}\left[\widetilde{D}(T)\,|\,\mathcal{F}_t\right] \le E_{\widetilde{Q}_1}\left[\widetilde{D}(T)\,|\,\mathcal{F}_t\right],$$

i.e.

$$E_{\widetilde{Q}_1}\left[\widetilde{D}(T)\,|\,\mathcal{F}_t\right] = E_{\widetilde{Q}_2}\left[\widetilde{D}(T)\,|\,\mathcal{F}_t\right].$$

Especially, we have for each $\widetilde{D}(T) := 1_A$, $A \in \mathcal{F}_T$, and $t := 0$

$$\widetilde{Q}_1(A) = E_{\widetilde{Q}_1}\left[\widetilde{D}(T)\right] = E_{\widetilde{Q}_2}\left[\widetilde{D}(T)\right] = \widetilde{Q}_2(A),$$

i.e. $\widetilde{Q}_1 = \widetilde{Q}_2$. Thus, since $\widetilde{Q}_1, \widetilde{Q}_2 \in I\!M$ were arbitrary but fixed, we conclude that $|I\!M| = 1$, i.e. the set $I\!M$ consisits of one point only. The second part of the statement is an immediate consequence of Theorem 3.19b). $\quad\square$

For more details on the opposite direction of Theorem 3.21, i.e. the implications of a unique martingale measure, see, e.g., Harrison and Pliska [HP83], Øksendal [Øks98], p. 266, or Bingham and Kiesel [BK98], p. 180.

3.4 Pricing and Hedging Contingent Claims

In this section we are interested in the price process of a (European) contingent claim $D = D(T)$ with maturity T. If we are given an equivalent martingale measure $\widetilde{Q} \in I\!M$ we know that the discounted price process of the primary traded assets is a $\widetilde{Q}$−martingale. So we can simply calculate the price of a (discounted) primary traded asset at time $t \in [0, T]$ to be the expected value of any corresponding and adequately discounted future market price, where the conditional expectation at time t is taken with respect to $\widetilde{Q}$. We therefore focus on the conditional expectation of the discounted value $\widetilde{D}(T)$ at time T and give the following definition.

Definition 3.22 (Contingent Claim Prices) *Let* $I\!\!M \neq \emptyset, \widetilde{Q} \in I\!\!M$, *and* $D = D(T)$ *be a (European) contingent claim with maturity* T. *The* ***expected-value process*** *of* D *under* $\widetilde{Q}$ *is given by the risk-neutral valuation formula*

$$V_D^{\widetilde{Q}}(t) := P_0(t) \cdot E_{\widetilde{Q}}\left[\widetilde{D}(T) | \mathcal{F}_t\right], \, t \in [0, T].$$

We call $V_D(t) := V_D^{\widetilde{Q}}(t)$ *the* ***price of the contingent claim*** D *if it is unique in* $I\!\!M$. *The price process* $V(\varphi) = (V(\varphi, t))_{t \in [0, T]}$ *of any* $\varphi \in \Phi(Q)$ *which replicates* D *is called the* ***arbitrage price process*** *of* D *in* $\mathcal{M}(Q)$ *under* φ.

Note that $V_D^{\widetilde{Q}}(t)$ is unique in $I\!\!M$ if the financial market $\mathcal{M}(Q)$ is complete or if σ has a left inverse $\lambda \otimes Q - a.s.$ This is straightforward from Theorem 3.21 since $|I\!\!M| = 1$ in this case. A little more generally, the following lemma holds.

Lemma 3.23 *Let* $I\!\!M \neq \emptyset$, $\left(\widetilde{M3}\right)$ *be satisfied, and let the contingent claim* $D = D(T)$ *with maturity* T *be* Φ^*−*attainable. Then the expected-value process is unique in* $I\!\!M$, *i.e. the price of the contingent claim* D *is given by*

$$V_D(t) = P_0(t) \cdot E_{\widetilde{Q}}\left[\widetilde{D}(T) | \mathcal{F}_t\right], \, t \in [0, T], \quad \text{for all } \widetilde{Q} \in I\!\!M.$$

Furthermore, the expected-value process and the arbitrage price process coincide under any replicating portfolio $\varphi \in \Phi^*(Q)$, *i.e.*

$$V(\varphi, t) = V(\varphi^*, t) = V_D(t) \quad \text{for all } t \in [0, T], \, \varphi, \varphi^* \in \Phi^*(Q).$$

Proof. Let $\widetilde{Q} \in I\!\!M$ and $D = D(T)$ be a contingent claim with maturity T, attainable by a trading strategy $\varphi \in \Phi^*(Q)$. Then $V(\varphi, T) = D(T)$, and, since $\widetilde{V}(\varphi)$ is a $\widetilde{Q}$−martingale by Lemma 3.10a),

$$\begin{aligned} V(\varphi, t) &= P_0(t) \cdot \widetilde{V}(\varphi, t) = P_0(t) \cdot E_{\widetilde{Q}}\left[\widetilde{V}(\varphi, T) | \mathcal{F}_t\right] \\ &= P_0(t) \cdot E_{\widetilde{Q}}\left[\widetilde{D}(T) | \mathcal{F}_t\right] = V_D^{\widetilde{Q}}(t). \end{aligned}$$

As this is true for any $\widetilde{Q} \in I\!\!M$ the mapping $\widetilde{Q} \rightarrow V_D^{\widetilde{Q}}(t)$ is constant on $I\!\!M$ for all $t \in [0, T]$. Also, the mapping $\varphi \rightarrow V(\varphi, t)$ is constant on $\Phi^*(Q)$ for all $t \in [0, T]$, which completes the proof. $\qquad\qquad\square$

Suppose we are offered a guaranteed payment of $D(T)$ at time $t = T$ for a (European) contingent claim $D = D(T)$ with maturity T. How much should we be willing to pay for such a guarantee at time t, $0 \leq t \leq T$? There are two ways to look at this question: If we are the buyer of such a guarantee and pay the price p for it at time t, our initial fortune (debt) at this time is $-p$. Making this deal is interesting only if we can get a profit out of it, i.e. if there is an admissible strategy $\varphi \in \Phi^*(Q)$ such that the corresponding price process satisfies $V(\varphi, t) = -p$ and $V(\varphi, T) + D(T) \geq 0$ $Q - a.s.$ Thus, the maximum price we would agree to, the so-called buyers or *bid price* $D_B(t)$ at time t, is given by

$$D_B(t) = \sup \left\{ \begin{array}{l} p \in \mathbb{R} : V(\varphi, t) = -p, \; V(\varphi, T) + D(T) \geq 0 \; Q - a.s. \\ \text{for some } \varphi \in \Phi^*(Q) \end{array} \right\}.$$

If, on the other hand, we are the seller of such a guarantee, we could use the initial payment p at time t to start with a trading strategy which should give us a final portfolio price at time $t = T$ not less than the amount $D(T)$ we are obliged to pay. Thus, we will make the deal if there is an admissible trading strategy $\varphi \in \Phi^*(Q)$ such that the corresponding price process satisfies $V(\varphi, t) = p$ and $V(\varphi, T) \geq D(T)$ $Q - a.s.$ The minimum price we would agree to, the so-called sellers or *ask price* $D_A(t)$ at time t, is given by

$$D_A(t) = \inf \left\{ \begin{array}{l} p \in \mathbb{R} : V(\varphi, t) = p, \; V(\varphi, T) \geq D(T) \; Q - a.s. \\ \text{for some } \varphi \in \Phi^*(Q) \end{array} \right\}.$$

Clearly, we expect the bid price for a (European) contingent claim to be less than the ask price. The difference $D_A(t) - D_B(t)$ of the bid and ask price at time t is called the *bid-ask spread* for the contingent claim D at time t. The major question is under which conditions the bid-ask spread equals zero.

Theorem 3.24   a) *Suppose that there exists a m-dimensional progressively measurable stochastic process γ such that the conditions (NA) and (NV) hold. Furthermore, let $\left(\widetilde{M3}' \right)$ be satisfied and the probability measure $\widetilde{Q}$ on $(\Omega, \mathcal{F}_T)$ be defined as in Lemma 3.9. Then for any (European) contingent claim $D = D(T)$ with maturity T we have*

$$D_B(t) \leq V_D^{\widetilde{Q}}(t) = P_0(t) \cdot E_{\widetilde{Q}} \left[\widetilde{D}(T) | \mathcal{F}_t \right] \leq D_A(t).$$

b) *Let the assumptions of part a) be satisfied. If the (primary) financial market $\mathcal{M}(Q)$ is Φ^*-complete, the price at time t of any (European) contingent claim D with maturity T is given by*

$$D_B(t) = V_D(t) = D_A(t).$$

Proof. a) Let $p \in \mathbb{R}$ and suppose there exists a trading strategy $\varphi \in \Phi^*(Q)$ such that $V(\varphi,t) = -p$ and $V(\varphi,T) + D(T) \geq 0$. Then since φ is self-financing,

$$
\begin{aligned}
V(\varphi,T) &= V(\varphi,0) + \int_0^T \varphi(s)' \, dP(s) \\
&= V(\varphi,0) + \int_0^t \varphi(s)' \, dP(s) + \int_t^T \varphi(s)' \, dP(s) \\
&= V(\varphi,t) + \int_t^T \varphi(s)' \, dP(s) \\
&= -p + \int_t^T \varphi(s)' \, dP(s) \geq -D(T) \quad Q-a.s.
\end{aligned}
$$

Following Lemma 3.4 this is equivalent to

$$
\begin{aligned}
\tilde{V}(\varphi,T) &= V(\varphi,0) + \int_0^T \varphi(s)' \, d\tilde{P}(s) \\
&= \tilde{V}(\varphi,t) + \int_t^T \varphi(s)' \, d\tilde{P}(s) \\
&= -\tilde{p}(t) + \int_t^T \varphi(s)' \, d\tilde{P}(s) \geq -\tilde{D}(T) \quad Q-a.s.
\end{aligned}
$$

with $\tilde{p}(t) := p \cdot P_0^{-1}(t)$. By means of Lemma 3.12, this is equivalent to

$$
\tilde{V}(\varphi,T) = -\tilde{p}(t) + \int_t^T \sum_{i=1}^n \varphi_i(s) \cdot \tilde{\sigma}_i(s) \, d\widetilde{W}(s) \geq -\tilde{D}(T) \quad Q-a.s.
$$

or equivalently, $Q - a.s.$,

$$
P_0(t) \cdot \tilde{V}(\varphi,T) = -p + P_0(t) \cdot \int_t^T \sum_{i=1}^n \varphi_i(s) \cdot \tilde{\sigma}_i(s) \, d\widetilde{W}(s) \geq -P_0(t) \cdot \tilde{D}(T)
$$

where $\widetilde{W}$ is as defined in Lemma 3.12. Since φ_i is $\lambda \otimes Q - a.s.$ bounded and because of $\left(\widetilde{M3'}\right)$ we know that $\int_0^t \varphi_i(s) \cdot \tilde{\sigma}_i(s) \, d\widetilde{W}(s)$ is a $\tilde{Q}$–martingale for each $i \in \{1,...,n\}$. Thus,

$$
E_{\tilde{Q}}\left[P_0(t) \cdot \tilde{V}(\varphi,T) | \mathcal{F}_t\right] = -p \geq E_{\tilde{Q}}\left[-P_0(t) \cdot \tilde{D}(T) | \mathcal{F}_t\right] = -V_D^{\tilde{Q}}(t),
$$

i.e. $p \leq V_D^{\tilde{Q}}(t)$. Hence,

$$
\begin{aligned}
D_B(t) &= \sup \left\{ \begin{array}{c} p \in \mathbb{R} : V(\varphi,t) = -p, \ V(\varphi,T) + D(T) \geq 0 \ Q-a.s. \\ \text{for some } \varphi \in \Phi^*(Q) \end{array} \right\} \\
&\leq V_D^{\tilde{Q}}(t).
\end{aligned}
$$

If no such pair (p, φ) exists we get $D_B(t) = -\infty \leq V_D^{\tilde{Q}}(t)$ which proves the first inequality. The same arguments are used to prove the second inequality.

b) Now, in addition to the assumptions of a), let $\mathcal{M}(Q)$ be Φ^*−complete. Furthermore, let the random variable $\tilde{D}_k = \tilde{D}_k(T)$ for $k \in I\!\!N$ be defined by

$$\tilde{D}_k(\omega) = \begin{cases} k, & if \ \tilde{D}(\omega) \geq k \\ \tilde{D}(\omega), & if \ \tilde{D}(\omega) < k. \end{cases}$$

Then $\tilde{D}_k(T)$ is a bounded contingent claim with maturity T and

$$D_k(T) := P_0(T) \cdot \tilde{D}_k(T) \leq P_0(T) \cdot \tilde{D}(T) = D(T).$$

Let $t \in [0, T]$ be arbitrary but fixed and

$$D_B^k(t) := \sup \left\{ \begin{array}{c} p \in I\!\!R : V(\varphi, t) = -p, \ V(\varphi, T) + D_k(T) \geq 0 \ Q - a.s. \\ \text{for some } \varphi \in \Phi^*(Q) \end{array} \right\}.$$

Since $D_k(T) \leq D(T)$, each pair (p, φ), $p \in I\!\!R$, $\varphi \in \Phi^*(Q)$, with $V(\varphi, t) = -p$, and $V(\varphi, T) + D_k(T) \geq 0 \ Q - a.s.$ also holds $V(\varphi, t) = -p$, $V(\varphi, T) + D(T) \geq 0 \ Q - a.s.$, i.e.

$$D_B^k(t) \leq D_B(t) \ \text{ for all } k \in I\!\!N. \tag{3.16}$$

Since $\mathcal{M}(Q)$ is Φ^*−complete so is $\widetilde{\mathcal{M}}(Q)$ and thus there exists a trading strategy $\varphi^k \in \Phi^*(Q)$ such that $\tilde{V}_k(\varphi^k, T) = \tilde{D}_k(T)$, i.e.

$$\begin{aligned} \tilde{V}_k(\varphi^k, T) &= V_k(\varphi^k, 0) + \int_0^T \sum_{i=1}^n \varphi_i^k(s) \cdot \tilde{\sigma}_i(s) \, d\widetilde{W}(s) \\ &= \tilde{V}_k(\varphi^k, t) + \int_t^T \sum_{i=1}^n \varphi_i^k(s) \cdot \tilde{\sigma}_i(s) \, d\widetilde{W}(s) \\ &= \tilde{D}_k(T) \quad Q - a.s. \end{aligned}$$

Setting $p_k := V_k(\varphi^k, t)$ and $\tilde{\varphi}_i^k(t) := -\varphi_i^k(t)$ we see that

$$\tilde{V}_k\left(\tilde{\varphi}^k, T\right) = -\tilde{V}_k(\varphi^k, T) = -\tilde{D}_k(T) \, Q - a.s.,$$

i.e.

$$V_k\left(\tilde{\varphi}^k, T\right) + D_k(T) = 0 \ \ Q - a.s. \ \text{ and } \ V_k\left(\tilde{\varphi}^k, t\right) = -V_k(\varphi^k, t) = -p_k.$$

Hence,

$$D_B^k(t) \geq p_k \ \text{ for all } k \in I\!\!N. \tag{3.17}$$

Furthermore,

$$
\begin{aligned}
\tilde{V}_k \left(\tilde{\varphi}^k, T \right) &= -\tilde{V}_k \left(\varphi^k, t \right) + \int_t^T \sum_{i=1}^n \tilde{\varphi}_i^k \left(s \right) \cdot \tilde{\sigma}_i \left(s \right) d\widetilde{W} \left(s \right) \\
&= -P_0^{-1} \left(t \right) \cdot p_k + \int_t^T \sum_{i=1}^n \tilde{\varphi}_i^k \left(s \right) \cdot \tilde{\sigma}_i \left(s \right) d\widetilde{W} \left(s \right) \\
&= -\tilde{D}_k \left(T \right) \quad Q - a.s.
\end{aligned}
$$

Taking expectations we get

$$
-P_0^{-1} \left(t \right) \cdot p_k = E_{\tilde{Q}} \left[-\tilde{D}_k \left(T \right) | \mathcal{F}_t \right] \quad \text{or} \quad p_k = P_0 \left(t \right) \cdot E_{\tilde{Q}} \left[\tilde{D}_k \left(T \right) | \mathcal{F}_t \right].
\tag{3.18}
$$

Combining equations (3.16),(3.17), and (3.18) we get

$$
D_B \left(t \right) \geq D_B^k \left(t \right) \geq P_0 \left(t \right) \cdot E_{\tilde{Q}} \left[\tilde{D}_k \left(T \right) | \mathcal{F}_t \right] \quad \text{for all } k \in I\!N.
$$

Notice that the mapping $k \longmapsto \tilde{D}_k \left(T \right)$ is increasing with $\lim_{k \to \infty} \tilde{D}_k \left(T \right) = \tilde{D} \left(T \right)$. By monotone convergence (see, e.g. Theorem 19.1 in Hinderer [Hin85]) we get

$$
\begin{aligned}
D_B \left(t \right) &\geq \lim_{k \to \infty} P_0 \left(t \right) \cdot E_{\tilde{Q}} \left[\tilde{D}_k \left(T \right) | \mathcal{F}_t \right] = P_0 \left(t \right) \cdot E_{\tilde{Q}} \left[\lim_{k \to \infty} \tilde{D}_k \left(T \right) | \mathcal{F}_t \right] \\
&= P_0 \left(t \right) \cdot E_{\tilde{Q}} \left[\tilde{D} \left(T \right) | \mathcal{F}_t \right] = V_D \left(t \right).
\end{aligned}
$$

Together with part a) we conclude that

$$
D_B \left(t \right) = V_D \left(t \right).
$$

For proving the right equation, let the random variable $\tilde{D}_{kl} = \tilde{D}_{kl} \left(T \right)$ for $k, l \in I\!N$ be defined by

$$
\tilde{D}_{kl} \left(\omega \right) = \begin{cases} k, & if \ \tilde{D} \left(\omega \right) \geq k \\ D \left(\omega \right), & if \ -l < \tilde{D} \left(\omega \right) < k \\ -l, & if \ \tilde{D} \left(\omega \right) \leq -l. \end{cases}
$$

Then $\tilde{D}_{kl} \left(T \right)$ is a bounded contingent claim with maturity T and

$$
D_{kl} \left(T \right) := P_0 \left(T \right) \cdot \tilde{D}_{kl} \left(T \right) \geq P_0 \left(T \right) \cdot \tilde{D}_k \left(T \right) = D_k \left(T \right).
$$

Let $t \in [0, T]$ be arbitrary but fixed and

$$
D_A^k \left(t \right) := \inf \left\{ \begin{array}{l} p \in I\!R : V \left(\varphi, t \right) = p, \ V \left(\varphi, T \right) \geq D_k \left(T \right) \ Q - a.s. \\ \text{for some } \varphi \in \Phi^* \left(Q \right) \end{array} \right\}
$$

and

$$D_A^{kl}(t) := \inf \left\{ \begin{array}{l} p \in I\!R : V(\varphi,t) = p, \ V(\varphi,T) \geq D_{kl}(T) \ Q - a.s. \\ \text{for some } \varphi \in \Phi^*(Q) \end{array} \right\}.$$

Because of $D_{kl}(T) \geq D_k(T)$, each pair (p,φ), $p \in I\!R$, $\varphi \in \Phi^*(Q)$, with $V(\varphi,t) = p$, $V(\varphi,T) \geq D_{kl}(T)$ $Q - a.s.$ also holds $V(\varphi,t) = p$, $V(\varphi,T) \geq D_k(T)$ $Q - a.s.$, i.e.

$$D_A^k(t) \leq D_A^{kl}(t) \quad \text{for all } k,l \in I\!N. \tag{3.19}$$

Since $\widetilde{\mathcal{M}}(Q)$ is Φ^*−complete there exists a trading strategy $\varphi^{kl} \in \Phi^*(Q)$ such that $\tilde{V}_{kl}(\varphi^{kl},T) = \tilde{D}_{kl}(T)$, i.e.

$$\tilde{V}_{kl}(\varphi^{kl},T) = V_{kl}(\varphi^{kl},0) + \int_0^T \sum_{i=1}^n \varphi_i^{kl}(s) \cdot \tilde{\sigma}_i(s) \, d\widetilde{W}(s) = \tilde{D}_{kl}(T) \ Q - a.s.$$

Setting $p_{kl} := V_{kl}(\varphi^{kl},t)$ we know that

$$p_{kl} \geq D_A^{kl}(t) \quad \text{for all } k,l \in I\!N \tag{3.20}$$

by definition of $D_A^{kl}(t)$. Furthermore,

$$\tilde{V}_{kl}(\varphi^{kl},T) = \tilde{p}_{kl} + \int_t^T \sum_{i=1}^n \varphi_i^{kl}(s) \cdot \tilde{\sigma}_i(s) \, d\widetilde{W}(s) = \tilde{D}_{kl}(T) \quad Q - a.s.$$

with $\tilde{p}_{kl} := P_0^{-1}(t) \cdot p_{kl}$. Taking expectations we get

$$p_{kl} = P_0(t) \cdot E_{\tilde{Q}}\left[\tilde{D}_{kl}(T) | \mathcal{F}_t \right]. \tag{3.21}$$

Combining equations (3.19), (3.20), and (3.21) we conclude that

$$D_A^k(t) \leq D_A^{kl}(t) \leq P_0(t) \cdot E_{\tilde{Q}}\left[\tilde{D}_{kl}(T) | \mathcal{F}_t \right] \quad \text{for all } k,l \in I\!N.$$

Since the mapping $l \longmapsto \tilde{D}_{kl}(T)$ is decreasing with $\lim_{l \to \infty} \tilde{D}_{kl}(T) = \tilde{D}_k(T)$, we get, by monotone convergence,

$$D_A^k(t) \leq \lim_{l \to \infty} P_0(t) \cdot E_{\tilde{Q}}\left[\tilde{D}_{kl}(T) | \mathcal{F}_t \right] = P_0(t) \cdot E_{\tilde{Q}}\left[\lim_{l \to \infty} \tilde{D}_{kl}(T) | \mathcal{F}_t \right]$$

$$= P_0(t) \cdot E_{\tilde{Q}}\left[\tilde{D}_k(T) | \mathcal{F}_t \right] \quad \text{for all } k \in I\!N.$$

Now, since the mapping $k \longmapsto \tilde{D}_k(T)$ is increasing with $\lim_{k \to \infty} \tilde{D}_k(T) = \tilde{D}(T)$ we know that

$$D_A^k(t) \leq P_0(t) \cdot E_{\tilde{Q}}\left[\tilde{D}_k(T) | \mathcal{F}_t \right] \leq P_0(t) \cdot E_{\tilde{Q}}\left[\tilde{D}(T) | \mathcal{F}_t \right] \quad \text{for all } k \in I\!N.$$

Since $D_k(T) \le D(T)$ each pair (p, φ), $p \in \mathbb{R}$, $\varphi \in \Phi^*(Q)$, with $V(\varphi, t) = p$, $V(\varphi, T) \ge D(T)$ $Q - a.s.$ also holds $V(\varphi, t) = p$, $V(\varphi, T) \ge D_k(T)$ $Q - a.s.$, i.e.

$$D_A^k(t) \le D_A(t), \text{ for all } k \in \mathbb{N}.$$

Thus,

$$D_A(t) = \sup_{k \in \mathbb{N}} D_A^k(t) \le P_0(t) \cdot E_{\widetilde{Q}}\left[\widetilde{D}(T)|\mathcal{F}_t\right] = V_D(t).$$

Together with part a) we conclude that

$$D_A(t) = P_0(t) \cdot E_{\widetilde{Q}}\left[\widetilde{D}(T)|\mathcal{F}_t\right] = V_D(t)$$

which completes the proof. □

The following lemma shows under which conditions we can find a replicating or hedging strategy for a contingent claim $D(T)$ with maturity T.

Lemma 3.25 *Let the financial market $\mathcal{M}(Q)$ be complete. Suppose that there exists a $m-$dimensional progressively measurable stochastic process γ such that (NA) and (NV) hold. Let $\left(\widetilde{M3}'\right)$ be satisfied and $\widetilde{Q}$ and $\widetilde{W}$ be defined as in Lemma 3.12. Furthermore, let $\phi = (\phi(t))_{t \in [0,T]}$ be a progressively measurable stochastic process with $\int_0^T \|\phi(t)\|^2\, dt < \infty$ $Q - a.s.$ such that*

$$\widetilde{D}(T) = E_{\widetilde{Q}}\left[\widetilde{D}(T)\right] + \int_0^T \phi(s)'\, d\widetilde{W}(s) \tag{3.22}$$

for the discounted contingent claim $\widetilde{D}(T)$ with maturity T. Then $\varphi = (\varphi_0, ..., \varphi_n)$ defined by

$$(\varphi_1(t), ..., \varphi_n(t)) := P_0(t) \cdot \phi(t)'\, \sigma^{-1}(t), \tag{3.23}$$

where σ^{-1} denotes the left-inverse of σ, and φ_0 is defined as in Lemma 3.5 with $v = E_{\widetilde{Q}}\left[\widetilde{D}(T)\right]$, is a self-financing trading strategy with

$$V(\varphi, T) = D(T) := P_0(T) \cdot \widetilde{D}(T)$$

if $(\varphi_1, ..., \varphi_n)$ is $\lambda \otimes Q - a.s.$ bounded.

Proof. Let φ_0 be as defined in Lemma 3.5 and $(\varphi_1, ..., \varphi_n)$ be $\lambda \otimes Q - a.s.$ bounded. Then φ is self-financing with

$$\widetilde{V}(\varphi, t) = \widetilde{V}(\varphi, 0) + \sum_{i=1}^{n} \int_0^t \varphi_i(s)\, d\widetilde{P}_i(s)$$

and $\widetilde{V}(\varphi, 0) = E_{\widetilde{Q}}\left[\widetilde{D}(T)\right]$. Using Lemma 3.12 we know that, under (NA), (NV), and $\left(\widetilde{M3}'\right)$, $d\widetilde{P}_i(t) = \widetilde{\sigma}_i(t) \, d\widetilde{W}(t)$, $i = 1, ..., n$, which gives us

$$
\begin{aligned}
\widetilde{V}(\varphi, T) &= \widetilde{V}(\varphi, 0) + \sum_{i=1}^{n} \int_0^T \varphi_i(s) \cdot \widetilde{\sigma}_i(t) \, d\widetilde{W}(t) \\
&= E_{\widetilde{Q}}\left[\widetilde{D}(T)\right] + \int_0^T P_0(t) \cdot \phi(t)' \cdot \sigma^{-1}(t) \cdot \widetilde{\sigma}(t) \, d\widetilde{W}(t) \\
&\quad \text{using (3.23)} \\
&= E_{\widetilde{Q}}\left[\widetilde{D}(T)\right] + \int_0^T \phi(t)' \, d\widetilde{W}(t) \\
&= \widetilde{D}(T) \text{ using (3.22).}
\end{aligned}
$$

$\square$

Note that the left-inverse of σ exists since $\mathcal{M}(Q)$ is complete (see Theorem 3.19). Furthermore, the self-financing trading strategy φ in Lemma 3.25 is a replicating or hedging strategy if the price processes $V(\varphi)$ and $\widetilde{V}(\varphi)$ are lower bounded.

3.5   The Generalized Black-Scholes Model

As an important example for our financial market model we specialize to a market which consists of only two traded securities, i.e. $n = 1, m = 1$. One of these securites is the bank account or numéraire, indexed by $i = 0$, the other is specified by setting

$$
\mu_1(t) := \mu(t) \cdot P_1(t), \ \sigma_1(t) := \sigma(t) \cdot P_1(t) \quad \text{and} \quad W_1(t) := W(t)
$$

for all $t \in [0, T]$, with progressively measurable stochastic processes μ and $\sigma > 0$ on $[0, T]$ such that the numéraire conditions and the financial market conditions $\left(\widehat{M1}\right)$ and $\left(\widehat{M2}\right)$ of Section 3.1 are satisfied with $P_1(0) > 0$. This special financial market model is a generalization of the model considered by Black and Scholes [BS73] in their pioneering paper on the pricing of contingent claims and will therefore be referred to as the Black-Scholes model $\mathcal{M}^{BS} = \mathcal{M}^{BS}(Q)$. We can easily see that

$$
P_1(t) = P_1(0) \cdot e^{\int_0^t \sigma(s) dW(s) + \int_0^t \left(\mu(s) - \frac{1}{2} \cdot \sigma^2(s)\right) ds} > 0, \ t \in [0, T].
$$

Furthermore, condition (NA) gets the form

$$
\mu(t) \cdot P_1(t) - \sigma(t) \cdot P_1(t) \cdot \gamma(t) = r(t) \cdot P_1(t) \quad \lambda \otimes Q - a.s.
$$

or equivalently

$$\gamma(t) = \frac{\mu(t) - r(t)}{\sigma(t)} \quad \lambda \otimes Q - a.s. \qquad (NA^{BS})$$

condition (NV) is given by

$$E_Q\left[e^{\frac{1}{2}\int_0^T \left(\frac{\mu(s)-r(s)}{\sigma(s)}\right)^2 ds}\right] < \infty. \qquad (NV^{BS})$$

Now let $\widetilde{Q}$ and $\widetilde{W}$ be defined as in Lemma 3.9. If the Novikov condition holds for σ, i.e.

$$E_{\widetilde{Q}}\left[e^{\frac{1}{2}\int_0^T \sigma^2(s)ds}\right] < \infty, \qquad (\widetilde{M3}^{BS})$$

we know by Theorem 2.42c) and Lemma 3.12 that $\widetilde{Q}$ is an equivalent martingale measure and that the considered Black-Scholes market contains no arbitrage opportunities in $\Phi^*(Q)$ (see Theorem 3.11 for the last statement). Moreover, since $\sigma_1(t) := \sigma(t) \cdot P_1(t) > 0$ on $[0, T]$, we know that σ_1 is invertible. Hence, Theorem 3.21 tells us that $\widetilde{Q}$ is unique. The actual situation is summarized in the followig theorem.

Theorem 3.26 (Generalized Black-Scholes) *Suppose that the primary traded assets with prices P_0 and P_1 are given by*

$$\begin{aligned} dP_0(t) &= r(t) \cdot P_0(t)\, dt, \quad P_0(0) = 1, \\ dP_1(t) &= \mu(t) \cdot P_1(t)\, dt + \sigma(t) \cdot P_1(t)\, dW(t), \quad P_1(0) > 0 \end{aligned}$$

with $\sigma > 0$ such that $(NA^{BS}), (NV^{BS})$ and $\left(\widetilde{M3}^{BS}\right)$ are satisfied. Then $\mathcal{M}^{BS}$ contains no arbitrage opportunities in $\Phi^*(Q)$, and the price process of any (European) contingent claim $D = D(T)$ with maturity T is given by

$$V_D(t) = P_0(t) \cdot E_{\widetilde{Q}}\left[\widetilde{D}(T)|\mathcal{F}_t\right] = E_{\widetilde{Q}}\left[e^{-\int_t^T r(s)ds} \cdot D(T)|\mathcal{F}_t\right] \qquad (3.24)$$

for $t \in [0, T]$, $\widetilde{Q} \in M$.

If we suppose that r and σ are deterministic, we know by Theorem 2.43 that, at time t, the random variable

$$Y_0 := \int_t^T \sigma(s)\, d\widetilde{W}(s) = \int_0^\infty 1_{[t,T]}(s) \cdot \sigma(s)\, d\widetilde{W}(s)$$

is normally distributed under $\widetilde{Q}$ with expectation $E_{\widetilde{Q}}^{y,t}[Y_0] = 0$ and variance $E_{\widetilde{Q}}^{y,t}[Y_0^2] = \int_t^T \sigma^2(s)\, ds$. Under (NA^{BS}), $P_1(T)$ is given by

$$\begin{aligned} P_1(T) &= P_1(t) \cdot e^{\int_t^T \sigma(s)dW(s) + \int_t^T \left(\mu(s) - \frac{1}{2}\cdot\sigma^2(s)\right)ds} \\ &= P_1(t) \cdot e^{\int_t^T \sigma(s)d\widetilde{W}(s) + \int_t^T \left(\mu(s) - \sigma(s)\cdot\gamma(s) - \frac{1}{2}\cdot\sigma^2(s)\right)ds} \\ &= P_1(t) \cdot e^{\int_t^T \sigma(s)d\widetilde{W}(s) + \int_t^T \left(r(s) - \frac{1}{2}\cdot\sigma^2(s)\right)ds}. \end{aligned}$$

Hence, using the Feynman-Kac representation, where we assume that the regularity conditions necessary to apply the Feynman-Kac representation and stated in Section 2.6 hold, we know that for any contingent claim D with $D(T) = f(P_1(T))$ equation (3.24) transforms to

$$
\begin{aligned}
V_D(t) &= e^{-\int_t^T r(s)ds} \cdot E_{\widetilde{Q}}^{P_1,t}\left[f(P_1(T))\right] \\
&= e^{-\int_t^T r(s)ds} \cdot E_{\widetilde{Q}}^{P_1,t}\left[f\left(P_1(t) \cdot e^{\int_t^T \sigma(s)d\widetilde{W}(s)+\int_t^T \left(r(s)-\frac{1}{2}\cdot\sigma^2(s)\right)ds}\right)\right] \\
&= e^{-\int_t^T r(s)ds} \cdot \int_{-\infty}^{\infty} f(e^y) \cdot n\left(y|\mu_Y,\sigma_Y^2\right) dy,
\end{aligned}
$$

where $n\left(y|\mu_Y,\sigma_Y^2\right)$ denotes the normal density function with expectation

$$
\mu_Y = \mu_Y(t,T) := \ln(P_1(t)) + \int_t^T \left(r(s) - \frac{1}{2}\cdot\sigma^2(s)\right) ds \qquad (3.25)
$$

and standard deviation

$$
\sigma_Y = \sigma_Y(t,T) := \sqrt{\int_t^T \sigma^2(s)\, ds} \qquad (3.26)
$$

evaluated at point $y \in I\!R$. Thus, we get the following corollary.

Corollary 3.27 *Let the assumptions of Theorem 3.26 be satisfied. Furthermore, let r and σ be deterministic. Then, under sufficient regularity conditions on f, the price at time $t \in [0,T]$ of any (European) contingent claim D with $D(T) = f(P_1(T))$ and maturity T is given by*

$$
V_D(t) = e^{-\int_t^T r(s)ds} \cdot \int_{-\infty}^{\infty} f(e^y) \cdot n\left(y|\mu_Y,\sigma_Y^2\right) dy \qquad (3.27)
$$

with μ_Y and σ_Y as defined in (3.25) and (3.26).

Let us apply this corollary to the function

$$
D(T) = f(P_1(T)) := \max\{P_1(T) - X, 0\}
$$

with $X \in [0, \infty)$ which is the terminal pay-off of a so-called (European) call option and can be shown to hold the Feynman-Kac regularity conditions. Correspondingly, a (European) put option is given by the terminal pay-off $\max\{X - P_1(T), 0\}$. For the (European) call option we get

$$
\begin{aligned}
V_D(t) &= e^{-\int_t^T r(s)ds} \cdot \int_{-\infty}^{\infty} f(e^y) \cdot n\left(y|\mu_Y,\sigma_Y^2\right) dy \\
&= e^{-\int_t^T r(s)ds} \int_{-\infty}^{\infty} \max\{e^y - X, 0\} \cdot n\left(y|\mu_Y,\sigma_Y^2\right) dy \\
&= e^{-\int_t^T r(s)ds} \cdot \int_{\ln X}^{\infty} (e^y - X) \cdot n\left(y|\mu_Y,\sigma_Y^2\right) dy \\
&= e^{-\int_t^T r(s)ds} \cdot (I_1 - I_2)
\end{aligned}
$$

with

$$
\begin{aligned}
I_1 &= \int_{\ln X}^{\infty} e^y \cdot n\left(y|\mu_Y, \sigma_Y^2\right) dy \\
&= \frac{1}{\sigma_Y \cdot \sqrt{2\pi}} \int_{\ln X}^{\infty} e^y \cdot e^{-\frac{1}{2\sigma_Y^2}\cdot(y-\mu_Y)^2} dy \\
&= \frac{1}{\sigma_Y \cdot \sqrt{2\pi}} \int_{\ln X}^{\infty} e^{-\frac{1}{2\sigma_Y^2}\cdot\left(y^2 - 2\left(\mu_Y + \sigma_Y^2\right)\cdot y + \mu_Y^2\right)} dy \\
&= \frac{1}{\sigma_Y \cdot \sqrt{2\pi}} \int_{\ln X}^{\infty} e^{-\frac{1}{2\sigma_Y^2}\cdot\left(\left[y-\left(\mu_Y + \sigma_Y^2\right)\right]^2 - \left(\mu_Y + \sigma_Y^2\right)^2 + \mu_Y^2\right)} dy \\
&= \frac{1}{\sigma_Y \cdot \sqrt{2\pi}} \int_{\ln X}^{\infty} e^{-\frac{1}{2\sigma_Y^2}\cdot\left(\left[y-\left(\mu_Y + \sigma_Y^2\right)\right]^2 - \left(2\cdot\mu_Y\cdot\sigma_Y^2 + \sigma_Y^4\right)\right)} dy \\
&= \frac{e^{\mu_Y + \frac{1}{2}\cdot\sigma_Y^2}}{\sigma_Y \cdot \sqrt{2\pi}} \int_{\ln X}^{\infty} e^{-\frac{1}{2\sigma_Y^2}\cdot\left(y-\left(\mu_Y + \sigma_Y^2\right)\right)^2} dy \\
&= e^{\mu_Y + \frac{1}{2}\cdot\sigma_Y^2} \cdot \left[1 - \mathcal{N}\left(\frac{\ln X - \left(\mu_Y + \sigma_Y^2\right)}{\sigma_Y}\right)\right] \\
&= P_1(t) \cdot e^{\int_t^T r(s)ds} \cdot \mathcal{N}\left(\frac{\ln\left(\frac{P_1(t)}{X}\right) + \int_t^T r(s)\, ds + \frac{1}{2}\cdot\sigma_Y^2}{\sigma_Y}\right)
\end{aligned}
$$

where $\mathcal{N}(y)$ denoted the standard normal distribution function evaluated at point $y \in \mathbb{R}$ and

$$
\begin{aligned}
I_2 &= \int_{\ln X}^{\infty} X \cdot n\left(y|\mu_Y, \sigma_Y^2\right) dy \\
&= X \cdot \left[1 - \mathcal{N}\left(\frac{\ln X - \mu_Y}{\sigma_Y}\right)\right] \\
&= X \cdot \mathcal{N}\left(\frac{\ln\left(\frac{P_1(t)}{X}\right) + \int_t^T r(s)\, ds - \frac{1}{2}\sigma_Y^2}{\sigma_Y}\right).
\end{aligned}
$$

This result is summarized in the following corollary which generalizes the results of Black and Scholes [BS73].

Corollary 3.28 (Generalized Black-Scholes Option Prices)
Let the assumptions of Corollary 3.27 be satisfied. Then the price of a (European) call option at time $t \in [0, T]$ is given by

$$
Call^{BS}(t, T, X) = P_1(t) \cdot \mathcal{N}(d_1) - e^{-\int_t^T r(s)ds} \cdot X \cdot \mathcal{N}(d_2)
$$

with

$$
d_1 := \frac{\ln\left(\frac{P_1(t)}{X}\right) + \int_t^T r(s)\, ds + \frac{1}{2}\sigma_Y^2}{\sigma_Y}, \quad d_2 := d_1 - \sigma_Y,
$$

and

$$\sigma_Y := \sqrt{\int_t^T \sigma^2(s)\, ds}.$$

Correspondingly, the price of a (European) put option is given by

$$Put^{BS}(t,T,X) = e^{-\int_t^T r(s)ds} \cdot X \cdot \mathcal{N}(-d_2) - P_1(t) \cdot \mathcal{N}(-d_1).$$

As another special case we consider the function[8]

$$F(t,T) := e^{\int_t^T r(s)ds} \cdot P_1(t) = P_0(T) \cdot \tilde{P}_1(t),\, t \in [0,T].$$

The idea behind this function is the following: Let us, for a moment of exposition, think in time steps of size Δt with $T - t = n \cdot \Delta t$ and let us consider an agreement $A_F^{\Delta t}$ which says that at the end of each time step the price $F(t + (i+1) \cdot \Delta t, T)$ is compared with the price $F(t + i \cdot \Delta t, T)$ at the beginning of the time step and the difference is put on a cash account, i.e. an account which gives us a continuous interest of $r(t + i \cdot \Delta t)$ for each time-period between $t+i\cdot\Delta t$ and $t+(i+1)\cdot\Delta t$, $i = 0, ..., n-1$. If Δt is equal to one trading day we call this procedure a *daily settlement*. Using Lemma 2.36, the price process $(F(t,T))_{t\in[0,T]}$ satisfies the stochastic differential equation

$$
\begin{aligned}
d_t F(t,T) &= P_0(T)\, d\left[P_0^{-1}(t) \cdot P_1(t)\right]\\
&= P_0(T) \cdot \left[P_1(t)\, dP_0^{-1}(t) + P_0^{-1}(t)\, dP_1(t)\right]\\
&= -r(t) \cdot P_0(T) \cdot P_0^{-1}(t) \cdot P_1(t)\, dt\\
&\quad + P_0(T) \cdot P_0^{-1}(t) \cdot \left[\mu(t) \cdot P_1(t)\, dt + \sigma(t) \cdot P_1(t)\, dW(t)\right]\\
&= -r(t) \cdot F(t,T)\, dt + \mu(t) \cdot F(t,T)\, dt\\
&\quad + \sigma(t) \cdot F(t,T)\, dW(t)\\
&= \mu_F(t) \cdot F(t,T)\, dt + \sigma(t) \cdot F(t,T)\, dW(t),
\end{aligned}
$$

with

$$\mu_F(t) := \mu(t) - r(t).$$

Since

$$\gamma(t) = \frac{\mu(t) - r(t)}{\sigma(t)} = \frac{\mu_F(t)}{\sigma(t)} \quad \lambda \otimes Q - \text{a.s.}$$

we see that

$$d_t F(t,T) = \sigma(t) \cdot \gamma(t) \cdot F(t,T)\, dt + \sigma(t) \cdot F(t,T)\, dW(t).$$

---

[8] Note, that we do assume that there are no cash payments with respect to P_1 for this analysis. Also note that $F(T,T) = P_1(T)$.

Hence, the price process for F under the martingale measure $\widetilde{Q}$ is given by

$$d_t F(t,T) = \sigma(t) \cdot F(t,T) \, d\widetilde{W}(t).$$

Under condition $\left(\widetilde{M3}^{BS}\right)$, $(F(t,T))_{t \in [0,T]}$ is a continuous $\widetilde{Q}$–martingale. So we know that for all $0 \leq t \leq t' \leq T$ we have

$$F(t,T) = E_{\widetilde{Q}}[F(t',T)|\mathcal{F}_t],$$

especially, setting $t' := T$,

$$F(t,T) = E_{\widetilde{Q}}[P_1(T)|\mathcal{F}_t].$$

Remark. *If interest rates are deterministic, the value of each element of agreement* $A_F^{\Delta t}$ *at time* t, $0 \leq t \leq t_1 \leq t_1 + \Delta t \leq T$, *can be derived by*

$$e^{-\int_t^{t_1+\Delta t} r(s)ds} \cdot E_{\widetilde{Q}}[F(t_1 + \Delta t, T) - F(t_1, T)|\mathcal{F}_t]$$

$$= e^{-\int_t^{t_1+\Delta t} r(s)ds} \cdot \left(E_{\widetilde{Q}}[F(t_1 + \Delta t, T)|\mathcal{F}_t] - E_{\widetilde{Q}}[F(t_1, T)|\mathcal{F}_t] \right)$$

$$= e^{-\int_t^{t_1+\Delta t} r(s)ds} \cdot (F(t,T) - F(t,T))$$

$$= 0.$$

$\square$

What we have seen so far is that the financial contract corresponding to the agreement $A_F^{\Delta t}$ is supposed to be traded at an exchange which does a regular settlement (daily, if $\Delta t = 1 \ day$) and will supply us with a terminal value equal to the underlying primary traded asset P_1. It costs nothing to enter into this agreement, i.e. to buy the correponding financial contract. In some sense, the daily settlement corrects the daily price-changes. Therefore, if signed at time t, this financial contract can be considered to be an agreement to buy or sell an asset at a certain time in the future for a certain price $F(t,T) = E_{\widetilde{Q}}[P_1(T)|\mathcal{F}_t]$, called the *futures price*[9], and is therefore known as *futures contract*[10]. On the other side, the forward price $Forward(t,T)$ of a financial instrument for a future time T, called maturity or delivery time, evaluated at time t is defined to be that delivery price X, fixed at time t, which would make this contract, called a *forward*

---

[9] Within other models, we will use the conditional expectation of a final payment or value under the martingale measure $\widetilde{Q}$ given $\mathcal{F}_t$ directly as the definition of the futures price at time $t \in [0,T]$.

[10] For a detailed discussion on Treasury bill and LIBOR or EURIBOR futures as well as on futures contracts and forward agreements on coupon bonds see Sections 5.3 and 5.4.

agreement, have zero value. Unlike a futures contract, a forward contract is usually not traded at an exchange therefore having no daily settlement. The forward price can be easily evaluated by the following arbitrage argument. Consider the two portfolios set up at time t:

A: One long position in the forward contract with delivery price X and maturity time T plus a zero-coupon bond[11] with time to maturity T and a notional amount of X.

B: One unit of the underlying primary traded asset with price P_1.

The sale of the maturing zero-coupon bond of portfolio A at time T gives us exactly the amount of money X to buy the underlying via the forward contract. Denoting the price at time t of a zero-coupon bond with maturity time T and a notional amount of 1 by $P(t,T)$ and the value of the forward contract by $V_{Forward}(t,T,X)$, this leads us to the following value table for the portfolios A and B at time t and at maturity time T:

Portfolio	t	T
A	$V_{Forward}(t,T,X) + X \cdot P(t,T)$	$P_1(T)$
B	$P_1(t)$	$P_1(T)$

Having the same terminal value, under the absence of arbitrage opportunities, the two portfoilos must have the same value at time t, i.e.

$$P_1(t) = V_{Forward}(t_0,T,X) + X \cdot P(t,T).\qquad(3.28)$$

Inserting the forward price $Forward(t,T)$ for X, we get

$$\begin{aligned}
P_1(t) &= V_{Forward}(t,T,Forward(t,T))\\
&\quad + Forward(t,T) \cdot P(t,T)\\
&= 0 + Forward(t,T) \cdot P(t,T),
\end{aligned}$$

or equivalently

$$Forward(t,T) = P^{-1}(t,T) \cdot P_1(t).\qquad(3.29)$$

To compare the futures and the forward price let us set up a few trading strategies where we assume that *interest rates are deterministic.* The first strategy $\varphi_A^{\Delta t}$ we consider is to successively buy $e^{\sum_{k=0}^{i} r(k\cdot\Delta t)\cdot\Delta t}$ units of

---

[11] A zero-coupon bond is a financial contract which pays out its notional amount at maturity and nothing in between. Zero-coupon bonds will be studied in more detail in Chapter 4.

agreement $A_F^{\Delta t}$ at time $t = i \cdot \Delta t$, $i = 0, ..., n-1$. This gives us a value at time T of

$$V\left(\varphi_A^{\Delta t}, T\right) = \sum_{i=0}^{n-1} e^{\sum_{k=0}^{i} r(k \cdot \Delta t) \cdot \Delta t}$$
$$\cdot \left(F\left((i+1) \cdot \Delta t, T\right) - F\left(i \cdot \Delta t, T\right)\right) \cdot e^{\sum_{k=i+1}^{n-1} r(k \cdot \Delta t) \cdot \Delta t}$$

$$= \sum_{i=0}^{n-1} \left(F\left((i+1) \cdot \Delta t, T\right) - F\left(i \cdot \Delta t, T\right)\right) \cdot e^{\sum_{k=0}^{n-1} r(k \cdot \Delta t) \cdot \Delta t}$$

$$= e^{\sum_{k=0}^{n-1} r(k \cdot \Delta t) \cdot \Delta t} \cdot \left(F\left(T, T\right) - F\left(0, T\right)\right)$$

$$= e^{\sum_{k=0}^{n-1} r(k \cdot \Delta t) \cdot \Delta t} \cdot \left(P_1\left(T\right) - F\left(0, T\right)\right).$$

If we define the strategy $\varphi_A := \varphi_A^0$ to continuously buy $e^{\int_0^t r(s)ds}$ units of agreement $A_F := A_F^0$ which itself is continuously settled with the settled amount being put on a cash account, we get

$$V\left(\varphi_A, t\right) = \left(F\left(t, T\right) - F\left(0, T\right)\right) \cdot e^{\int_0^t r(s)ds} \quad \text{for all } t \in [0, T].$$

Because of $F\left(t, T\right) = e^{\int_t^T r(s)ds} \cdot P_1\left(t\right)$ by definition this gives us

$$V\left(\varphi_A, t\right) = e^{\int_0^T r(s)ds} \cdot P_1\left(t\right) - F\left(0, T\right) \cdot e^{\int_0^t r(s)ds} \quad \text{for all } t \in [0, T].$$

So φ_A is equivalent to buying $e^{\int_0^T r(s)ds}$ units of the underlying primary traded asset P_1 plus borrowing an amount of $F\left(0, T\right)$ at time $t = 0$, both together at an initial cost of $V\left(\varphi_A, 0\right) = 0$. Since interest rates are supposed to be deterministic, we know that φ_A is progressively measurable and self-financing.

Let us now set up the two trading strategies φ_B and φ_C. Strategy φ_B is to buy, at time $t = 0$, a zero coupon bond with a maturity time T and a notional amount of $e^{\int_0^T r(s)ds} \cdot Forward\left(0, T\right)$ and to take a long position in a number of $e^{\int_0^T r(s)ds}$ forward contracts. This guarantees an amount of $e^{\int_0^T r(s)ds}$ units of the underlying primary traded asset P_1 paid at time T by the money we get from the maturing zero coupon bond. Using (3.28) the value of this portfolio at time $t \in [0, T]$ is given by

$$V\left(\varphi_B, t\right) = e^{\int_0^T r(s)ds} \cdot Forward\left(0, T\right) \cdot P\left(t, T\right)$$
$$+ e^{\int_0^T r(s)ds} \cdot V_{Forward}\left(t, T, Forward\left(0, T\right)\right)$$
$$= e^{\int_0^T r(s)ds} \cdot Forward\left(0, T\right) \cdot P\left(t, T\right)$$
$$+ e^{\int_0^T r(s)ds} \cdot \left(P_1\left(t\right) - Forward\left(0, T\right) \cdot P\left(t, T\right)\right)$$
$$= e^{\int_0^T r(s)ds} \cdot P_1\left(t\right).$$

Strategy φ_C is nothing other than φ_A plus an investment at time $t = 0$ into a zero coupon bond with maturity time T and a notional amount of $F(0,T) \cdot e^{\int_0^T r(s)ds}$. This gives us to the following value table for the strategies φ_B and φ_C at time $t = 0$ and time $t \in (0,T]$:

Strategy	$t = 0$	t
φ_B	$e^{\int_0^T r(s)ds} \cdot Forward(0,T) \cdot P(0,T)$	$P_1(t) \cdot e^{\int_0^T r(s)ds}$
φ_C	$e^{\int_0^T r(s)ds} \cdot P_1(0)$	$P_1(t) \cdot e^{\int_0^T r(s)ds}$

Since interest rates are supposed to be deterministic, we know that φ_B and φ_C are progressively measurable and self-financing. Furthermore, $V(\varphi_B, t)$ and $V(\varphi_C, t)$ are bounded below for all $t \in [0,T]$. Hence, φ_B and φ_C are admissible trading strategies. Having the same terminal value, under the absence of arbitrage opportunities, the two portfolios must have the same value at $t = 0$, i.e.

$$
\begin{aligned}
e^{\int_0^T r(s)ds} \cdot Forward(0,T) \cdot P(0,T) &= e^{\int_0^T r(s)ds} \cdot P_1(0) \\
&= e^{\int_0^T r(s)ds} \cdot F(0,T) \cdot P(0,T)
\end{aligned}
$$

where we used that $P_1(0) = F(0,T) \cdot P(0,T)$ because of equation 3.29. Hence,

$$Forward(0,T) = F(0,T),$$

i.e. the forward price and the price for the futures contract are identical. The previous results are summarized in the following theorem.

Theorem 3.29 (Futures and Forward Prices) *Let the assumptions of Theorem 3.26 be satisfied. Then the futures price at time $t \in [0,T]$ is given by*

$$F(t,T) = E_{\widetilde{Q}}[P_1(T) | \mathcal{F}_t].$$

On the other hand, the forward price at time $t \in [0,T]$ is given by

$$Forward(t,T) = P^{-1}(t,T) \cdot P_1(t).$$

Furthermore, if interest rates are deterministic, the forward and the futures price are identical and it costs nothing to enter into a futures contract or a forward-rate agreement.

Let us continue with deriving the price of contingent claims based on the traded futures prices described above. We already know that for all $t' \in [t,T]$, under the assumptions of Theorem 3.26,

$$
\begin{aligned}
F(t',T) &= F(t,T) \cdot e^{\int_t^{t'} \sigma(s)dW(s) + \int_t^{t'} \left(\mu_F(s) - \frac{1}{2} \cdot \sigma^2(s)\right)ds} \\
&= F(t,T) \cdot e^{\int_t^{t'} \sigma(s)d\widetilde{W}(s) + \int_t^{t'} \left(\mu_F(s) - \gamma(s) \cdot \sigma(s) - \frac{1}{2} \cdot \sigma^2(s)\right)ds} \\
&= F(t,T) \cdot e^{\int_t^{t'} \sigma(s)d\widetilde{W}(s) - \frac{1}{2} \int_t^{t'} \sigma^2(s)ds}.
\end{aligned}
$$

If we suppose that $r(t)$ and $\sigma(t)$ *are deterministic*, we already know by Theorem 2.43 that, at time t, the random variable

$$Y_0 := \int_t^T \sigma(s) \, d\widetilde{W}(s) = \int_0^\infty 1_{[t,T]}(s) \cdot \sigma(s) \, d\widetilde{W}(s)$$

is normally distributed under $\widetilde{Q}$ with expectation $E_{\widetilde{Q}}^{y,t}[Y_0] = 0$ and variance $E_{\widetilde{Q}}^{y,t}[Y_0^2] = \sigma_Y^2 := \int_t^T \sigma^2(s) \, ds$. Hence, using the Feyman-Kac representation, for any contingent claim D with $D(T) = f(F(T,T))$, equation (3.24) transforms to

$$
\begin{aligned}
V_D(t) &= e^{-\int_t^T r(s)ds} \cdot E_{\widetilde{Q}}^{F,t}[f(F(T,T))] \\
&= e^{-\int_t^T r(s)ds} \cdot \int_{-\infty}^\infty f(e^y) \cdot n\left(y|\mu_{Y_F}, \sigma_Y^2\right) dy,
\end{aligned}
$$

where $n\left(y|\mu_{Y_F}, \sigma_Y^2\right)$ denotes the normal density function with expectation

$$\mu_{Y_F} := \ln(F(t,T)) - \frac{1}{2} \cdot \sigma_Y^2.$$

Using these results we get the following analogue to the generalized Black-Scholes formula which is due to Black [Bla76].

Corollary 3.30 (Generalized Black) *Let the assumptions of Corollary 3.27 be satisifed. Then the price at time $t = 0$ of any (European) contingent claim D with $D(T) = f(F(T,T))$ and maturity T is given by*

$$V_D(t) = e^{-\int_t^T r(s)ds} \cdot \int_{-\infty}^\infty f(e^y) \cdot n\left(y|\mu_{Y_F}, \sigma_Y^2\right) dy$$

with μ_{Y_F} and σ_Y as defined above. Furthermore, let

$$D(T) := \max\{F(T,T) - X, 0\}$$

and $X \in [0, \infty)$ which is the terminal pay-off of a (European) call option on a financial instrument characterized by the price process $(F(t,T))_{t \in [0,T]}$, i.e. a futures option. Then the price of this (European) futures call option at time t is given by

$$Call_F^{Black}(t, T, X) = e^{-\int_t^T r(s)ds} \cdot [F(t,T) \cdot \mathcal{N}(d_1) - X \cdot \mathcal{N}(d_2)],$$

with

$$d_1 := \frac{\ln\left(\frac{F(t,T)}{X}\right) + \frac{1}{2} \cdot \sigma_Y^2}{\sigma_Y} \quad and \quad d_2 := d_1 - \sigma_Y.$$

Correspondingly, the price of a (European) futures put option is given by

$$Put_F^{Black}(t,T,X) = e^{-\int_t^T r(s)ds} \cdot [X \cdot \mathcal{N}(-d_2) - F(t,T) \cdot \mathcal{N}(-d_1)]$$
$$= Call_F^{BS}(t,T,X) + e^{-\int_t^T r(s)ds} \cdot (X - F(t,T)).$$

The last line is called the put-call parity for (European) futures options.

3.6 Change of Numéraire

Up to now, we considered the safe process P_0 to be our numéraire. However, it may make sense to change the scaling or unit price to simplify the pricing of a specific derivative or the evaluation of the corresponding (conditional) expected value. We therefore relax the assumptions $(N1)$ and $(N2)$ and return to the more general case of $(M1)$ and $(M2)$. As in Definition 3.1 we define a numéraire in $\mathcal{M} = \mathcal{M}(Q)$ in general to be a strictly positive Itô process $X = (X(t))_{t\in[0,T]}$. Discounting with respect to X is denoted by a tilde and the upper index X, i.e. $\widetilde{P}^X(t) = X^{-1}(t) \cdot P(t)$ are the primary traded asset prices discounted with respect to X. Correspondingly, the notations $\mathcal{M}^X = \mathcal{M}^X(Q)$ and $\widetilde{\mathcal{M}}^X = \widetilde{\mathcal{M}}^X(Q)$ show the usage of X as a numéraire. As already defined, we relax the upper index X for the special case $X := P_0$. The set of all trading strategies in $\mathcal{M}^X$ (and $\widetilde{\mathcal{M}}^X$) is defined such that $\int_0^t \varphi_i(s) dP_i(s)$ as well as $\int_0^t \varphi_i(s) d\widetilde{P}_i^X(s)$ exist as local Q-martingales for all $t \in [0,T]$ and for all $i \in \{0,...,n\}$. The following theorem shows that we do not have to change the trading strategy as a consequence of a numéraire change. But if we do carefully adjust the probability measure, the newly discounted price process will keep it's martingale property with respect to the changed probability measure. This is an essential characteristic, and makes the change of numéraire technique one of the most powerful tools for pricing financial derivatives. We will use this property extensively in Section 3.7 as well as in Section 4.7.

Theorem 3.31 (Change of Numéraire) *Let $X = (X(t))_{t\in[0,T]}$ be a non-dividend-paying numéraire in $\mathcal{M} = \mathcal{M}(Q)$ and $\widetilde{Q} \in \mathbb{M}$.*

  a) *The trading strategy φ is self-financing in $\mathcal{M}$ if and only if φ is self-financing in $\widetilde{\mathcal{M}}^X$, i.e. a self-financing trading strategy remains self-financing after a numéraire change. Furthermore, if X is $\lambda \otimes Q$-a.s. bounded, the trading strategy φ is admissible in $\mathcal{M}$ if and only if φ is admissible in $\mathcal{M}^X$. Especially, if a contingent claim D with maturity T is (Φ^*-) attainable in $\mathcal{M}$ it is also (Φ^*-) attainable in $\mathcal{M}^X$ and the replicating portfolio is the same.*

b) *If the discounted numéraire process* $\tilde{X} = \left(\tilde{X}(t)\right)_{t \in [0,T]}$ *with* $\tilde{X}(t) := P_0^{-1}(t) \cdot X(t)$, $t \in [0,T]$, *is a* $\tilde{Q}-$*martingale, then there exists a probability measure* Q^X *on* $(\Omega, \mathcal{F})$, *defined by its Radon-Nikodým derivative* $L(T)$ *with respect to* $\tilde{Q}$,

$$L(t) = \frac{dQ^X}{d\tilde{Q}}|_{\mathcal{F}_t} = \frac{X(t)}{X(0) \cdot P_0(t)}, \ t \in [0,T],$$

such that the discounted primary traded asset prices $\tilde{P}_i^X$, $i = 1, ..., n$, *are* Q^X-*martingales. Furthermore, the expected-value process of a contingent claim* $D = D(T)$ *with maturity* T *under* $\tilde{Q}$ *and numéraire* P_0 *coincides with expected-value process of* D *under* Q^X *and numéraire* X, *i.e.*

$$P_0(t) \cdot E_{\tilde{Q}}\left[\tilde{D}(T)|\mathcal{F}_t\right] = X(t) \cdot E_{Q^X}\left[\tilde{D}^X(T)|\mathcal{F}_t\right]$$

for all $t \in [0,T]$.

Proof. a) Let $\varphi = (\varphi_0, ..., \varphi_n)$ be a self-financing trading strategy in $\mathcal{M}$. Then, using the product rule (Lemma 2.36), we get

$$
\begin{aligned}
d\tilde{V}^X(\varphi, t) &= d\left(X^{-1}(t) \cdot V(\varphi, t)\right) \\
&= V(\varphi, t)dX^{-1}(t) + X^{-1}(t)\, dV(\varphi, t) \\
&\quad + d < V(\varphi, t), X^{-1}(t) > (t) \\
&= V(\varphi, t)dX^{-1}(t) + X^{-1}(t) \cdot \varphi(t)'dP(t) \\
&\quad + d < V(\varphi, t), X^{-1}(t) > (t) \quad \text{using (3.8)} \\
&= \sum_{i=0}^{n} \varphi_i(t) \cdot \left[\begin{array}{c} P_i(t)dX^{-1}(t) + X^{-1}(t)\, dP_i(t) \\ + d < P_i(t), X^{-1}(t) > (t) \end{array}\right] \\
&= \varphi(t)'d\tilde{P}^X(t),
\end{aligned}
$$

i.e. $\varphi = (\varphi_0, ..., \varphi_n)$ is a self-financing trading strategy in $\tilde{\mathcal{M}}^X$. The opposite direction is proved by similar arguments. Furthermore,

$$\tilde{V}^X(\varphi, t) = \varphi(t)'\tilde{P}^X(t) = X^{-1}(t) \cdot \varphi(t)'P(t) = X^{-1}(t) \cdot V(\varphi, t)$$

for all $t \in [0,T]$. Thus, since $X^{-1} > 0$, we know that $\tilde{V}^X(\varphi) \geq 0$ if and only if $V(\varphi) \geq 0$. Furthermore, since X is $\lambda \otimes Q-$a.s. bounded, $\tilde{V}^X(\varphi)$ is $\lambda \otimes Q-$a.s. lower bounded if and only if $V(\varphi)$ is $\lambda \otimes Q-$a.s. lower bounded. Hence, the admissible trading strategies in $\mathcal{M}$ and $\mathcal{M}^X$ coincide. Consequently, the same is true for the replicating strategies in $\mathcal{M}$ and $\mathcal{M}^X$ which proves statement a).

b) Let the discounted numéraire process $\tilde{X} = \left(\tilde{X}(t)\right)_{t\in[0,T]}$ be a $\tilde{Q}$-martingale and Q^X be as defined in statement b). Since the discounted primary traded asset prices $\tilde{P}_i$ are $\tilde{Q}$-matingales for all $i = 1, ..., n$, we get, using the Bayes formula (Theorem 2.7) with $\mathcal{F}_s = \mathcal{F}$, $\mathcal{G} = \mathcal{F}_s$, $\tilde{Q} = Q^X$, $Q = \tilde{Q}$, $X = \tilde{P}_i^X(t)$, and $f = L(t)$,

$$
\begin{aligned}
E_{Q^X}\left[\tilde{P}_i^X(t)\,|\mathcal{F}_s\right] &= \frac{1}{E_{\tilde{Q}}[L(t)\,|\mathcal{F}_s]} \cdot E_{\tilde{Q}}\left[\frac{P_i(t)}{X(t)} \cdot L(t)\,|\mathcal{F}_s\right] \\
&= \frac{1}{L(s)} \cdot E_{\tilde{Q}}\left[\frac{P_i(t)}{X(t)} \cdot L(t)\,|\mathcal{F}_s\right] \\
&= \frac{X(0) \cdot P_0(s)}{X(s)} \cdot E_{\tilde{Q}}\left[\frac{P_i(t)}{X(0) \cdot P_0(t)}|\mathcal{F}_s\right] \\
&= \frac{P_0(s)}{X(s)} \cdot E_{\tilde{Q}}\left[\tilde{P}_i(t)\,|\mathcal{F}_s\right] \\
&= \frac{P_0(s)}{X(s)} \cdot \tilde{P}_i(s) \\
&= \tilde{P}_i^X(s) \quad \text{for all } i = 1, ..., n,\ 0 \le s \le t \le T.
\end{aligned}
$$

Hence, the discounted primary traded asset prices $\tilde{P}_i^X$ are Q^X-martingales for all $i = 1, ..., n$ and so is any portfolio φ of these. For the second part, let $D = D(T)$ be a (Φ^*-) attainable contingent calim in $\mathcal{M}$ with maturity T. Then there exists an admissible portfolio $\varphi \in \Phi(Q)$ $[\Phi^*(Q)]$ such that $V_\varphi(T) = D(T)$ which is equivalent to

$$
\tilde{V}_\varphi^X(T) = X^{-1}(T) \cdot V_\varphi(T) = X^{-1}(T) \cdot D(T) = \tilde{D}^X(T).
$$

Using the Bayes formula (Theorem 2.7) with $\mathcal{G} = \mathcal{F}_t$, $\tilde{Q} = Q^X$, $Q = \tilde{Q}$, $X = \tilde{D}^X(T)$, and $f = L(T)$, we get for all $t \in [0,T]$

$$
\begin{aligned}
E_{Q^X}\left[\tilde{D}^X(T)\,|\mathcal{F}_t\right] &= E_{Q^X}\left[\frac{D(T)}{X(T)}|\mathcal{F}_t\right] \\
&= \frac{1}{E_{\tilde{Q}}[L(T)\,|\mathcal{F}_t]} \cdot E_{\tilde{Q}}\left[\frac{D(T)}{X(T)} \cdot L(T)\,|\mathcal{F}_t\right] \\
&= \frac{1}{L(t)} \cdot E_{\tilde{Q}}\left[\frac{D(T)}{X(T)} \cdot L(T)\,|\mathcal{F}_t\right] \\
&= \frac{X(0) \cdot P_0(t)}{X(t)} \cdot E_{\tilde{Q}}\left[\frac{D(T)}{X(0) \cdot P_0(T)}|\mathcal{F}_t\right] \\
&= \frac{P_0(t)}{X(t)} \cdot E_{\tilde{Q}}\left[\tilde{D}(T)\,|\mathcal{F}_t\right],
\end{aligned}
$$

i.e.

$$
P_0(t) \cdot E_{\tilde{Q}}\left[\tilde{D}(T)\,|\mathcal{F}_t\right] = X(t) \cdot E_{Q^X}\left[\tilde{D}^X(T)\,|\mathcal{F}_t\right].
$$

Hence, the expected-value process of D under $\widetilde{Q}$ and numéraire P_0 coincides with expected-value process of D under Q^X and numéraire X. □

As an immediate consequence of the previous theorem we get the following corollary.

Corollary 3.32 (Change of Numéraire Formula)
Let $X = (X(t))_{t\in[0,T]}$ and $Y = (Y(t))_{t\in[0,T]}$ be two $\lambda \otimes Q$-a.s. bounded numéraires in $\mathcal{M} = \mathcal{M}(Q)$ such that the discounted numéraire processes $\widetilde{X}$ and $\widetilde{Y}$ are $\widetilde{Q}$-martingales, and let $D = D(T)$ be a contingent claim with maturity T. Then the expected-value processes of D under Q^X and Q^Y with numéraires X and Y, respectively, coincide, i.e.

$$X(t) \cdot E_{Q^X}\left[\widetilde{D}^X(T)|\mathcal{F}_t\right] = Y(t) \cdot E_{Q^Y}\left[\widetilde{D}^Y(T)|\mathcal{F}_t\right] \quad \text{for all } t \in [0,T],$$

and the Radon-Nikodým derivative $\frac{dQ^X}{dQ^Y} = \frac{dQ^X}{dQ^Y}|_{\mathcal{F}_T}$ of Q^X with respect to Q^Y is given by

$$\frac{dQ^X}{dQ^Y}|_{\mathcal{F}_t} = \frac{X(t) \cdot Y(0)}{X(0) \cdot Y(t)} \quad \text{for all } t \in [0,T].$$

3.7   The T-Forward Measure

Having discussed the effects of a numéraire change for the expected-value process, we now define a measure due to a specific choice of numéraire to expose the power of this technique. For the fixed time T let $P(t,T)$ denote the price at time $t \in [0,T]$ of a zero-coupon bond[12] with maturity T and a notional of 1. As we have already seen, the forward price $P_i^F(t,T)$ at time t of the primary traded asset with price process P_i is given by

$$P_i^F(t,T) := Forward_i(t,T) = P^{-1}(t,T) \cdot P_i(t), \, i = 1,...,n.$$

Since $P(t,T) > 0$ for all $t \in [0,T]$ it is quite natural to consider the zero-coupon bond price process $(P(t,T))_{t\in[0,T]}$ as a numéraire, i.e.

$$X(t) := P(t,T), \, t \in [0,T].$$

If the discounted numéraire process $\left(\widetilde{P}(t,T)\right)_{t\in[0,T]}$ with

$$\widetilde{P}(t,T) := P_0^{-1}(t) \cdot P(t,T), \, t \in [0,T],$$

---

[12] For more details on zero coupon bonds see Section 4.1.

is a $\widetilde{Q}$−martingale, then the forward price processes

$$\left(P_i^F(t,T)\right)_{t\in[0,T]} = \widetilde{P}_i^{P(\cdot,T)}, \; i = 1, ..., n,$$

are $Q^{P(\cdot,T)}$−martingales if we define the probability measure $Q^{P(\cdot,T)}$ on $(\Omega, \mathcal{F})$ via its Radon-Nikodým derivative $L(T)$ with respect to $\widetilde{Q}$ by

$$L(t) = \frac{dQ^{P(\cdot,T)}}{d\widetilde{Q}}\bigg|_{\mathcal{F}_t} = \frac{P(t,T)}{P(0,T)\cdot P_0(t)}, \; t \in [0,T],$$

i.e.[13]

$$L(T) = \frac{dQ^{P(\cdot,T)}}{d\widetilde{Q}}\bigg|_{\mathcal{F}_T} = \frac{1}{P(0,T)\cdot P_0(T)}.$$

Usually, the probability measure $Q^{P(\cdot,T)}$ is called the T−*forward measure* and simply denoted by Q^T. Furthermore, using the change of numéraire formula (Corollary 3.32), the expected-value process of a contingent claim $D = D(T)$ with maturity T is given by

$$\begin{aligned}
P_0(t) \cdot E_{\widetilde{Q}}\left[\widetilde{D}(T)|\mathcal{F}_t\right] &= P(t,T) \cdot E_{Q^T}\left[\widetilde{D}^X(T)|\mathcal{F}_t\right] \\
&= P(t,T) \cdot E_{Q^T}[D(T)|\mathcal{F}_t].
\end{aligned}$$

Example. The price $Call(t,T,K)$ at time $t = 0$ of a European call option on P_i, $i \in \{1, ..., n\}$, with maturity T and exercise price K is given by

$$\begin{aligned}
Call(0,T,K) &= P(0,T) \cdot E_{Q^T}[\max\{P_i(T) - K, 0\}] \\
&= P(0,T) \cdot \left(E_{Q^T}[P_i(T) \cdot 1_A] - K \cdot Q^T(A)\right)
\end{aligned}$$

with

$$A := \{\omega \in \Omega : P_i(\omega, T) > K\}.$$

Using Corollary 3.32 with P_i as a numéraire process we get

$$\begin{aligned}
Call(0,T,K) &= P(0,T) \cdot E_{Q^T}[P_i(T) \cdot 1_A] - P(0,T) \cdot K \cdot Q^T(A) \\
&= P_i(0) \cdot E_{Q^{P_i}}\left[\frac{P_i(T) \cdot 1_A}{P_i(T)}\right] - P(0,T) \cdot K \cdot Q^T(A) \\
&= P_i(0) \cdot Q^{P_i}(A) - P(0,T) \cdot K \cdot Q^T(A).
\end{aligned}$$

---

[13] If we are given a so-called short rate process $(r(t))_{t\in[0,T]}$ with $P_0(t) = e^{\int_0^t r(s)ds}$ for all $t \in [0,T]$ as it will be the case for most of the interest rate market models in Section 4, this is equivalent to

$$L(T) = \frac{1}{P(0,T)} \cdot e^{-\int_0^T r(s)ds}.$$

We could also calculate the price of an option to exchange the two primary traded assets with price processes P_1 and P_2 according to the exchange agreement $K \cdot P_2$ against P_1 at time T giving the owner of this so-called *exchange call option* the right to receive the cash flow $D(T) = (P_1(T) - K \cdot P_2(T))^+$. Using P_2 as a numéraire process, the price $ExCall(t, T, K)$ at time $t = 0$ of the exchange call option is given by

$$
\begin{aligned}
ExCall(0, T, K) &= P_2(0) \cdot E_{Q^{P_2}}\left[\tilde{D}^{P_2}(T)\right] \\
&= P_2(0) \cdot E_{Q^{P_2}}\left[\left(\frac{P_1(T)}{P_2(T)} - K\right)^+\right] \\
&= P_2(0) \cdot \left(E_{Q^{P_2}}\left[\frac{P_1(T)}{P_2(T)} \cdot 1_A\right] - K \cdot Q^{P_2}(A)\right)
\end{aligned}
$$

with
$$
A := \{\omega \in \Omega : P_1(\omega, T) > K \cdot P_2(\omega, T)\}.
$$

If we now apply Corollary 3.32 with P_1 as a numéraire we get

$$
\begin{aligned}
ExCall(0, T, K) &= P_2(0) \cdot E_{Q^{P_2}}\left[\frac{P_1(T)}{P_2(T)} \cdot 1_A\right] - P_2(0) \cdot K \cdot Q^{P_2}(A) \\
&= P_1(0) \cdot Q^{P_1}(A) - P_2(0) \cdot K \cdot Q^{P_2}(A).
\end{aligned}
$$

$\square$

Part II

Modelling and Pricing in Interest-Rate Markets

4

Interest-Rate Markets

One dollar today is better than one dollar tomorrow and one dollar tomorrow is certainly better than one dollar in one year. In other words, time is money. But what should be paid today or tomorrow or, more generally at time t, for a guaranteed cash payment of one dollar at a time T, $T \geq t$, in the far future? This is one of the questions which will be answered in the interest-rate market. Up to today this market is one of the most important financial markets trading instruments such as coupon bonds, forward-rate agreements on coupon bonds, interest-rate futures and swaps as well as standard or exotic interest-rate options. Because the interest-rate market is a specific financial market, we will apply the results of Chapter 3 to embed it in the general framework of Section 3.1. We start by defining the general interest-rate market model in Section 4.1. No-arbitrage and completeness conditions in the interest-rate market model are given in Section 4.2, while Section 4.3 deals with the pricing of interest-rate-related contingent claims. Because there are infinitely many possible maturity times when we could get an invested amount of money back from a bank or the market, there are also infinitely many interest rates representing the possible investment horizons and changing their values over time. It has therefore been and still is a specific challenge to find factors which sufficiently well describe the behaviour of all interest rates over time. In Section 4.4 we discuss one of the most general platforms for pricing interest-rate derivatives, the Heath-Jarrow-Morton framework. It deals with an infinite number of so-called forward short rates, specified today for an infinitely short time-period at some future point in time, to describe the movement of interest rates. If the forward short rate is specified for today, the resulting rate is called the short rate. There is a huge class of one-factor models, the famous short-rate models, dealing with the short rate as the only driving factor. Even if their concentration on one factor is rather restrictive, the attraction of these models is the technical tractability and very often the numerical advantage of closed-form solutions for the prices of contingent claims. One-factor models are discussed in Section 4.5 including an excursion to the Gaussian models which easily fit into the Heath-Jarrow-Morton framework. As a representative of the Heath-Jarrow-Morton and especially of the Gaussian models, we will pick out the Hull-White model, mainly for three reasons. First, it is general enough to allow for arbitrage-free prices and closed-form solutions. Second, it allows for a detailed derivation of distributional results not only for the interest rates but also for the zero-coupon bond prices. And third, it is possible to explicitly derive Green's function for this model. The latter will help us to show a specific application of this concept to calculate directly the prices of derivatives, even those of exotic options. A brief overview of multi-factor models can be found in Section 4.6.

Up to now there is no dominating model within the interest-rate market. Very often it even depends on the specific derivative which model is used. A pretty new candidate for becoming a benchmark model is the LIBOR market model, describing the behaviour of market rates rather than that of the short or forward short rates. It is presented in Section 4.7 and extensively uses the change of numéraire technique and the forward measure derived in Sections 3.6 and 3.7.

All of the previous models do not deal with the possibility of zero-rate changes because of defaults in the financial market. In Section 4.8 we therefore give a brief overview of the most famous credit-risk models.

If you are already familiar with the modelling of interest-rate markets you may immediately switch to Chapter 5 for a more practically oriented description and pricing of interest-rate derivatives. If you are interested in further insights with respect to tests and implementations of interest-rate models you may refer to Brigo and Mercurio [BM01], Pelsser [Pel00], or Rebonato [Reb96].

4.1   The Interest-Rate Market Model

The guaranteed cash payment of one dollar in the future is one of the basic or primary assets in the interest-rate market and is called a zero-coupon bond or discount bond. At an evaluation time t we have to pay a price of $P(t,T)$ to get one dollar back at the maturity date T, $T \geq t$ of the zero-coupon bond. The supplement zero-coupon indicates that there will be no exchange of payments, i.e. no coupon payments, during the lifetime of this contract. The time $T - t$ of the period from t to T is called the time to maturity of the zero-coupon bond. Note that two zero-coupon bonds with the same time to maturity but evaluated at different points in time may quite well have different prices as the interest-rate market may change over time. Also note that the price of a specific zero-coupon bond will change over time even if the interest-rate market doesn't change since the time to maturity of that zero-coupon bond will decrease until it finally satisfies the condition $P(T,T) = 1$ at its maturity date $t = T$. This effect is called the time to maturity effect of a zero-coupon bond. If we plot, for a fixed point in time t, the (market) prices $P(t,T)$ of the zero-coupon bonds with different maturity dates $T \geq t$ depending on their time to maturity $T - t$, we get a chart which is called the zero-coupon bond or discount curve at time t. The time to maturity effect is pictured by the starting value of 1. Usually the discount curve is decreasing and random market movements are represented by varying discount curves at different points in time t.

In practice, the zero-coupon bond prices $P(t,T)$ are usually translated into an implicit rate of return $R(t,T)$, the so-called (continuous) zero rate or spot rate at time t for the maturity time T or for the time to maturity $T - t$. This is done by the relation

$$P(t,T) = e^{-R(t,T) \cdot (T-t)} \quad \text{or} \quad R(t,T) = -\frac{\ln P(t,T)}{T - t} \,.$$

If, as with the discount curve, we plot, for a fixed point in time t, the zero rates $R(t,T)$ for different maturity dates $T \geq t$ dependent on their time to maturity $T - t$, we get the so-called zero-rate curve at time t. According to the relation of the logarithmic zero-coupon bond prices and the time to maturity the zero-rate curve may have different shapes such as an increasing or decreasing curve. We call this the term structure of the zero-rate curve. As the discount curve the zero-rate curve as well as the term structure of the zero-rate curve may change over time due to changes of the interest-rate market. In some sense the zero-rate curve expresses the market view on how the zero rates may evolve over time. If, for example, the market expects the zero rates to increase, the zero-coupon bonds with a longer time to maturity will have a higher zero rate than those with a shorter time to maturity. The resulting increasing zero-rate curve is called normal. If interest rates are expected to fall the corresponding zero-rate curve will be decreasing. We call this an inverse zero-rate curve.

If we let the time to maturity approach a value of zero the corresponding limit of the zero rate is of special interest, especially from a modelling point of view. Formally we define

$$r(t) := R(t,t) := -\lim_{\Delta t \to 0} \frac{\ln P(t, t + \Delta t)}{\Delta t} = -\frac{\partial}{\partial T} \ln P(t,t)$$

where $\frac{\partial}{\partial T} \ln P(t,t)$ is defined to be the partial derivative $\frac{\partial}{\partial T} \ln P(t,T)$ evaluated at $T = t$. We hereby and in the sequel assume that all derivatives, integrals, and limits exist as they appear. Hence, $r(t)$ is the zero rate for an infinitesimal time to maturity at time point t and is therefore called the (instantaneous) short rate at time t. Many of the commonly applied yield models use the short rate to describe interest-rate movements. Nevertheless, it should be mentioned that the short rate, due to the restriction to an infinitesimal time to maturity, does not depend on the maturity time T any longer. We therefore lose information if we move from a whole zero-rate curve to a representative short rate. In general we will not be able to derive the whole zero-rate curve by knowing the short rate at time t only.

Considering infinitesimal time-periods but still keeping the dependence on the time to maturity T is possible via the definition of the so-called (continuous) forward rates. To do this imagine a contract (forward zero-coupon bond) in which, at time t, two parties agree at no cost to exchange at a future time $T_1 \geq t$ a zero-coupon bond with time to maturity $T_2 - T_1$, $T_1 \leq T_2$, for a cash payment of $P(t, T_1, T_2)$. To what price $P(t, T_1, T_2)$ should the two parties agree at time $t \leq T_1$? The answer is easy: If we sell a number of $P(t, T_1, T_2)$ zero-coupon bonds with time to maturity T_1 at time t and at the same time agree to invest the amount of cash $P(t, T_1, T_2)$ we receive at time T_1 in a zero-coupon bond with time to maturity $T_2 - T_1$, this portfolio is identical to a zero-coupon bond with maturity T_2. The price for this portfolio must equal the price of the zero-coupon bond with time to maturity T_2 if the interest-rate market is arbitrage-free, i.e.

$$P(t, T_1, T_2) \cdot P(t, T_1) = P(t, T_2) \quad \text{or} \quad P(t, T_1, T_2) = \frac{P(t, T_2)}{P(t, T_1)}.$$

If we denote the forward zero rate corresponding to the forward zero-coupon bond with $R(t, T_1, T_2)$, we get

$$P(t, T_1, T_2) = e^{-R(t, T_1, T_2) \cdot (T_2 - T_1)} \quad \text{or} \quad R(t, T_1, T_2) = -\frac{\ln P(t, T_1, T_2)}{T_2 - T_1}.$$

Therefore, the forward zero rate is given by

$$R(t, T_1, T_2) = -\frac{\ln P(t, T_2) - \ln P(t, T_1)}{T_2 - T_1}.$$

If we let the time to maturity $T_2 - T_1$ of the forward zero-rate approach zero we get for $T_1 = T$ the so-called forward short rate or briefly forward rate $f(t, T)$, i.e.

$$
\begin{aligned}
f(t, T) \quad &:= \quad R(t, T, T) := -\lim_{\Delta t \to 0} \frac{\ln P(t, T + \Delta t) - \ln P(t, T)}{\Delta t} \\
&= \quad -\frac{\partial}{\partial T} \ln P(t, T).
\end{aligned}
$$

$f(t, T)$ represents the interest rate for an infinitesimal time-period at time T, derived at time point t. Especially, for $T = t$ we get

$$f(t, t) = r(t).$$

Unlike the short rate, the forward short rate keeps the dependence on T. Compared to the zero-coupon bond prices we will not lose information concentrating on the forward short rates. In other words we can derive the zero-coupon bond prices or the zero-rate curve if we know all forward short rates by

$$P(t,T) = e^{-\int_t^T f(t,s)ds} \quad \text{or} \quad R(t,T) = \frac{1}{T-t} \int_t^T f(t,s)ds\,.$$

Note that we are dealing with an infinite number of forward short rates to derive the zero-coupon bond prices or zero rates. If we plot at a given point in time t the forward rates $f(t,T)$ for different time horizons $T \geq t$ dependent on the difference $T-t$, we get the so-called forward curve at time t. Since we have the following relation between the forward rate $f(t,T)$ and the corresponding zero rate $R(t,T)$:

$$
\begin{aligned}
f(t,T) &= -\frac{\partial}{\partial T} \ln P(t,T) = \frac{\partial}{\partial T}\left[R(t,T) \cdot (T-t)\right] \\
&= R(t,T) + (T-t) \cdot \frac{\partial}{\partial T} R(t,T)\,,
\end{aligned}
$$

the forward curve will lie above (below) the zero-rate curve if the latter is normal (inverse). For $T = t$ both curves coincide at a value of $r(t)$. For fixed T the forward short rate is a stochastic process in t with a final value of $f(T,T) = r(T)$ at time T.

Another security will be very useful for the evaluation of contingent claims or derivatives, the so-called cash account. It will be the numéraire to define our discounted primary assets and therefore the discounted interest-rate market. If at time t_0 we invest one dollar in a zero-coupon bond with maturity time t we get a final payment at time t of $P^{-1}(t_0,t)$ with

$$-\ln P(t_0,t) = \int_{t_0}^t f(t_0,s)ds = \lim_{\substack{n \to \infty \\ t = t_0 + n \cdot \Delta s_n}} \sum_{k=0}^n f(t_0, t_0 + k \cdot \Delta s_n) \cdot \Delta s_n\,.$$

From todays point of view (time t_0) we can derive the final value of this investment with time to maturity t by investing one dollar at time t_0 into infinitely many forward zero-coupon bonds with infinitesimal time to maturity such that one maturity date is the starting date of the next forward zero-coupon bond and all zero-coupon bonds together span the time-period from t_0 to t. The number of forward zero-coupon bonds in which we have to reinvest equals the final payment of the previous investment. If the investment in these forward zero-coupon bonds doesn't take place under the actual market conditions at time t_0 but successively in time, the conditions and prices at which the forward zero-coupon bonds can be bought are not already fixed but stochastic. Therefore, we have to replace the forward short rates by the corresponding future short rates to get the so-called cash account with a final random payment of $P_0(t)$ with

$$\ln P_0(t) := \lim_{\substack{n \to \infty \\ t = t_0 + n \cdot \Delta s_n}} \sum_{k=0}^{n} r(t_0 + k \cdot \Delta s_n) \cdot \Delta s_n = \int_{t_0}^{t} r(s)ds.$$

The corresponding differential equation has the form

$$dP_0(t) = r(t) \cdot P_0(t)dt$$

with $P_0(t_0) = 1$. Hence, $P_0^{-1}(t)$ follows the differential equation

$$dP_0^{-1}(t) = -r(t) \cdot P_0^{-1}(t)\, dt,$$

with $P_0(t_0) = P_0^{-1}(t_0) = 1$.

We have already defined the basic financial instruments of an interest-rate market, the zero-coupon bonds as well as the cash account which we will choose to be our numéraire. As in Section 3.1 we suppose that we are dealing with a frictionless security market where investors are allowed to trade continuously up to some fixed finite horizon T^*. Uncertainty in the interest-rate market is modelled by a complete probability space $(\Omega, \mathcal{F}, Q)$ where all prices are driven by a $m-$dimensional Wiener process $W = (W(t))_{t \in [t_0, T^*]}$ and $t_0 \in [0, T^*]$ denotes an arbitrary point in time. We assume that the probability space is filtered by the natural filtration $\mathbb{F} = \mathbb{F}(W)$ with $\mathcal{F} = \mathcal{F}_{T^*} = \mathcal{F}_{T^*}(W)$ and $\mathcal{F}_{t_0} = \mathcal{F}_{t_0}(W) = \{\emptyset, \Omega\}$. The prices of the zero-coupon bonds with different maturities are described by non-negative Itô processes $(P(t, T))_{t \in [t_0, T]}$ with

$$\begin{aligned} d_t P(t, T) &= \mu_P(t, T)\, dt + \sigma_P(t, T)\, dW(t) \\ &= \mu_P(t, T)\, dt + \sum_{j=1}^{m} \sigma_{P,j}(t, T)\, dW_j(t) \end{aligned}$$

for all $t \in [t_0, T]$, with progressively measurable stochastic processes μ_P and σ_P such that for all $T \in \mathcal{T} := [t_0, T^*]$

$$\int_{t_0}^{T} |\mu_P(s, T)|\, ds < \infty \quad Q - a.s. \tag{M1^{IRM}}$$

and

$$\int_{t_0}^{T} \sigma_{P,j}^2(s, T)\, ds < \infty \quad Q - a.s. \text{ for all } j = 1, ..., m.$$

As in Section 3.1 we furthermore assume for technical convenience that for all $T \in \mathcal{T}$ and for all $j \in \{1, ..., m\}$

$$E_Q\left[\int_{t_0}^{T} \sigma_{P,j}^2(s, T)\, ds\right] < \infty \tag{M2^{IRM}}$$

and refer to this model under conditions $\left(M1^{IRM}\right)$ and $\left(M2^{IRM}\right)$ as the (primary) *interest-rate market* $\mathcal{M}^{IRM} = \mathcal{M}^{IRM}\left(Q,\mathcal{T}\right)$. Also, we assume that the numéraire P_0 satisfies conditions $(N1)$ and $(N2)$ of Section 3.1 with time 0 replaced by t_0. According to equation (3.4) the discounted or normalized zero-coupon bond prices

$$\widetilde{P}\left(t,T\right) = P_0^{-1}(t) \cdot P\left(t,T\right), \, t_0 \leq t \leq T \leq T^*,$$

are described by

$$d\widetilde{P}\left(t,T\right) = \widetilde{\mu}_P\left(t,T\right) dt + \widetilde{\sigma}_P\left(t,T\right) dW\left(t\right) \tag{4.1}$$

with

$$\widetilde{\mu}_P\left(t,T\right) = \left(\mu_P\left(t,T\right) - r\left(t\right)\cdot P(t,T)\right)\cdot P_0^{-1}\left(t\right)$$

and

$$\widetilde{\sigma}_{P,j}\left(t,T\right) = \sigma_{P,j}\left(t,T\right)\cdot P_0^{-1}\left(t\right)$$

for all $j = 1,...,m, t_0 \leq t \leq T \leq T^*$. Correspondingly, we set $\widetilde{\mu}_0 \equiv 0$ and $\widetilde{\sigma}_{0j} \equiv 0$, $j = 1,...,m$ for the discounted numéraire process.

The *discounted* or *normalized* (primary) *interest-rate market* $\widetilde{\mathcal{M}}^{IRM} = \widetilde{\mathcal{M}}^{IRM}\left(Q,\mathcal{T}\right)$ is defined by the discounted zero-coupon bond prices and the conditions

$$\int_{t_0}^{T} \left|\widetilde{\mu}_P\left(s,T\right)\right| ds < \infty \quad Q-a.s. \qquad (\widetilde{M1}^{IRM})$$

and

$$E_Q\left[\int_{t_0}^{T} \widetilde{\sigma}_{P,j}^2\left(s,T\right) ds\right] < \infty \qquad (\widetilde{M2}^{IRM})$$

for all $j = 1,...,m$ and for all $T \in \mathcal{T}$.

With respect to the pricing of specific contingent claims $D = D\left(T_D\right)$ maturing at time $T_D \in \mathcal{T}$, we consider T_D to be the arbitrary but fixed planning horizon in our interest-rate markets $\mathcal{M}^{IRM}$ and $\widetilde{\mathcal{M}}^{IRM}$.

4.2 No-Arbitrage and Completeness

Because there are infinitely many zero-coupon bonds corresponding to the different maturity dates we have to carefully define what we consider to be a trading strategy or portfolio process. By Theorem 3.11 the absence of arbitrage opportunities in a financial market is basically guaranteed by the existence of an equivalent martingale measure. Under this measure,

the discounted price processes of the primary traded assets are, by definition, martingales. For the interest-rate market we consider the zero-coupon
bonds as our primary traded assets. Our interest is therefore focused on
the discounted zero-coupon bond prices $\widetilde{P}(t,T)$ with $t_0 \leq t \leq T \leq T^*$.
However, in extension to our financial market model, we have to deal with
an infinite number of discounted zero-coupon bond prices which we want to
be martingales with respect to a specific measure $\widetilde{Q}$. According to practical
trading restrictions we concentrate on portfolios consisting of an arbitrary
but finite number of assets. For any set $\mathcal{T}_n := \{T_1, ..., T_n\} \subset \mathcal{T}$, $t_0 < T_D \leq$
$T_1 < \cdots < T_n = T^*$, $n \in \mathbb{N}$, of maturity times we consider the finite
interest-rate market $\mathcal{M}^{IRM}(\mathcal{T}_n) := \mathcal{M}^{IRM}(\mathcal{T}_n, Q)$ consisting of the zero-
coupon bonds with maturity times in $\mathcal{T}_n$, i.e. $P_i(t) = P(t, T_i)$, $i = 1, ..., n$,
and the planning horizon T_D[1]. Hence, conditions $\left(M1^{IRM}\right)$, $\left(M2^{IRM}\right)$ and
$\left(\widetilde{M1}^{IRM}\right)$, $\left(\widetilde{M2}^{IRM}\right)$ restricted on all $T \in \mathcal{T}_n$ are the corresponding assumptions to $(M1), (M2)$ and $\left(\widetilde{M1}\right), \left(\widetilde{M2}\right)$ of Section 3.1 for $\mathcal{M}^{IRM}(\mathcal{T}_n)$.
Assumed for all $T \in \mathcal{T}$, they ensure that all integrals exist in any finite
interest-rate market. With respect to finite interest-rate markets we are
therefore playing on familiar ground. We denote the set of all admissible
trading strategies in $\mathcal{M}^{IRM}(\mathcal{T}_n)$ by $\Phi(\mathcal{T}_n, Q)$ and the set of equivalent martingale measures to Q in $\mathcal{M}^{IRM}(\mathcal{T}_n)$ by $\mathbb{M}(\mathcal{T}_n) = \mathbb{M}(Q, \mathcal{T}_n)$ and give the
following definition.

Definition 4.1 *The interest-rate market* $\mathcal{M}^{IRM}$ *is called* **arbitrage-free**
if any finite interest-rate market $\mathcal{M}^{IRM}(\mathcal{T}_n)$ *with* $\mathcal{T}_n := \{T_1, ..., T_n\}$,
$n \in \mathbb{N}$, *(and planning horizon* $T_D \leq T_1$*) is arbitrage-free.*

Using Theorem 3.11 we know that $\mathcal{M}^{IRM}$ contains no $\lambda \otimes Q - a.s.$
bounded arbitrage opportunitites if $\mathbb{M}(\mathcal{T}_n) \neq \emptyset$ and the martingale condition

$$E_{\widetilde{Q}}\left[\widetilde{P}^2(t,T)\right] < \infty \quad \text{for all } t \in [t_0, T], \, T \in \mathcal{T}_n$$

is satisfied for all finite interest-rate markets $\mathcal{M}^{IRM}(\mathcal{T}_n)$, $n \in \mathbb{N}$. Necessary
and sufficient conditions for each single market were given in Lemma 3.12.
If there has to be an equivalent martingale measure in any finite interest-
rate market, the natural question which follows is if we can find a single
probability measure $\widetilde{Q}$ on $(\Omega, \mathcal{F})$ being an equivalent martingale measure in

---

[1] Note that the assumption $T_D \leq T_1 = \min\{T : T \in \mathcal{T}_n\}$ is done for the ease of
exposition. Basically we could also allow for $T_1 < T_D \leq T^*$ and synthetically extend the
maturity time of all discount bonds with maturity $T_i < T_D$ by setting $P(t, T_i) = 0$ for
all $t \in (T_i, T^*]$. Additionally we then set $\varphi_i(t) = 0$ for all $t \in (T_i, T^*]$ and each trading
strategy $(\varphi_1, ..., \varphi_n)$. This is to say that after the maturity of a synthetically extended
discount bond, i.e. depending on t, the market is considered to be reduced by this asset
and the corresponding conditions like those on completeness have to be fullfilled in the
reduced market only.

every finite interest-rate market. We therefore slightly extend our definition of an equivalent martingale measure.

Definition 4.2 (Equivalent Martingale Measure) *A probability measure $\widetilde{Q}$ on $(\Omega, \mathcal{F})$ is called an equivalent martingale measure to Q if:*

(i) *$\widetilde{Q}$ is equivalent to Q.*

(ii) *The discounted price process $\left(\widetilde{P}(t,T)\right)_{t \in [t_0, T]}$ is a $\widetilde{Q}$-martingale for all $T \in \mathcal{T}$.*

The **set of equivalent martingale measures** *to Q is denoted by $I\!M = I\!M(Q)$.*

Using this definition the following lemma is straightforward.

Lemma 4.3 *The following statements are equivalent:*

a) *$I\!M \neq \emptyset$.*

b) *There is an equivalent martingale measure $\widetilde{Q}$ to Q such that for any finite set $\mathcal{T}_n := \{T_1, ..., T_n\}$, $n \in I\!N$, $t_0 < T_1 < \cdots < T_n = T^*$, the discounted price processes $\left(\widetilde{P}(t,T)\right)_{t \in [t_0, T_1]}$ are $\widetilde{Q}$-martingales for all $T \in \mathcal{T}_n$.*

The following theorem tells us, under which conditions an equivalent martingale measure exists.

Theorem 4.4   a) *Suppose that there exists a m-dimensional progressively measurable stochastic process γ such that:*

(i) *The following Novikov condition holds for γ:*

$$E_Q\left[e^{\frac{1}{2}\int_{t_0}^{T^*} \|\gamma(s)\|^2 ds}\right] < \infty. \qquad (NV^{IRM})$$

(ii) *$E_Q[L(\gamma, T^*)] = 1$ with*

$$L(\gamma, t) := e^{-\int_{t_0}^{t} \gamma(s)' dW(s) - \frac{1}{2}\int_{t_0}^{t} \|\gamma(s)\|^2 ds}.$$

(iii) *The no-arbitrage condition*

$$\mu_P(t,T) - \sigma_P(t,T)\gamma(t) = r(t) \cdot P(t,T) \qquad (NA^{IRM})$$

holds for all $t_0 \leq t \leq T \leq T^*$.

Furthermore, let the probability measure $\widetilde{Q}$ *on* $(\Omega, \mathcal{F}_{T^*}) = (\Omega, \mathcal{F})$ *be defined by* $\widetilde{Q} := Q_{L(\gamma, T^*)}$. *Then the stochastic process* $\widetilde{W} = \left(\widetilde{W}(t) \right)_{t \in [t_0, T^*]}$ *defined by*

$$d\widetilde{W}(t) := \gamma(t)\, dt + dW(t)\, , \, t \in [t_0, T^*]\, ,$$

is a $\widetilde{Q}-$ *Wiener process and the discounted price processes* $\left(\widetilde{P}(t, T) \right)_{t \in [t_0, T]}$ *have the following representation in terms of* $\widetilde{W}$:

$$d\widetilde{P}_0(t) = 0 \text{ and } d\widetilde{P}(t, T) = \tilde{\sigma}_P(t, T) d\widetilde{W}(t)\, , \, t_0 \le t \le T \le T^*.$$

Furthermore,

$$dP(t, T) = r(t) \cdot P(t, T)\, dt + \sigma_P(t, T) d\widetilde{W}(t)\, .$$

If the martingale condition

$$E_{\widetilde{Q}}\left[\int_{t_0}^{T^*} \|\tilde{\sigma}_P(s, T)\|^2\, ds \right] < \infty \quad \text{for all } t_0 \le T \le T^* \quad (\widetilde{M3}^{IRM})$$

is satisfied, then $\widetilde{Q} \in \mathbb{M}$, *i.e.* $\mathbb{M} \ne \emptyset$.

b) Let $\mathbb{M} \ne \emptyset$. *Then for each* $\widetilde{Q} \in \mathbb{M}$ *there exists a* $m-$*dimensional progressively measurable stochastic process* γ *such that conditions (i') and (ii) hold with*

(i') $\displaystyle\int_{t_0}^{T^*} \|\gamma(s)\|^2\, ds < \infty \; Q - a.s.$

If, in extension to (i') Novikov's condition (i) holds, then condition (iii) is satisfied.

Proof. Suppose there exists an equivalent martingale measure $\widetilde{Q} \in \mathbb{M}$. Then, using Lemma 3.9, there exists a $m-$dimensional progressively measurable stochastic process γ such that $\int_{t_0}^{T^*} \|\gamma(s)\|^2\, ds < \infty \; Q - a.s.$ and

$$\widetilde{Q}(A) = E_Q\left[1_A \cdot L(\gamma, T^*) \right] \quad \text{for all } A \in \mathcal{F}_{T^*} = \mathcal{F}$$

with

$$L(\gamma, t) := e^{-\int_{t_0}^{t} \gamma(s)'\, dW(s) - \frac{1}{2} \int_{t_0}^{t} \|\gamma(s)\|^2\, ds}.$$

Since $\Omega \in \mathcal{F}$ we know that

$$E_Q\left[L(\gamma, T^*) \right] = E_Q\left[1_\Omega \cdot L(\gamma, T^*) \right] = \widetilde{Q}(\Omega) = 1.$$

So conditions (i') and (ii) are satisfied. Now let (i) be satisfied. Then $\widetilde{W} = \left(\widetilde{W}(t)\right)_{t \in [t_0, T^*]}$, defined by $d\widetilde{W}(t) := \gamma(t)\,dt + dW(t)$ on $[t_0, T^*]$, is a $\widetilde{Q}$–Wiener process on $(\Omega, \mathcal{F})$. Thus, using equation (4.1), we get

$$
\begin{aligned}
d_t \widetilde{P}(t,T) &= P_0^{-1}(t) \cdot [(\mu_P(t,T) - r(t) \cdot P(t,T))\,dt + \sigma_P(t,T)dW(t)] \\
&= P_0^{-1}(t) \cdot (\mu_P(t,T) - r(t) \cdot P(t,T) - \sigma_P(t,T)\gamma(t))\,dt \\
&\quad + \widetilde{\sigma}_P(t,T)d\widetilde{W}(t) \text{ for all } t \in [t_0, T].
\end{aligned}
$$

Hence, since the discounted zero-coupon bond prices are $\widetilde{Q}$–martingales for all $t_0 \leq T \leq T^*$ we know by Theorem 2.42 that for all $t_0 \leq t \leq T \leq T^*$,

$$
\mu_P(t,T) - r(t) \cdot P(t,T) - \sigma_P(t,T)\gamma(t) = 0. \tag{4.2}
$$

For the opposite direction, suppose that $\widetilde{Q} := Q_{L(\gamma,T^*)}$ is defined by

$$
\widetilde{Q}(A) = E_Q\left[1_A \cdot L(\gamma,T^*)\right] \text{ for all } A \in \mathcal{F}_{T^*} = \mathcal{F}.
$$

Then $\widetilde{Q}$ is a probability measure on $(\Omega, \mathcal{F})$ because of (ii). Furthermore, $\widetilde{W} = \left(\widetilde{W}(t)\right)_{t \in [t_0, T^*]}$, defined by $d\widetilde{W}(t) := \gamma(t)\,dt + dW(t)$ on $[t_0, T^*]$, is a $\widetilde{Q}$–Wiener process on $(\Omega, \mathcal{F})$ because of (i). Using equation (4.1), we get

$$
\begin{aligned}
d_t \widetilde{P}(t,T) &= P_0^{-1}(t) \cdot [(\mu_P(t,T) - r(t) \cdot P(t,T))\,dt + \sigma_P(t,T)dW(t)] \\
&= P_0^{-1}(t) \cdot [\sigma_P(t,T)\gamma(t)\,dt + \sigma_P(t,T)dW(t)] \text{ using (iii)} \\
&= \widetilde{\sigma}_P(t,T)d\widetilde{W}(t).
\end{aligned}
$$

Since $\widetilde{W}$ is a $\widetilde{Q}$–Wiener process we know, following Theorem 2.42, that $\left(\widetilde{P}(t,T)\right)_{t \in [t_0, T]}$ is a $\widetilde{Q}$–martingale for all $t_0 \leq T \leq T^*$ because of $\left(\widetilde{M3}^{IRM}\right)$, i.e. $\widetilde{Q}$ is an equivalent martingale measure for Q. Using the product rule (Lemma 2.36), we get

$$
\begin{aligned}
dP(t,T) &= d\left(P_0(t) \cdot \widetilde{P}(t,T)\right) = P_0(t)\,d\widetilde{P}(t,T) + \widetilde{P}(t,T)\,dP_0(t) \\
&= P_0(t) \cdot \widetilde{\sigma}_P(t,T)d\widetilde{W}(t) + \widetilde{P}(t,T) \cdot r(t) \cdot P_0(t)\,dt \\
&= r(t) \cdot P(t,T)\,dt + \sigma_P(t,T)d\widetilde{W}(t)
\end{aligned}
$$

which completes the proof. $\square$

The next question we have to deal with is the completeness of our interest-rate market. Again, we start with extending the definition of attainability and completeness to the case of infinitely many primary traded assets (discount bonds).

Definition 4.5 *Let the random variable* $D = D(T_D)$ *on* $(\Omega, \mathcal{F}_{T_D})$ *be a (European) contingent claim with maturity* T_D. *If there exists a finite set* $\mathcal{T}_n := \{T_1, ..., T_n\}, n \in \mathbb{N}, t_0 < T_D \leq T_1 < \cdots < T_n = T^*$, *and an admissible trading strategy* $\varphi \in \Phi(\mathcal{T}_n, Q)$ *with* $V(\varphi, T_D) = D(T_D)$ *we call* φ *a* **replicating** *or* **hedging strategy** *for* D *in* $\mathcal{M}^{IRM}$ *or* $\mathcal{M}^{IRM}(\mathcal{T}_n)$ *respectively. The contingent claim* $D = D(T_D)$ *is called* (Φ^*-) **attainable** *in* $\mathcal{M}^{IRM}$ *if there exists a finite interest-rate market* $\mathcal{M}^{IRM}(\mathcal{T}_n)$ *with* $\mathcal{T}_n := \{T_1, ..., T_n\}, n \in \mathbb{N}, t_0 < T_D \leq T_1 < \cdots < T_n = T^*$, *such that* D *is* $(\Phi^*(\mathcal{T}_n)-)$ *attainable in* $\mathcal{M}^{IRM}(\mathcal{T}_n)$. *The interest-rate market* $\mathcal{M}^{IRM}$ *is said to be* (Φ^*-) **complete** *if any contingent claim* $D = D(T_D)$, *such that* $\widetilde{D}(T_D) := P_0^{-1}(T_D) \cdot D(T_D)$ *is bounded, is* (Φ^*-) *attainable in* $\mathcal{M}^{IRM}$.

Thus, the completeness of $\mathcal{M}^{IRM}$ is closely related to completeness of finite interest-rate markets. The following theorem is an immediate consequence of Theorem 3.19 and Theorem 3.21.

Theorem 4.6 (Completeness) *Let* $\mathcal{T}_n := \{T_1, ..., T_n\}, n \in \mathbb{N}$, *with* $t_0 < T_D \leq T_1 < \cdots < T_n = T^*$. *Suppose there exists a* $m-$*dimensional progressively measurable stochastic process* γ *such that assumptions* $(i), (ii)$, *and* (iii) *of Theorem 4.4 are satisfied.*

a) *Let the matrix* $\sigma_P(t, \mathcal{T}_n) := (\sigma_{P,j}(t, T_i))_{\substack{i=1,...,n \\ j=1,...,m}}$ *have a left inverse* $\lambda \otimes Q - a.s.$ *on* $\mathcal{B}([t_0, T_D]) \otimes \mathcal{F}_{T_D}$. *Then* $|\mathbb{M}(\mathcal{T}_n)| = 1$ *on* $\mathcal{F}_{T_D}{}^2$.

b) *Let the matrix* $\sigma_P(t, \mathcal{T}_n)$ *have a left inverse* $\lambda \otimes Q - a.s.$ *on* $\mathcal{B}([t_0, T_D]) \otimes \mathcal{F}_{T_D}$ *such that* $\widetilde{\sigma}_P^{-1}(t, \mathcal{T}_n) := \sigma_P^{-1}(t, \mathcal{T}_n) \cdot P_0(t)$ *is bounded* $\lambda \otimes Q - a.s.$ *on* $\mathcal{B}([t_0, T_D]) \otimes \mathcal{F}_{T_D}$. *Then the finite interest-rate market* $\mathcal{M}^{IRM}(\mathcal{T}_n)$ *is complete.*

c) *Let* $\mathcal{M}^{IRM}(\mathcal{T}_n)$ *be complete. Then* $\sigma_P(t, \mathcal{T}_n)$ *has a left inverse* $\lambda \otimes Q - a.s.$ *on* $\mathcal{B}([t_0, T_D]) \otimes \mathcal{F}_{T_D}$. *Especially,* $|\mathbb{M}(\mathcal{T}_n)| = 1$ *on* $\mathcal{F}_{T_D}$.

The following theorem gives an answer on the question when the equivalent martingale measure is unique.

Theorem 4.7 (Unique Martingale Measure) *Let* $n = m$ *and let us suppose that for each* $\widetilde{Q} \in \mathbb{M}$ *there exists a* $n-$*dimensional progressively measurable stochastic process* γ *such that assumptions* $(i), (ii)$, *and* (iii) *of Theorem 4.4 as well as* $\left(\widetilde{M3'}^{IRM}\right)$ *are satisfied. Then the following conditions are equivalent:*

a) $|\mathbb{M}| = 1$.

---

[2]Note: $|\mathbb{M}(\mathcal{T}_n)| = 1$ on $\mathcal{F}_{T_D}$ means that for each pair $\widetilde{Q}_1, \widetilde{Q}_2 \in \mathbb{M}(\mathcal{T}_n)$ we have $\widetilde{Q}_1(A) = \widetilde{Q}_2(A)$ for all $A \in \mathcal{F}_{T_D}$.

b) For each $T_1 \in [t_0, T^*]$ there exists a finite set $\mathcal{T}_n := \{T_1, ..., T_n\}$, $t_0 < T_1 < \cdots < T_n = T^*$, such that the matrix $\sigma_P(t, \mathcal{T}_n)$ is invertible $\lambda \otimes Q - a.s.$ on $\mathcal{B}([t_0, T_1]) \otimes \mathcal{F}_{T_1}$.

Proof. From the assumptions we know that $I\!M \neq \emptyset$ and that each $\widetilde{Q} \in I\!M$ satisfies condition $\left(NA^{IRM}\right)$, i.e.

$$\mu_P(t, T) - \sigma_P(t, T)\gamma(t) = r(t) \cdot P(t, T) \quad \text{for all } t_0 \leq t \leq T \leq T^*.$$

Let $\mathcal{T}_n := \{T_1, ..., T_n\}$, $\mu_P(t, \mathcal{T}_n) := (\mu_P(t, T_i))_{i=1,...,n}$, $P(t, \mathcal{T}_n) := (P(t, T_i))_{i=1,...,n}$, and $\sigma_P(t, \mathcal{T}_n) := (\sigma_P(t, T_i))_{i=1,...,n}$. Then

$$\mu_P(t, \mathcal{T}_n) - r(t) \cdot P(t, \mathcal{T}_n) = \sigma_P(t, \mathcal{T}_n)\gamma(t) \quad \text{for all } t_0 \leq t \leq T_1. \quad (4.3)$$

Now equation (4.3) has a unique solution $\gamma(t)$ on $[t_0, T_1]$ if and only if $\sigma_P(t, \mathcal{T}_m)$ is invertible. Since T_1 was an arbitrary element of $[t_0, T^*]$, $\gamma(t)$ is unique on $[t_0, T^*]$, i.e., by Theorem 4.4b), there is only one equivalent martingale measure to Q, if and only if condition b) holds.   □

4.3 Pricing Contingent Claims

For the pricing of a (European) contingent claim $D = D(T_D)$ with maturity T_D we already defined the *expected-value process* of D under a martingale measure $\widetilde{Q} \in I\!M \neq \emptyset$ by the risk-neutral valuation formula

$$V_D^{\widetilde{Q}}(t) := P_0(t) \cdot E_{\widetilde{Q}}\left[\widetilde{D}(T_D)|\mathcal{F}_t\right], \ t \in [t_0, T_D]$$

and called $V_D(t) := V_D^{\widetilde{Q}}(t)$ the *price of the contingent claim* D if it is unique in $I\!M$. This is, of course, true if $|I\!M| = 1$ as it was discussed in Theorem 4.7. For a more general statement let

$$\Phi(Q) = \Phi(Q, T_D) := \bigcup_{\substack{\mathcal{T}_n = \{T_1, ..., T_n\}: \\ T_D \leq T_1 < \cdots < T_n = T^*, \\ n \in I\!N}} \Phi(\mathcal{T}_n, Q).$$

Then $\Phi(Q)$ is the set of all admissible trading strategies using zero-coupon bonds which mature not before T_D[3]. As above, $\Phi^*(Q) = \Phi^*(Q, T_D)$ is defined to be the set of all $\lambda \otimes Q - a.s.$ bounded trading strategies $\varphi \in \Phi(Q)$. Furthermore, let the *bid price* at time $t \in [t_0, T_D]$ be defined by

$$D_B(t) = \sup\left\{\begin{array}{c} p \in I\!R : V(\varphi, t) = -p, \ V(\varphi, T_D) + D(T_D) \geq 0 \ Q - a.s. \\ \text{for some } \varphi \in \Phi^*(Q) \end{array}\right\}$$

---

[3] As already noted, we could also define $\Phi(Q) := \bigcup_{\substack{\mathcal{T}_n = \{T_1, ..., T_n\}: \\ t_0 < T_1 < \cdots < T_n = T^*, \\ n \in I\!N}} \Phi(\mathcal{T}_n, Q)$ with after maturity definitions as discussed earlier.

and the *ask price* by

$$D_A(t) = \inf \left\{ \begin{array}{c} p \in I\!\!R : V(\varphi, t) = p, \ V(\varphi, T_D) \geq D(T_D) \ Q - a.s. \\ \text{for some } \varphi \in \Phi^*(Q) \end{array} \right\}.$$

Using Lemma 3.23 and Theorem 3.24 we then get the following corollary.

Corollary 4.8 (Pricing Contingent Claims) *Let us suppose that there exists a $m-$dimensional progressively measurable stochastic process γ such that assumptions (i), (ii), and (iii) of Theorem 4.4 are satisfied. Furthermore, let the probability measure $\widetilde{Q}$ on $(\Omega, \mathcal{F})$ be defined as in Theorem 4.4 and let $\left(\widetilde{M3}^{IRM}\right)$ be satisfied.*

a) *The expected-value process of any Φ^*-attainable (European) contingent claim $D = D(T_D)$ with maturity T_D is unique in $I\!\!M$, i.e. the price of the contingent claim D is given by*

$$V_D(t) = V_D^{\widetilde{Q}}(t) = P_0(t) \cdot E_{\widetilde{Q}}\left[\widetilde{D}(T_D)|\mathcal{F}_t\right] \ \text{for all } t \in [t_0, T_D], \widetilde{Q} \in I\!\!M.$$

Furthermore, the expected-value process and the arbitrage price process of the contingent claim D coincide under any replicating portfolio $\varphi \in \Phi^*(Q)$, i.e.

$$V(\varphi, t) = V(\varphi^*, t) = V_D(t) \quad \text{for all } t \in [t_0, T_D], \ \varphi, \varphi^* \in \Phi^*(Q).$$

b) *For any (European) contingent claim $D = D(T_D)$ with maturity T_D, we have*

$$D_B(t) \leq V_D^{\widetilde{Q}}(t) \leq D_A(t) \ \text{for all } t \in [t_0, T_D].$$

c) *If the interest-rate market $\mathcal{M}^{IRM}$ is Φ^*-complete, the price at time t of any (European) contingent claim $D = D(T_D)$ with maturity T_D is given by*

$$D_B(t) = V_D(t) = D_A(t) \quad \text{for all } t \in [t_0, T_D].$$

To avoid too many technicalities within each of the upcoming specific interest-rate models, we generally assume that there always exists an equivalent matingale measure $\widetilde{Q} \in I\!\!M$ and that the expected-value process is unique in $I\!\!M$ for all the contingent claims we will consider. This is to say that we will concentrate our interest on the expected-value process if we are to determine the price of a contingent claim.

4.4   The Heath-Jarrow-Morton Framework

As we have learned, there are infinitely many zero-coupon bonds corresponding to different maturity dates. It has been one of the main challenges
in interest-rate theory to find the driving factors of these zero-coupon bond
prices. As a consequence different interest-rate models appeared. One of
the most general platforms is the Heath-Jarrow-Morton (HJM) framework
which will be discussed in this section. The model is defined in Section 4.4.1,
where we also derive the stochastic differential equations for the short rate
and the zero-coupon bond prices within this model. We then embed the
HJM model in the general setup of Section 4.1 to derive the conditions for
an absence of arbitrage and completeness as well as the HJM arbitrage-
free price system. This is done in Sections 4.4.2 and 4.4.3. Transfering the
results of Section 3.7, we will also discuss the forward measures within the
HJM model and derive the price dynamics of a forward zero-coupon bond.
This is the essential link to the newly developed LIBOR market models of
Section 4.7.

4.4.1   The Heath-Jarrow-Morton Model

In 1992 Heath, Jarrow and Morton (see [HJM92]) published a new framework for evaluating interest-rate derivatives. The starting point of this
model is the forward short rates $f(t, T)$ at time $t \geq t_0$ with fixed maturity T, $t \leq T \leq T^*$, where we interpret T^* to be the maximum time
horizon or trading period and t_0 to be the current time. The dynamics of
the forward short rates are given by the Itô processes

$$d_t f(t, T) = \mu(t, T)\, dt + \sigma(t, T) dW(t). \qquad (4.4)$$

It is assumed that $W(t)$ is a $m-$dimensional $(Q-)$ Brownian motion
and that the so-called drift $\mu = (\mu(t, T))_{t \in [t_0, T]}$ and volatility
$\sigma = (\sigma(t, T))_{t \in [t_0, T]} = (\sigma_1(t, T), ..., \sigma_m(t, T))_{t \in [t_0, T]}$ are $m-$dimensional
progressively measurable stochastic processes satisfying the conditions

$(M1^{HJM})$ $\displaystyle\int_{t_0}^{T} |\mu(s, T)|\, ds < \infty \; Q - a.s.$ for all $t_0 \leq T \leq T^*$

$(M2^{HJM})$ $\displaystyle\int_{t_0}^{T} \sigma_j^2(s, T) ds < \infty \; Q - a.s.$ for all $t_0 \leq T \leq T^*, j = 1, ..., m$

$(M3^{HJM})$ $\displaystyle\int_{t_0}^{T^*} |f(t_0, u)|\, ds < \infty \; Q - a.s.$

$(M4^{HJM})$ $\displaystyle\int_{t_0}^{T^*} \int_{t_0}^{u} |\mu(s, u)|\, ds du < \infty \; Q - a.s.$

Then the forward short rate $f(t, T)$ is given by

$$f(t, T) = f(t_0, T) + \int_{t_0}^{t} \mu(s, T) ds + \int_{t_0}^{t} \sigma(s, T) dW(s). \qquad (4.5)$$

Under assumption $\left(M2^{HJM} \right)$ we know that for all $t_0 \leq T \leq T^*$ the stochastic integral $J \left(\sigma_j \left(\cdot, T \right) \right)_{t \in [t_0, T]}$ with $J_t \left(\sigma_j \left(\cdot, T \right) \right) = \int_{t_0}^{t} \sigma_j(s, T) dW(s)$, $t \in [t_0, T]$, is a continuous local martingale on $[t_0, T]$ for all $j = 1, ..., m$. It is a continuous martingale if $E_Q \left[\int_{t_0}^{T} \sigma_j^2(s, T) ds \right] < \infty$ for all $j = 1, ..., m$. Furthermore, we get for all $t_0 \leq t \leq T \leq T' \leq T^*$

$$f(t, T') - f(t, T) = f(t_0, T') - f(t_0, T) + \int_{t_0}^{t} \left(\mu(s, T') - \mu(s, T) \right) ds$$

$$+ \int_{t_0}^{t} \left(\sigma(s, T') - \sigma(s, T) \right) dW(s).$$

We assume that the processes are smooth enough to allow for differentiation and certain operations involving the change of order of integration and differentiation. Since these conditions are rather technical[4] we refer the reader to Heath, Jarrow and Morton [HJM92] or Protter [Pro92], p. 159ff for more details and proofs. Nevertheless, we note that they are satisfied if $(f(t_0, T))_{T \in [t_0, T^*]}$ as well as $\mu(t, T)$ and $\sigma(t, T)$ are bounded, differentiable in T, and if each σ_j is of the form $\sigma_j(t, T) = \sigma_{1j}(t) \cdot \sigma_{2j}(T)$ for all $t_0 \leq t \leq T \leq T^*$ with integrable real, positive functions σ_{1j} and σ_{2j} such that $\int_{t_0}^{T^*} \sigma_{1j}^2(s) ds < \infty$ $Q - a.s.$, $j = 1, ..., m$, as it will be the case for the Gaussian one-factor models with exponential volatility structure (see Definition 4.21). Under these regularity conditions we have

$$f_T(t, T) = f_T(t_0, T) + \int_{t_0}^{t} \left(\mu_T(s, T) \right) ds + \int_{t_0}^{t} \left(\sigma_T(s, T) \right) dW(s), \qquad (4.6)$$

---

[4] To be precise, the HJM regularity conditions are as follows

(R1) $E_Q \left[\int_{t_0}^{T} \left(\int_{s}^{T} \sigma_j(s, u) du \right)^2 ds \right] < \infty$ for all $t_0 \leq T \leq T^*$, $j = 1, ..., m$

(R2) $\int_{t_0}^{t} \left(\int_{t}^{T} \sigma_j(s, u) du \right)^2 ds < \infty$ $Q - a.s.$ for all $t_0 \leq t \leq T \leq T^*$, $j = 1, ..., m$

(R3) $t \longmapsto \int_{t}^{T} \left(\int_{t_0}^{t} \sigma_j(s, u) dW_j(s) \right) du < \infty$ is continuous $Q - a.s.$ for all
$t_0 \leq T \leq T^*$, $j = 1, ..., m$.

For the ease of exposition we furthermore assume that $\mu(t, T)$, $\sigma(t, T)$, and $f(t_0, T)$ are differentiable in T for all $t_0 \leq t \leq T \leq T^*$.

where the index T denotes the derivative of the corresponding function with respect to T. The short-rate process $r(t) = f(t, t)$ is given by

$$r(t) = f(t_0, t) + \int_{t_0}^t \mu(s, t) ds + \int_{t_0}^t \sigma(s, t) dW(s). \tag{4.7}$$

Using Theorem 2.43 we get the following lemma.

Lemma 4.9 *Let μ and σ be deterministic functions. Then the short rate and the forward short rates are normally distributed. Furthermore, if σ is independent of T, the difference $f(t, T') - f(t, T)$ of the forward short rates is a deterministic function in t, T and T' for all $t_0 \leq t \leq T \leq T' \leq T^*$. In this case the forward short rates are perfectly correlated.*

We can also prove the following lemma which gives us the stochastic differential equation for the short-rate process $(r(t))_{t \in [t_0, T^*]}$.

Lemma 4.10 (Short-Rate Dynamics) *Let $\mu(t) := \mu(t, t)$ and $\sigma(t) := \sigma(t, t)$. Then the short-rate dynamics are given by*

$$dr(t) = (f_T(t, t) + \mu(t)) \, dt + \sigma(t) dW(t).$$

Proof. Using the equations $\mu(s, t) = \mu(s) + \int_s^t \mu_T(s, u) du$ and $\sigma(s, t) = \sigma(s) + \int_s^t \sigma_T(s, u) du$ we conclude from (4.7) that

$$
\begin{aligned}
r(t) &= f(t_0, t) + \int_{t_0}^t \mu(s, t) \, ds + \int_{t_0}^t \sigma(s, t) dW(s) \\
&= f(t_0, t) + \int_{t_0}^t \mu(s) ds + \int_{t_0}^t \int_s^t \mu_T(s, u) du \, ds \\
&\quad + \int_{t_0}^t \sigma(s) dW(s) + \int_{t_0}^t \int_s^t \sigma_T(s, u) du \, dW(s) \\
&= f(t_0, t) + \int_{t_0}^t \mu(s) ds + \int_{t_0}^t \sigma(s) dW(s) \\
&\quad + \int_{t_0}^t \left(\int_{t_0}^u \mu_T(s, u) ds + \int_{t_0}^u \sigma_T(s, u) dW(s) \right) du \\
&= f(t_0, t) + \int_{t_0}^t \mu(s) ds + \int_{t_0}^t \sigma(s) dW(s) \\
&\quad + \int_{t_0}^t (f_T(u, u) - f_T(t_0, u)) \, du \quad \text{because of (4.6)} \\
&= \int_{t_0}^t (f_T(s, s) + \mu(s)) \, ds + \int_{t_0}^t \sigma(s) dW(s) \\
&\quad + f(t_0, t) - (f(t_0, t) - f(t_0, t_0)) \\
&= r(t_0) + \int_{t_0}^t (f_T(s, s) + \mu(s)) \, ds + \int_{t_0}^t \sigma(s) dW(s).
\end{aligned}
$$

Thus,

$$\int_{t_0}^{t} dr\,(s) = r(t) - r\,(t_0) = \int_{t_0}^{t} \left(f_T(s,s) + \mu(s) \right) ds + \int_{t_0}^{t} \sigma(s) dW(s).$$

which completes the proof.   □

In this model, the cash account as our natural numéraire satisfies the equation

$$\ln P_0(t) \;=\; \int_{t_0}^{t} r(u) du \ \text{ which is, using } (4.7)\,,$$

$$= \int_{t_0}^{t} f(t_0, u) du + \int_{t_0}^{t} \int_{t_0}^{u} \mu(s, u) ds du + \int_{t_0}^{t} \int_{t_0}^{u} \sigma(s, u) dW(s) du$$

$$= \int_{t_0}^{t} f(t_0, u) du + \int_{t_0}^{t} \int_{s}^{t} \mu(s, u) du ds + \int_{t_0}^{t} \int_{s}^{t} \sigma(s, u) du dW(s)\,.$$

For ease of simplicity, we define for fixed $t \in [t_0, T]$ the progressively measurable stochastic processes $m = (m(t,T))_{t\in[t_0,T]}$ and $v = (v(t,T))_{t\in[t_0,T]}$ by

$$m(t,T) = \int_{t}^{T} \mu(t,u) du \ \text{ and } \ v(t,T) = \int_{t}^{T} \sigma(t,u) du\,.$$

This leads us to

$$\ln P_0(t) = \int_{t_0}^{t} f(t_0, u) du + \int_{t_0}^{t} m(s,t) ds + \int_{t_0}^{t} v(s,t) dW(s)\,.$$

The primary traded assets within our general setup of Section 4.1 have been the zero-coupon bonds with maturity T and time to maturity $T - t$. To embed the HJM model we therefore have to derive the stochastic differential equation for these instruments. To do this, let $T \in [t, T^*]$ be arbitrary but fixed. Within the HJM model the price $P(t,T)$ of the corresponding zero-coupon bond satisfies the equation

$$-\ln P(t,T) \;=\; \int_{t}^{T} f(t,u) du \ \text{ and therefore, using } (4.5)\,,$$

$$= \int_{t}^{T} f(t_0, u) du + \int_{t}^{T} \int_{t_0}^{t} \mu(s, u) ds du$$

$$+ \int_{t}^{T} \int_{t_0}^{t} \sigma(s, u) dW(s) du$$

$$= \int_{t}^{T} f(t_0, u) du + \int_{t_0}^{t} \int_{t}^{T} \mu(s, u) du ds$$

$$+ \int_{t_0}^{t} \int_{t}^{T} \sigma(s, u) du dW(s).$$

For the discounted zero-coupon bond prices $\left(\widetilde{P}(t,T)\right)_{t\in[t_0,T]}$ with $\widetilde{P}(t,T) = P_0^{-1}(t) \cdot P(t,T)$ for all $T \in [t_0,T^*]$ we thus get

$$
\begin{aligned}
-\ln \widetilde{P}(t,T) &= \ln P_0(t) - \ln P(t,T) \\
&= \int_{t_0}^t r(u)du + \int_t^T f(t,u)du \\
&= \int_{t_0}^T f(t_0,u)du + \int_{t_0}^t \int_s^T \mu(s,u)duds \\
&\quad + \int_{t_0}^t \int_s^T \sigma(s,u)dudW(s) \\
&= \int_{t_0}^T f(t_0,u)du + \int_{t_0}^t m(s,T)ds + \int_{t_0}^t v(s,T)dW(s) \\
&= -\ln P(t_0,T) + \int_{t_0}^t m(s,T)ds + \int_{t_0}^t v(s,T)dW(s).
\end{aligned}
$$

Hence,

$$
\ln P(t,T) = \ln P(t_0,T) + \int_{t_0}^t \left(r(s) - m(s,T)\right)ds - \int_{t_0}^t v(s,T)dW(s). \quad (4.8)
$$

Lemma 4.11 (Zero-Coupon Bond Price Dynamics)
For all $t \in [t_0,T]$, the zero-coupon bond price dynamics are given by

$$
\begin{aligned}
d_t P(t,T) &= -P(t,T) \cdot \left(m(t,T) - r(t) - \frac{1}{2} \cdot \|v(t,T)\|^2\right) dt \\
&\quad -P(t,T) \cdot v(t,T)dW(t).
\end{aligned}
$$

Furthermore, the discounted zero-coupon bond price dynamics are given by

$$
d_t \widetilde{P}(t,T) = -\widetilde{P}(t,T) \cdot \left[\left(m(t,T) - \frac{1}{2} \cdot \|v(t,T)\|^2\right) dt + v(t,T)dW(t)\right].
$$

Proof. Using (4.8) we know that

$$
d_t \ln P(t,T) = (r(t) - m(t,T))\, dt - v(t,T)dW(t).
$$

Thus, applying Ito's lemma (Theorem 2.34) with $X(t) := \ln P(t,T)$ and $G(x,t) = e^x$ we get

$$
\begin{aligned}
dG(X(t),t) &= d_t P(t,T) \\
&= \left(P(t,T) \cdot (r(t) - m(t,T)) + \frac{1}{2} \cdot P(t,T) \cdot \|v(t,T)\|^2\right) dt \\
&\quad -P(t,T) \cdot v(t,T)dW(t).
\end{aligned}
$$

Now let us apply Ito's lemma to $X(t) := -\ln \widetilde{P}(t,T)$ and $G(x,t) = e^{-x}$ with

$$dX(t) = m(t,T)dt + v(t,T)dW(t).$$

Doing this we get

$$
\begin{aligned}
dG\left(X\left(t\right),t\right) &= d_t \widetilde{P}\left(t,T\right) \\
&= \left(-\widetilde{P}(t,T) \cdot m(t,T) + \frac{1}{2} \cdot \widetilde{P}(t,T) \cdot \|v(t,T)\|^2\right) dt \\
&\quad -\widetilde{P}(t,T) \cdot v(t,T) dW\left(t\right).
\end{aligned}
$$

$\square$

4.4.2 No-Arbitrage and Completeness within the HJM Model

Using Lemma 4.11 we can easily see how the Heath-Jarrow-Morton framework fits in our general interest-rate market model. Conditions $\left(M1^{HJM}\right)-\left(M4^{HJM}\right)$ and $(R1)-(R3)$ ensure that the assumptions of the interest-rate market model are satisfied[5]. The following theorem tells us under which conditions an equivalent martingale measure exists.

Theorem 4.12   a) *Suppose that there exists a $m-$dimensional progressively measurable stochastic process γ such that:*

(i) *The following Novikov condition holds for γ:*

$$E_Q\left[e^{\frac{1}{2}\int_{t_0}^{T^*}\|\gamma(s)\|^2 ds}\right] < \infty. \qquad (NV^{HJM})$$

(ii) $E_Q\left[L\left(\gamma,T^*\right)\right] = 1$ *with*

$$L\left(\gamma,t\right) := e^{-\int_{t_0}^t \gamma(s)'dW(s)-\frac{1}{2}\int_{t_0}^t \|\gamma(s)\|^2 ds}.$$

(iii) *The no-arbitrage condition*

$$m(t,T) - \frac{1}{2} \cdot \|v(t,T)\|^2 = v(t,T)\gamma(t) \qquad (NA^{HJM})$$

holds for all $t_0 \leq t \leq T \leq T^*$. This condition is equivalent to the condition that

$$\mu(t,T) = \sigma(t,T) \cdot \left(v(t,T)' + \gamma(t)\right) \qquad (HJM^{drift})$$

is satisfied for all $t_0 \leq t \leq T \leq T^*$.

---

[5] To be more precise, it is sufficient to consider the assumptions corresponding to conditions $\left(\widehat{M1}\right)$ and $\left(\widehat{M2}\right)$ of Section 4.1.

Furthermore, let the probability measure $\widetilde{Q}$ on $(\Omega, \mathcal{F}_{T^*}) = (\Omega, \mathcal{F})$ be defined by $\widetilde{Q} := Q_{L(\gamma, T^*)}$. Then the stochastic process $\widetilde{W} = \left(\widetilde{W}(t) \right)_{t \in [t_0, T^*]}$, defined by

$$d\widetilde{W}(t) := \gamma(t)\, dt + dW(t), \ t \in [t_0, T^*],$$

is a $\widetilde{Q}$–Wiener process and the discounted price processes $\left(\widetilde{P}(t, T) \right)_{t \in [t_0, T]}$ have the following representation in terms of $\widetilde{W}$:

$$d\widetilde{P}_0(t) = 0 \text{ and } d\widetilde{P}(t, T) = -\widetilde{P}(t, T) \cdot v(t, T)\, d\widetilde{W}(t)$$

for all $t_0 \leq t \leq T \leq T^*$. Furthermore,

$$d_t P(t, T) = P(t, T) \cdot \left[r(t)dt - v(t, T)d\widetilde{W}(t) \right].$$

If the martingale (Novikov) condition

$$E_{\widetilde{Q}} \left[e^{\frac{1}{2} \int_{t_0}^T \|v(s, T)\|^2 ds} \right] < \infty \text{ for all } t_0 \leq T \leq T^* \qquad (\widetilde{M3}^{HJM})$$

is satisfied, then $\widetilde{Q} \in \mathbb{M}$, i.e. $\mathbb{M} \neq \emptyset$.

b) *Let $\mathbb{M} \neq \emptyset$. Then for each $\widetilde{Q} \in \mathbb{M}$ there exists a m–dimensional progressively measurable stochastic process γ such that conditions (i') and (ii) hold with*

(i') $\displaystyle \int_{t_0}^{T^*} \|\gamma(s)\|^2\, ds < \infty \ Q - a.s.$

If, in extension to (i') Novikov's condition (i) holds, then condition (iii) is satisfied.

Proof. The proof is quite similar to that of Theorem 4.4. The only difference is the dynamics of the discounted zero-coupon bond prices which we get from Lemma 4.11 by

$$\begin{aligned}
d_t \widetilde{P}(t, T) &= -\widetilde{P}(t, T) \cdot \left[\left(m(t, T) - \frac{1}{2} \cdot \|v(t, T)\|^2 \right) dt + v(t, T) dW(t) \right] \\
&= -\widetilde{P}(t, T) \cdot \left[\left(m(t, T) - \frac{1}{2} \cdot \|v(t, T)\|^2 - v(t, T)\gamma(t) \right) dt \right. \\
&\quad \left. + v(t, T) d\widetilde{W}(t) \right] \text{ for all } t \in [t_0, T].
\end{aligned}$$

Hence, if the discounted zero-coupon bond prices are $\widetilde{Q}$–martingales for all $t_0 \leq T \leq T^*$ we know by Theorem 2.42 that for all $t_0 \leq t \leq T \leq T^*$:

$$m(t, T) - \frac{1}{2} \cdot \|v(t, T)\|^2 - v(t, T)\gamma(t) = 0. \qquad (4.9)$$

Differentiation of (4.9) with respect to T gives us

$$\mu(t,T) - \sigma(t,T)v(t,T) - \sigma(t,T)\gamma(t) = 0 \quad \text{for all } t_0 \le t \le T \le T^*.$$

On the other hand, we get

$$
\begin{aligned}
d_t \widetilde{P}(t,T) &= -\widetilde{P}(t,T) \cdot \left[\left(m(t,T) - \frac{1}{2} \cdot \|v(t,T)\|^2 \right) dt + v(t,T) dW(t) \right] \\
&= -\widetilde{P}(t,T) \cdot [v(t,T)\gamma(t)\, dt + v(t,T) dW(t)] \quad \text{using (iii)} \\
&= -\widetilde{P}(t,T) \cdot v(t,T) d\widetilde{W}(t).
\end{aligned}
$$

If $\widetilde{W}$ is a $\widetilde{Q}$–Wiener process we know, following Theorem 2.42, that $\left(\widetilde{P}(t,T) \right)_{t \in [t_0,T]}$ is a $\widetilde{Q}$–martingale for all $t_0 \le T \le T^*$ because of $\left(\widetilde{M3}^{HJM} \right)$. The stochastic differential equation for $P(t,T)$ is straightforward from Lemma 4.11 and $\left(NA^{HJM} \right)$. $\qquad \square$

The progressively measurable stochastic process $\gamma(t)$ of equation $\left(NA^{HJM} \right)$ is called the *market price of risk*. If we consider the case of a one-factor model, i.e. $m = 1$, with $\sigma > 0$ we get

$$\gamma(t) = \frac{m(t,T)}{v(t,T)} - \frac{1}{2} \cdot v(t,T) \tag{4.10}$$

which is, under $\left(NA^{HJM} \right)$, independent of T. The following theorem is the analogon to Theorem 4.7.

Theorem 4.13 (Unique Martingale Measure) *Let $n = m$ and let us suppose that for each $\widetilde{Q} \in \mathbb{M}$ there exists an n–dimensional progressively measurable stochastic process γ such that assumptions $(i), (ii)$, and (iii) of Theorem 4.12 as well as $\left(\widetilde{M3}^{HJM} \right)$ are satisfied. Then the following conditions are equivalent:*

a) $|\mathbb{M}| = 1$.

b) *For each $T_1 \in [t_0, T^*]$ there exists a finite set $\mathcal{T}_n := \{T_1, ..., T_n\}$, $t_0 < T_1 < \cdots < T_n = T^*$, such that the matrix $v(t, \mathcal{T}_n) := (v_j(t,T_i))_{i,j=1,...,n}$ is invertible $\lambda \otimes Q$–a.s. on $\mathcal{B}([t_0,T_1]) \otimes \mathcal{F}_{T_1}$.*

c) *For each $T_1 \in [t_0, T^*]$ there exists a finite set $\mathcal{T}_n := \{T_1, ..., T_n\}$, $t_0 < T_1 < \cdots < T_n = T^*$, such that the matrix $\sigma(t, \mathcal{T}_n) := (\sigma_j(t,T_i))_{i,j=1,...,n}$ is invertible $\lambda \otimes Q$–a.s. on $\mathcal{B}([t_0,T_1]) \otimes \mathcal{F}_{T_1}$.*

Proof. The proof is similar to that of Theorem 4.7. From the assumptions we know that $I\!M \neq \emptyset$ and that each $\widetilde{Q} \in I\!M$ satisfies condition (NA^{HJM}), i.e.

$$m(t,T) - \frac{1}{2} \cdot \|v(t,T)\|^2 = v(t,T)\gamma(t) \quad \text{for all } t_0 \leq t \leq T \leq T^*.$$

Let $\mathcal{T}_n := \{T_1, ..., T_n\}$, $m(t, \mathcal{T}_n) := (m(t,T_i))_{i=1,...,n}$ and $\|v(t,\mathcal{T}_n)\|^2 := \left(\|v(t,T_i)\|^2\right)_{i=1,...,n}$. Then

$$m(t, \mathcal{T}_n) - \frac{1}{2} \cdot \|v(t,\mathcal{T}_n)\|^2 = v(t,\mathcal{T}_n)\gamma(t) \quad \text{for all } t_0 \leq t \leq T_1. \qquad (4.11)$$

Now equation (4.11) has a unique solution $\gamma(t)$ on $[t_0, T_1]$ if and only if $v(t, \mathcal{T}_m)$ is invertible. Since T_1 was an arbitrary element of $[t_0, T^*]$, $\gamma(t)$ is unique on $[t_0, T^*]$, i.e., by Theorem 4.4b), there is only one equivalent martingale measure to Q, if and only if condition b) holds. Since (NA^{HJM}) and (HJM^{drift}) are equivalent the same arguments can be used to show that $a)$ is equivalent to $c)$. $\qquad \square$

4.4.3 The HJM Arbitrage-free Price System

Combining the results of Sections 4.4.1 and 4.4.2 we can now set up the arbitrage-free price system for the Heath-Jarrow-Morton model. From condition (HJM^{drift}) we directly conclude that

$$\mu(t) = \sigma(t) \cdot \gamma(t), \quad \int_{t_0}^t \mu(s,T)\,ds = \int_{t_0}^t \sigma(s,T)v(s,T)\,ds + \int_{t_0}^t \sigma(s,T)\gamma(s)\,ds$$

and

$$\int_{t_0}^t \mu_T(s,T)\,ds = \int_{t_0}^t \frac{\partial}{\partial T}\left(\sigma(s,T)\left(v(s,T) + \gamma(s)\right)\right)\,ds$$

$$= \int_{t_0}^t \sigma_T(s,T)\gamma(s)\,ds + \int_{t_0}^t \sigma_T(s,T)v(s,T)\,ds + \int_{t_0}^t \|\sigma(t,T)\|^2\,ds.$$

As a summary we get the following *arbitrage-free price system for the Heath-Jarrow-Morton framework*

$$d_t f(t,T) \quad = \quad \sigma(t,T)\left(v(t,T)' + \gamma(t)\right)dt + \sigma(t,T)dW(t),$$

$$f(t,T) \quad = \quad f(t_0,T) + \int_{t_0}^{t} \sigma(s,T)v(s,T)'ds + \int_{t_0}^{t} \sigma(s,T)\gamma(s)ds$$

$$+ \int_{t_0}^{t} \sigma(s,T)dW(s),$$

$$r(t) \quad = \quad f(t_0,t) + \int_{t_0}^{t} \sigma(s,t)v(s,t)'ds + \int_{t_0}^{t} \sigma(s,t)\gamma(s)ds$$

$$+ \int_{t_0}^{t} \sigma(s,t)dW(s),$$

$$dr(t) \quad = \quad \left(f_T(t_0,t) + \int_{t_0}^{t} \sigma_T(s,t)\gamma(s)ds + \int_{t_0}^{t} \sigma_T(s,t)v(s,t)'ds \right.$$

$$+ \int_{t_0}^{t} \|\sigma(s,T)\|^2\, ds + \sigma(t)\gamma(t)$$

$$\left. + \int_{t_0}^{t} \sigma_T(s,t)dW(s) \right) dt + \sigma(t)dW(t),$$

$$d_t P(t,T) \quad = \quad -P(t,T) \cdot [(-r(t) + v(t,T)\gamma(t,T))\,dt + v(t,T)dW(t)]$$

$$-\ln P(t,T) \quad = \quad \int_{t}^{T} f(t_0,u)du + \int_{t_0}^{t}\int_{t}^{T} \sigma(s,u)v(s,u)'duds$$

$$+ \int_{t_0}^{t}\int_{t}^{T} \sigma(s,u)\gamma(s)duds + \int_{t_0}^{t}\int_{t}^{T} \sigma(s,u)dudW(s).$$

Setting $\gamma(t) \equiv 0$ and replacing $W(t)$ by $\widetilde{W}(t)$ we get the corresponding properties under the (martingale) measure $\widetilde{Q}$.

4.4.4   Forward Measures within the HJM Model

In this section we would like to transfer the results of Section 3.7 and derive the forward measures within the HJM model as well as the price dynamics of a forward zero-coupon bond. As we will see later on, this is the essential link to the LIBOR market models of Section 4.7. Let us suppose within this section that the volatility process $\sigma = (\sigma\,(t,T))_{t\in[t_0,T]}$ of the forward rates is bounded for all $t_0 \leq T \leq T^*$. Furthermore, let the assumptions of Theorem 4.12 be satisfied and let the corresponding interest-rate market $\mathcal{M}$ be complete. We then know from this theorem that the discounted zero-coupon bond price process $\left(\widetilde{P}(t,T)\right)_{t\in[t_0,T]}$ follows the stochastic differential equation

$$d_t\widetilde{P}(t,T) = -\widetilde{P}(t,T) \cdot v(t,T)d\widetilde{W}\,(t)$$

under the martingale measure $\widetilde{Q}$ with $v(t,T)$ as defined in Section 4.4.1 and $\widetilde{W}$ being a $\widetilde{Q}$−Wiener process. Hence,

$$\begin{aligned}
\widetilde{P}(t,T) &= \widetilde{P}(t_0,T) \cdot e^{-\int_{t_0}^{t} v(s,T)d\widetilde{W}(s)-\frac{1}{2}\int_{t_0}^{t}\|v(s,T)\|^2 ds} \\
&= P(t_0,T) \cdot L\,(v(\cdot,T),t)
\end{aligned}$$

for all $t \in [t_0,T]$, $v(\cdot,T) := (v(t,T))_{t\in[t_0,T]}$, and $L\,(v(\cdot,T),t)$ being as defined in Lemma 3.9. This is equivalent to

$$P(t,T) = P_0\,(t) \cdot P(t_0,T) \cdot L\,(v(\cdot,T),t)$$

for all $t \in [t_0,T]$. Following the results of Section 3.7 with $t = 0$ replaced by $t = t_0$, the Radon-Nikodým derivative $L\,(t)$ of the T−forward measure Q^T is given by

$$L\,(t) = \frac{dQ^T}{d\widetilde{Q}}|_{\mathcal{F}_t} = \frac{P\,(t,T)}{P\,(t_0,T) \cdot P_0\,(t)} = L\,(v(\cdot,T),t)$$

for all $t \in [t_0,T]$, or consequently, $Q^T = \widetilde{Q}_{L(T)}$ with

$$L\,(T) = e^{-\int_{t_0}^{T} v(s,T)d\widetilde{W}(s)-\frac{1}{2}\int_{t_0}^{T}\|v(s,T)\|^2 ds}.$$

Since $\sigma = (\sigma\,(t,T))_{t\in[t_0,T]}$ is bounded for all $t_0 \leq T \leq T^*$, the Radon-Nikodým derivative $(L\,(t))_{t\in[t_0,T]} := (L\,(v(\cdot,T),t))_{t\in[t_0,T]}$ is a continuous $\widetilde{Q}$−martingale because of Lemma 2.40 and thus, using the Girsanov theorem (Theorem 2.41), the stochastic process $W^T = \left(W_1^T, ..., W_m^T\right)$, defined by

$$W_j^T\,(t) := \widetilde{W}_j\,(t) + \int_{t_0}^{t} v\,(s,T)\,ds,\ t \in [t_0,T]\,,\ j = 1,...,m,$$

is a m−dimensional Q^T−Wiener process on the measure space $(\Omega, \mathcal{F}_T)$.

Now−for $t_0 \leq t \leq T_1 \leq T_2 \leq T^*$ let $P^F\,(t,T_1,T_2) := P\,(t,T_1)\,/P\,(t,T_2)$ denote the forward price at time t for the forward maturity time T_2 of a zero-coupon bond with maturity time T_1. Furthermore, with $m\,(t,T) := \int_t^T \mu\,(t,u)\,du$, $t_0 \leq t \leq T \leq T^*$, let

$$m\,(t,T_1,T_2) := m\,(t,T_2) - m\,(t,T_1) = \int_{T_1}^{T_2} \mu\,(t,u)\,du$$

and[6]

$$v(t,T_1,T_2) := v(t,T_2) - v(t,T_1) = \int_{T_1}^{T_2} \sigma(t,u)\,du.$$

Lemma 4.14 (Forward Zero-Coupon Bond Price Dynamics)
Let $t_0 \le t \le T_1 \le T_2 \le T^*$. Then the forward zero-coupon bond price dynamics under Q are given by[7]

$$\frac{d_t P^F(t,T_1,T_2)}{P^F(t,T_1,T_2)} = \left(m(t,T_1,T_2) + \frac{1}{2} \cdot \|v(t,T_1,T_2)\|^2 \right) dt + v(t,T_1,T_2)\,dW(t)$$

where W is a $m-$dimensional Wiener process under Q. Equivalently,

$$d_t \ln P^F(t,T_1,T_2) = m(t,T_1,T_2)\,dt + v(t,T_1,T_2)\,dW(t).$$

Proof. Using (4.8) we know that

$$d_t \ln P(t,T) = (r(t) - m(t,T))\,dt - v(t,T)\,dW(t).$$

Thus, applying Ito's lemma (Theorem 2.34) with $X_1(t) := \ln P(t,T_1)$, $X_2(t) := \ln P(t,T_2)$, and $G(x_1,x_2,t) = x_1 - x_2$ we get

$$
\begin{aligned}
d_t \ln P^F(t,T_1,T_2) &= dG(X_1(t), X_2(t), t) = d_t \ln P(t,T_1) - d_t \ln P(t,T_2) \\
&= (m(t,T_2) - m(t,T_1))\,dt + (v(t,T_2) - v(t,T_1))\,dW(t) \\
&= m(t,T_1,T_2)\,dt + v(t,T_1,T_2)\,dW(t).
\end{aligned}
$$

---

[6] Since $\sigma = (\sigma(t,T))_{t\in[t_0,T]}$ is bounded for all $t_0 \le T \le T^*$, we know that

$$E_Q\left[e^{\frac{1}{2}\int_{t_0}^T \|v(s,T,T^*)\|^2\,ds} \right] < \infty$$

for all $T \in [t_0, T^*]$ and for all probability measures Q on $(\Omega, \mathcal{F})$. What we basically have to ensure is that this Novikov condition holds under the T^*-Forward measure, i.e. for $Q = Q^{T^*}$.

[7] Since

$$\|v(t,T_2) - v(t,T_1)\|^2 = \|v(t,T_2)\|^2 - 2 \cdot v(t,T_1)v(t,T_2)' + \|v(t,T_1)\|^2$$

this equation can also be written as

$$
\begin{aligned}
d_t P^F(t,T_1,T_2) &= P^F(t,T_1,T_2) \cdot [(\alpha(t,T_1) - \alpha(t,T_2) + v(t,T_2)(v(t,T_2)' - v(t,T_1)'))dt \\
&\quad + (v(t,T_2) - v(t,T_1))\,dW(t)]
\end{aligned}
$$

with

$$\alpha(t,T) := -\left(m(t,T) - r(t) - \frac{1}{2} \cdot \|v(t,T)\|^2 \right).$$

Now, if we apply Itô's lemma to $X(t) := \ln P^F(t,T_1,T_2)$ and $G(x,t) = e^x$, we get

$$
\begin{aligned}
\frac{d_t P^F(t,T_1,T_2)}{P^F(t,T_1,T_2)} &= \frac{dG(X(t),t)}{P^F(t,T_1,T_2)} \\
&= \left(m(t,T_1,T_2) + \frac{1}{2} \cdot \|v(t,T_1,T_2)\|^2 \right) dt \\
&\quad + v(t,T_1,T_2)\, dW(t).
\end{aligned}
$$

$\square$

For pricing purposes we are interested in finding a measure under which the forward zero-coupon bond process $\left(P^F(t,T_1,T_2) \right)_{t \in [t_0,T_1]}$ with $t_0 \le T_1 \le T_2 \le T^*$ is a martingale. The following theorem gives an answer to that question.

Theorem 4.15 (Change of Measure) *Let $t_0 \le t \le T_1 \le T_2 \le T^*$. Then the forward zero-coupon bond price dynamics under the T_2–forward measure are given by*

$$
d_t P^F(t,T_1,T_2) = P^F(t,T_1,T_2) \cdot v(t,T_1,T_2)\, dW^{T_2}(t),
$$

where W^{T_2} is a $m-$dimensional Wiener process under Q^{T_2}. In other words, $\left(P^F(t,T_1,T_2) \right)_{t \in [t_0,T_1]}$ is a $Q^{T_2}-$martingale.

Proof. For $t_0 \le t \le T_1 \le T_2 \le T^*$ we know that

$$
P^F(t,T_1,T_2) = \frac{\widetilde{P}(t,T_1)}{\widetilde{P}(t,T_2)} \quad \text{with} \quad \widetilde{P}(t,T) = \frac{P(t,T)}{P_0(t)},\ T \in [t,T^*].
$$

Since

$$
d\widetilde{P}(t,T) = -\widetilde{P}(t,T) \cdot v(t,T)\, d\widetilde{W}(t) \quad \text{for all } t_0 \le t \le T \le T^*
$$

under the equivalent martingale measure $\widetilde{Q}$, we get by using Itô's lemma (Theorem 2.34) with $X(t) := \widetilde{P}(t,T)$ and $G(x,t) = \ln x_1$

$$
d\ln \widetilde{P}(t,T) = -\frac{1}{2} \cdot \|v(t,T)\|^2\, dt - v(t,T)\, d\widetilde{W}(t).
$$

Hence, with $X_1(t) := \ln \widetilde{P}(t,T_1)$, $X_2(t) := \ln \widetilde{P}(t,T_2)$, and $G(x_1,x_2,t) = x_1 - x_2$ we get

$$
\begin{aligned}
d\ln P^F(t,T_1,T_2) &= d\ln \widetilde{P}(t,T_1) - d\ln \widetilde{P}(t,T_2) \\
&= \frac{1}{2} \cdot \left(\|v(t,T_2)\|^2 - \|v(t,T_1)\|^2 \right) dt \\
&\quad + v(t,T_1,T_2)\, d\widetilde{W}(t).
\end{aligned}
$$

Again, using Itô's lemma (Theorem 2.34) with $X(t) := \ln P^F(t, T_1, T_2)$ and $G(x, t) = e^x$, we get

$$
\begin{aligned}
\frac{dP^F(t, T_1, T_2)}{P^F(t, T_1, T_2)} &= \frac{1}{2} \cdot \left[\left(\|v(t, T_2)\|^2 - \|v(t, T_1)\|^2 \right) + \|v(t, T_1, T_2)\|^2 \right] dt \\
&\quad + v(t, T_1, T_2)\, d\widetilde{W}(t) \\
&= \left[\|v(t, T_2)\|^2 - v(t, T_1)\, v(t, T_2)' \right] dt \\
&\quad + v(t, T_1, T_2)\, d\widetilde{W}(t) \\
&= v(t, T_1, T_2) \cdot \left[v(t, T_2)'\, dt + d\widetilde{W}(t) \right]
\end{aligned}
$$

under the equivalent martingale measure $\widetilde{Q}$. Since $(L(v(\cdot, T_2), t))_{t \in [t_0, T_2]}$ is a $\widetilde{Q}$-martingale, the stochastic process $W^{T_2} = \left(W_1^{T_2}, ..., W_m^{T_2} \right)$, defined by

$$
W_j^{T_2}(t) := W_j(t) + \int_{t_0}^{t} v_j(s, T_2)\, ds,\ t \in [t_0, T_2],\ j = 1, ..., m,
$$

is a m-dimensional $(Q^{T_2}-)$Wiener process on the measure space $(\Omega, \mathcal{F}_{T_2})$ with $Q^{T_2} = \widetilde{Q}_{L(v(\cdot, T_2), T_2)}$ denoting the T_2-forward measure. Hence,

$$
dP^F(t, T_1, T_2) = P^F(t, T_1, T_2) \cdot v(t, T_1, T_2)\, dW^{T_2}(t),
$$

i.e. $\left(P^F(t, T_1, T_2) \right)_{t \in [t_0, T_1]}$ is a Q^{T_2}-martingale. $\qquad \square$

Most of the time contingent claims are rather based on interest rates than on zero-coupon bond prices as an underlying. Due to the usual market convention that these interest rates R_L are quoted as linear rates we use the subscript L. More precisely, $R_L(t, T_1, T_2)$ is the rate for the time-period $[T_1, T_2]$ quoted at time t giving a one dollar investment at time T_1 a final value of $1 + R_L(t, T_1, T_2) \cdot (T_2 - T_1)$ at time T_2. Because of Lemma 2.6 we immediately get the following corollary.

Corollary 4.16 *Let* $t_0 \le t \le T_1 \le T_2 \le T^*$. *Then the stochastic process* $R_L(T_1, T_2) = (R_L(t, T_1, T_2))_{t \in [t_0, T_1]}$ *defined by*

$$
R_L(t, T_1, T_2) := \frac{P^F(t, T_1, T_2) - 1}{T_2 - T_1} = \frac{P^F(t, T_1, T_2)}{T_2 - T_1} - \frac{1}{T_2 - T_1}
$$

is a martingale under the T_2-*forward measure* Q^{T_2}.

4.5   One-Factor Models

Starting in the late seventies there appeared a variety of one-factor models concentrating on the short rate as the only driving factor. Even if these models only deal with one factor, they gained a lot of popularity because of their mathematical tractability and very often the numerical advantage of closed-form solutions for the prices of contingent claims. In this chapter we give an overview of some of the most popular one-factor models. We use our general interest-rate model of Section 4.1 for the special case $m = 1$, i.e. the underlying Wiener process is one-dimensional. As already mentioned, we generally assume that there exists an equivalent matingale measure $\widetilde{Q} \in I\!M$ and that the expected-value process is unique on $I\!M$ for all the contingent claims we will consider.

In Section 4.5.1 we start with general models of the short rate and move on to models which lead to a Gaussian distribution of the forward and short rate in Section 4.5.2. We will show how the latter models can be easily derived as a special case of the Heath-Jarrow-Morton model. As a representative of the Heath-Jarrow-Morton model we will pick out one of the most commonly used one-factor Gaussian short-rate models, the Hull-White model. This is done in Section 4.5.3. We do this for the special reason that it is possible to explicitly derive Green's function for this model. We will use this result and therefore the Hull-White model in Section 5 to show how we can directly calculate the prices of derivatives, even those of exotic options, using Green's function.

4.5.1   Short-Rate Models

Most of the one-factor interest-rate models take the short rate as the basis for modelling the term structure of interest rates. For a study of these models we assume that the short-rate process under the equivalent martingale measure $\widetilde{Q}$ is given by[8]

$$dr(t) = \alpha\,(r,t)\,dt + \sigma(r,t)d\widetilde{W}(t) \qquad (4.12)$$

with $\sigma > 0$, that the contingent claims with maturity T we consider only depend on the short rate $r\,(T)$ at time $T \in [t_0, T^*]$, i.e. $D = D\,(r\,(T),T)$, and that D is a sufficiently smooth function (see Section 2.6 for more

---

[8] It can be shown that each short rate model is a specific HJM model derived from a specific choice of the volatility structure. However, a general proof is beyond the scope of this book. To get a first impression of this relation, set $\gamma\,(t) \equiv 0$ in the arbitrage-free price system of the HJM framework in Section 4.4.3. More details and conditions can be found in Baxter [BR96], p. 149-151. In Section 4.5.2 we will show how a more general version of the Hull-White model fits into the HJM framework.

details). Then the price

$$
\begin{aligned}
V_D\left(r,t\right) & = P_0\left(t\right) \cdot E_{\tilde{Q}}\left[\tilde{D}\left(r\left(T\right),T\right)|\mathcal{F}_t\right] \\
& = E_{\tilde{Q}}\left[e^{-\int_t^T r(u)du} \cdot D\left(r\left(T\right),T\right)|\mathcal{F}_t\right]
\end{aligned}
$$

of the contingent claim at time $t \in [t_0, T]$ is given by the Feynman-Kac representation

$$
V_D\left(r,t\right) = E_{\tilde{Q}}^{r,t}\left[e^{-\int_t^T r(u)du} \cdot D\left(r\left(T\right),T\right)\right],
$$

where $V_D\left(r,t\right)$ is a solution of the Cauchy problem (see Definition 2.47ff.)

$$
V_t\left(r,t\right) + \alpha\left(r,t\right) \cdot V_r\left(r,t\right) + \frac{1}{2} \cdot \sigma^2\left(r,t\right) \cdot V_{rr}\left(r,t\right) = r \cdot V\left(r,t\right) \quad (4.13)
$$

for all admissible $(r,t) \in \mathbb{R} \times [t_0, T]$, and

$$
V\left(r,T\right) = D\left(r,T\right) \quad \text{for all admissible } r \in \mathbb{R}. \quad (4.14)
$$

If D is a discount bond with maturity T, the terminal condition is given by

$$
V\left(r,T\right) = 1 \quad \text{for all admissible } r \in \mathbb{R}.
$$

Definition 4.17 (Affine Term Structure) *If the discount bond prices are given by*

$$
P\left(t,T\right) = P\left(r,t,T\right) = e^{\tilde{A}(t,T)-r \cdot B(t,T)} \quad (4.15)
$$

*for all admissible $r \in \mathbb{R}$, $t_0 \leq t \leq T \leq T^*$, with deterministic functions $\tilde{A}$ and B, we call $\mathcal{M}$ an interest-rate market with affine term structure (ATS) or, correspondingly, the interest-rate market model a **short-rate model with ATS**.*

Note that the characterization of an affine term structure is defined implicitly via the prices of discount bonds, since these are our primary traded assets. Nevertheless, if we take a look at the zero rates $R\left(r,t,T\right)$ we see that (4.15) is equivalent to

$$
R\left(t,T\right) = R\left(r,t,T\right) = -\frac{1}{T-t} \cdot \ln P\left(r,t,T\right) = -\frac{\tilde{A}\left(t,T\right)}{T-t} + \frac{B\left(t,T\right)}{T-t} \cdot r
$$

for all admissible $r \in \mathbb{R}$, $t_0 \leq t \leq T \leq T^*$, which is an affine function in r. Setting $A\left(t,T\right) := e^{\tilde{A}(t,T)}$ for all $t_0 \leq t \leq T \leq T^*$, equation (4.15) is equivalent to

$$
P\left(r,t,T\right) = A\left(t,T\right) \cdot e^{-r \cdot B(t,T)} \quad \text{for all admissible } r \in \mathbb{R}, t_0 \leq t \leq T \leq T^*,
$$

which is the most popular notation for discount bond prices in interest-rate markets with ATS. The following lemma gives conditions under which the interest-rate model has an affine term structure.

Lemma 4.18 (Models With ATS) *Let the stochastic differential equation for the short rate r under the equivalent martingale measure $\widetilde{Q}$ be given by (4.12) with*

$$\alpha(r,t) = \theta(t) - a(t) \cdot r, \ \sigma(r,t) = \sqrt{b(t) + c(t) \cdot r} \qquad (4.16)$$

for all admissible $(r,t) \in \mathbb{R} \times [t_0, T]$ and deterministic functions $\theta : [t_0, T] \to \mathbb{R}$, and $a, b, c : [t_0, T] \to [0, \infty)$ such that $\sigma > 0$ on $\mathbb{R} \times [t_0, T]$. Then $\mathcal{M}$ is an interest-rate market with affine term structure where A and B are solutions of the system of PDEs

$$
\begin{aligned}
\widetilde{A}_t(t,T) - \theta(t) \cdot B(t,T) + \tfrac{1}{2} \cdot b(t) \cdot B^2(t,T) &= 0, \quad \widetilde{A}(T,T) = 0, \\
1 + B_t(t,T) - a(t) \cdot B(t,T) - \tfrac{1}{2} \cdot c(t) \cdot B^2(t,T) &= 0, \quad B(T,T) = 0,
\end{aligned}
$$
(4.17)

for all $t \in [t_0, T]$.

Proof. From (4.13) and (4.14) of the Cauchy problem we know that for all admissible $(r,t) \in \mathbb{R} \times [t_0, T]$ we have

$$P_t(r,t,T) + \alpha(r,t) \cdot P_r(r,t,T) + \frac{1}{2} \cdot \sigma^2(r,t) \cdot P_{rr}(r,t,T) = r \cdot P(r,t,T) \tag{4.18}$$

with

$$P(r,T,T) = 1 \text{ for all admissible } r \in \mathbb{R}.$$

If $\mathcal{M}$ is an interest-rate market with affine term structure we know that for all admissible $(r,t) \in \mathbb{R} \times [t_0, T]$

$$
\begin{aligned}
P_t(r,t,T) &= \widetilde{A}_t(t,T) \cdot P(r,t,T) - r \cdot B_t(t,T) \cdot P(r,t,T), \\
P_r(r,t,T) &= -B(t,T) \cdot P(r,t,T), \text{ and} \\
P_{rr}(r,t,T) &= B^2(t,T) \cdot P(r,t,T).
\end{aligned}
$$

Hence, (4.18) becomes

$$\widetilde{A}_t(t,T) - r \cdot B_t(t,T) - \alpha(r,t) \cdot B(t,T) + \frac{1}{2} \cdot \sigma^2(r,t) \cdot B^2(t,T) = r.$$

Using (4.16) we get

$$
\begin{aligned}
\widetilde{A}_t(t,T) - r \cdot B_t(t,T) - (\theta(t) - a(t) \cdot r) \cdot B(t,T) & \\
+ \tfrac{1}{2} \cdot (b(t) + c(t) \cdot r) \cdot B^2(t,T) &= r
\end{aligned}
$$

or equivalently

$$
\begin{aligned}
\widetilde{A}_t(t,T) - \theta(t) \cdot B(t,T) + \tfrac{1}{2} \cdot b(t) \cdot B^2(t,T) & \\
- r \cdot \left(B_t(t,T) - a(t) \cdot B(t,T) - \tfrac{1}{2} \cdot c(t) \cdot B^2(t,T) + 1 \right) &= 0
\end{aligned}
$$

for all admissible $(r, t) \in \mathbb{R} \times [t_0, T]$. Comparing coefficients, this is equivalent to the system

$$\tilde{A}_t(t, T) - \theta(t) \cdot B(t, T) + \frac{1}{2} \cdot b(t) \cdot B^2(t, T) = 0,$$

$$B_t(t, T) - a(t) \cdot B(t, T) - \frac{1}{2} \cdot c(t) \cdot B^2(t, T) + 1 = 0$$

for all $t \in [t_0, T]$. Furthermore, the terminal condition

$$\ln P(r, T, T) = \tilde{A}(T, T) - r \cdot B(T, T) = \ln 1 = 0$$

for all admissible $r \in \mathbb{R}$ is equivalent to

$$\tilde{A}(T, T) = 0 \quad \text{and} \quad B(T, T) = 0.$$

$\square$

Note that the second equation of (4.17), i.e. the ordinary differential equation (ODE) for B is a Riccati equation which can be solved analytically (see, e.g. Ince [Inc44]). Using the solution for B, we find $\tilde{A}$ by integrating the first equation of (4.17). A first *example* for a short-rate model with ATS is the *Vasicek model* (see Vasicek [Vas77]) described by

$$dr(t) = (\theta - a \cdot r(t)) \, dt + \sigma d\widetilde{W}(t),$$

i.e. $\theta(t) \equiv \theta$, $a(t) \equiv a$, $b(t) \equiv \sigma^2$, and $c(t) \equiv 0$. Solving (4.17) for this model we get an ATS with

$$B(t, T) = \frac{1 - e^{-a \cdot (T-t)}}{a}$$

and

$$\tilde{A}(t, T) = \frac{[B(t, T) - (T - t)] \cdot [\theta \cdot a - \frac{1}{2} \cdot \sigma^2]}{a^2} - \frac{\sigma^2 \cdot B^2(t, T)}{4 \cdot a}$$

for all $t_0 \leq t \leq T \leq T^*$ (see, e.g. Hull [Hul00], p. 567-568). Another *example* for a short-rate model with ATS is the *Cox-Ingersoll-Ross model* (see Cox, Ingersoll, and Ross [CIR85]) given by

$$dr(t) = (\theta - a \cdot r(t)) \, dt + \sigma \cdot \sqrt{r(t)} d\widetilde{W}(t),$$

i.e. $\theta(t) \equiv \theta$, $a(t) \equiv a$, $b(t) \equiv 0$, and $c(t) \equiv \sigma^2$. For this model it can be shown (see, e.g. Hull [Hul00], p. 570) that we get an ATS with

$$A(t, T) = \left[\frac{2 \cdot c \cdot e^{\frac{1}{2} \cdot (a+c) \cdot (T-t)}}{(a + c) \cdot (e^{c \cdot (T-t)} - 1) + 2 \cdot c} \right]^{\frac{2 \cdot \theta}{\sigma^2}}$$

with $c = \sqrt{a^2 + 2 \cdot \sigma^2}$ and

$$B(t,T) = \frac{2 \cdot \left(e^{c \cdot (T-t)} - 1\right)}{(a+c) \cdot \left(e^{c \cdot (T-t)} - 1\right) + 2 \cdot c}$$

for all $t_0 \leq t \leq T \leq T^*$. The *Hull-White model* (see Hull and White [HW90]) with short rates described by

$$dr(t) = (\theta(t) - a \cdot r(t)) dt + \sigma d\widetilde{W}(t)$$

can be considered as a generalization of the Vasicek model with θ depen-dending on time as well as of the *Ho-Lee model* with $a = 0$ (see Ho and Lee [HL86]). We will give a more detailed discussion of the Hull-White model in Chapter 4.5.3. The most important feature of this model is the possibility of an exact fit to the initial term structure due to the time dependence of θ. A more general version of the Hull-White model assumes that

$$dr(t) = (\theta(t) - a(t) \cdot r(t)) dt + \sigma(t) \cdot r^{\beta} d\widetilde{W}(t)$$

for some deterministic functions θ, a, and $\sigma > 0$ and some constant $\beta \geq 0$ which has an ATS for $\beta = 0$ and $\beta = \frac{1}{2}$. The case $\beta = 0$ is known as the *generalized Vasicek model*, the case $\beta = \frac{1}{2}$ is called the *generalized Cox-Ingersoll-Ross model*. For more details and assumptions on these models see, e.g. Musiela and Rutkowski [MR97], p. 292-295.

The *lognormal model* goes back to Black, Derman, and Toy [BDT90] and is usually referred to as the *Black-Derman-Toy (BDT) model*. In this model, the short rate is described by

$$d\ln r(t) = (\theta(t) - a(t) \cdot \ln r(t)) dt + \sigma(t) d\widetilde{W}(t)$$

for some deterministic functions θ, a, and $\sigma > 0$. The continious time limit of the original BDT model is given by choosing $a = -\frac{\sigma'}{\sigma}$. For practical aspects and variations of this model see, e.g., Black and Karasinski [BK91]. Of course this is just a brief selection of the various existing short-rate models. However, there is a special class of models characterized by their distributional properties which follow a Gaussian law. These models will be the topic of the next section.

4.5.2   One-Factor Gaussian Models

Returning to our Heath-Jarrow-Morton framework we leave the underly-ing Wiener process at a dimension of $m = 1$ as in the previous section and assume that there exists a one-dimensional progressively measurable sto-chastic process γ such that conditions $(i), (ii)$, and (iii) of Theorem 4.12 are satisfied. We hereby let the martingale measure $\widetilde{Q}$ be defined as in Theorem 4.12, suppose that condition $\left(\widetilde{M3}^{HJM}\right)$ holds, and examine the arbitrage-free price system with the additional assumption that the volatility $\sigma > 0$ is a deterministic function and $\int_{t_0}^T \sigma^2(s,T)\,ds < \infty$ for all $t_0 \leq T \leq T^*$. Following Theorem 2.43 we know that under this assumption the forward rate $f(t,T)$, the short rate $r(t)$ and $\ln P(t,T)$ are normally distributed under the equivalent martingale measure $\widetilde{Q}$. Because of their Gaussian distribution we call these models Gaussian. If, in addition to this assumption, the drift μ or equivalently the market price of risk γ are deterministic functions we call this a one-factor Gaussian model with deterministic market price of risk (deterministic forward-rate drift). Following Lemma 4.9 and under the arbitrage-free price system of the Heath-Jarrow-Morton framework we can easily prove the following lemma.

Lemma 4.19 (Forward- and Short-Rate Distributions)
Let the market price of risk be a deterministic function. Then for all (t,T), $t_0 \leq t \leq T \leq T^*$, the forward rate $f(t,T)$ is normally distributed (under Q) with expected value

$$E_Q[f(t,T)|\mathcal{F}_{t_0}] = f(t_0,T) + \int_{t_0}^t \sigma(s,T) \cdot v(s,T)ds + \int_{t_0}^t \sigma(s,T) \cdot \gamma(s)ds$$

and variance

$$s_f^2(t_0,t,T) := Var_Q[f(t,T)|\mathcal{F}_{t_0}] = \int_{t_0}^t \sigma^2(s,T)ds\,.$$

Furthermore, for all $t \in [t_0,T]$, the short rate $r(t)$ is normally distributed (under Q) with expected value

$$E_Q[r(t)|\mathcal{F}_{t_0}] = f(t_0,t) + \int_{t_0}^t \sigma(s,t) \cdot v(s,t)\,ds + \int_{t_0}^t \sigma(s,t) \cdot \gamma(s)ds$$

and variance

$$s_r^2(t_0,t) := Var_Q[r(t)|\mathcal{F}_{t_0}] = \int_{t_0}^t \sigma^2(s,t)ds\,.$$

Setting $\gamma \equiv 0$ we get the corresponding properties under the (martingale) measure $\widetilde{Q}$ which hold for all one-factor Gaussian models.

We also state the stochastic behaviour of the discount or zero-coupon bond prices.

Lemma 4.20 (Zero-Coupon Bond Price Distributions) *Let the market price of risk be a deterministic function. Then for all (t,T), $t_0 \leq t \leq$*

$T \leq T^*$, *the zero-coupon bond prices are lognormally distributed (under Q),*
i.e. $\ln P(t, T)$ *is normally distributed (under Q) with expected value*

$$
E_Q[\ln P(t,T)|\mathcal{F}_{t_0}] = -\int_t^T f(t_0, u) du - \int_{t_0}^t \int_t^T \sigma(s, u) \cdot v(s, u) du ds
$$
$$
- \int_{t_0}^t \gamma(s) \int_t^T \sigma(s, u) du ds
$$

and variance

$$
s_P^2(t_0, t, T) := Var_Q[\ln P(t,T)|\mathcal{F}_{t_0}] = \int_{t_0}^t \left(\int_t^T \sigma(s, u) du \right)^2 ds.
$$

Setting $\gamma \equiv 0$ *we get the corresponding properties under the (martingale)*
measure $\widetilde{Q}$ *which hold for all one-factor Gaussian models.*

Another interesting subset of the one-factor Gaussian models is specified
in the following definition.

Definition 4.21 (Exponential Volatility Structure) *A Gaussian one-*
factor model is called a Gaussian one-factor model with exponential volatil-
ity structure if the forward-rate volatility function satisfies the condition

$$
\sigma(t,T) = \sigma(t) \cdot b(t,T) \quad with \quad b(t,T) = e^{-\int_t^T a(u) du} \qquad (EV)
$$

for all $t_0 \leq t \leq T \leq T^*$ *with*

$$
a(t) > 0 \quad and \quad \sigma(t) > 0 \text{ independent of } T.
$$

Note that, under (EV), we have a separable volatility structure in the
sense that there are functions $\sigma_1(t) := \sigma_1(t_0, t)$ and $\sigma_2(T) := \sigma_2(t_0, T)$
such that

$$
\sigma(t,T) = \sigma_1(t) \cdot \sigma_2(T) \quad \text{for all } t_0 \leq t \leq T \leq T^*.
$$

To be more precise,

$$
\sigma_1(t) = \sigma(t) \cdot e^{\int_{t_0}^t a(u) du} \quad and \quad \sigma_2(T) = b(t_0, T).
$$

We can easily see that under these assumptions we have

$$
\begin{aligned}
b(t,t) &= b(T,T) = 1, \sigma(s,T) = \sigma(s,t) \cdot b(t,T), \\
b_T(t,T) &:= \tfrac{\partial}{\partial T} b(t,T) = -a(T) \cdot b(t,T), \\
b_t(t,T) &:= \tfrac{\partial}{\partial t} b(t,T) = a(t) \cdot b(t,T) \quad and \\
\sigma_T(t,T) &= -a(T) \cdot \sigma(t,T).
\end{aligned}
$$

Using these equations we can specify the stochastic differential equations
for the forward rate and the short rate.

Lemma 4.22 *The stochastic differential equations for the forward rate and the short rate in an arbitrage-free Gaussian one-factor model with exponential volatility structure are given by*

$$dr(t) = (\theta(t_0, t) + \sigma(t) \cdot \gamma(t) - a(t) \cdot r(t)) \, dt + \sigma(t) dW(t) \qquad (4.19)$$

with

$$\theta(t_0, t) = f_T(t_0, t) + a(t) \cdot f(t_0, t) + s_r^2(t_0, t)$$

and

$$
\begin{aligned}
d_t f(t, T) \;=\; & [\sigma^2(t) \cdot b(t, T) \cdot \left(\frac{1}{a(t)} - \frac{b(t, T)}{a(T)} - \int_t^T \frac{a_t(u)}{a^2(u)} \cdot b(t, u) du \right) \\
& + \sigma(t) \cdot b(t, T) \cdot \gamma(t)] \, dt + \sigma(t) \cdot b(t, T) dW(t) .
\end{aligned}
$$

Furthermore, we get the following relation between the forward rate $f(t_0, T)$ and the short rate $r(t_0)$:

$$
\begin{aligned}
f(t_0, T) = & \; r(t_0) \cdot b(t_0, T) + \int_{t_0}^T \theta(t_0, u) \cdot b(u, T) du \\
& - \int_{t_0}^T \sigma^2(s) \cdot b(s, T) \cdot \left(\frac{1}{a(s)} - \frac{b(s, T)}{a(T)} - \int_s^T \frac{a_t(u)}{a^2(u)} \cdot b(s, u) \, du \right) ds .
\end{aligned}
$$

$$\text{(4.20)}$$

Proof. To derive the stochastic differential equation for the short rate in this model we use the arbitrage-free equation for $r(t)$ in the HJM arbitrage-free price system to derive

$$
\begin{aligned}
\int_{t_0}^t \sigma_T(s, t) dW(s) & = -a(t) \int_{t_0}^t \sigma(s, t) dW(s) \\
& = -a(t) \cdot \left(r(t) - f(t_0, t) - \int_{t_0}^t \sigma(s, t) \cdot v(s, t) \, ds - \int_{t_0}^t \sigma(s, t) \cdot \gamma(s) ds \right) \\
& = -a(t) \cdot r(t) + a(t) \cdot f(t_0, t) - \int_{t_0}^t \sigma_T(s, t) \cdot v(s, t) \, ds \\
& \quad - \int_{t_0}^t \sigma_T(s, t) \cdot \gamma(s) ds.
\end{aligned}
$$

So the stochastic differential equation for the short rate in the HJM arbitrage-free price system is given by

$$
\begin{aligned}
dr(t) &= \left(f_T(t_0, t) + \int_{t_0}^t \sigma_T(s,t) \cdot \gamma(s)ds + \int_{t_0}^t \sigma_T(s,t) \cdot v(s,t)ds \right.\\
&\quad \left. + \int_{t_0}^t \sigma^2(s,t)ds + \sigma(t) \cdot \gamma(t) + \int_{t_0}^t \sigma_T(s,t)dW(s) \right) dt \\
&\quad + \sigma(t)dW(t) \\
&= \left(f_T(t_0,t) - a(t) \cdot r(t) + a(t) \cdot f(t_0,t) + s_r^2(t_0,t) + \sigma(t) \cdot \gamma(t) \right) dt \\
&\quad + \sigma(t)dW(t),
\end{aligned}
$$

or equivalently

$$
dr(t) = \left(\theta(t_0,t) + \sigma(t) \cdot \gamma(t) - a(t) \cdot r(t) \right) dt + \sigma(t)dW(t)
$$

with

$$
\theta(t_0,t) = f_T(t_0,t) + a(t) \cdot f(t_0,t) + s_r^2(t_0,t) .
$$

Furthermore, we get for each $t_0 \le t \le t^* \le T$:

$$
\begin{aligned}
\int_{t^*}^T \sigma(t,u)\,du &= \int_{t^*}^T -\frac{1}{a(u)} \cdot \sigma_T(t,u)du \\
&= -\frac{\sigma(t,u)}{a(u)}\Big|_{t^*}^T - \int_{t^*}^T \frac{a_t(u)}{a^2(u)} \cdot \sigma(t,u)du \\
&= \frac{\sigma(t,t^*)}{a(t^*)} - \frac{\sigma(t,T)}{a(T)} - \int_{t^*}^T \frac{a_t(u)}{a^2(u)} \cdot \sigma(t,u)du
\end{aligned}
$$

or equivalently

$$
\int_{t^*}^T \sigma(t,u)du = \sigma(t) \cdot \left(\frac{b(t,t^*)}{a(t^*)} - \frac{b(t,T)}{a(T)} - \int_{t^*}^T \frac{a_t(u)}{a^2(u)} \cdot b(t,u)du \right). \quad (4.21)
$$

As a special case we conclude for $t^* = t$ that

$$
\begin{aligned}
v(t,T) &= \int_t^T \sigma(t,u)du \\
&= \sigma(t) \cdot \left(\frac{1}{a(t)} - \frac{b(t,T)}{a(T)} - \int_t^T \frac{a_t(u)}{a^2(u)} \cdot b(t,u)du \right).
\end{aligned}
$$

Hence, using the first equation of the HJM arbitrage-free price system, we get

$$
\begin{aligned}
d_t f(t,T) &= [\sigma^2(t) \cdot b(t,T) \cdot \left(\frac{1}{a(t)} - \frac{b(t,T)}{a(T)} - \int_t^T \frac{a_t(u)}{a^2(u)} \cdot b(t,u)du \right) \\
&\quad + \sigma(t) \cdot b(t,T) \cdot \gamma(t)]dt + \sigma(t) \cdot b(t,T)dW(t) .
\end{aligned}
$$

Another equation can be derived using (4.19):

$$\int_t^T \theta(t_0, u) \cdot b(u, T) du = \int_t^T f_T(t_0, u) \cdot b(u, T) + \underbrace{f(t_0, u) \cdot a(u) \cdot b(u, T)}_{=b_t(u, T)} du$$

$$+ \int_t^T b(u, T) \int_{t_0}^u \sigma^2(s, u) ds du$$

$$= [f(t_0, u) \cdot b(u, T)]_t^T + \int_t^T \int_{t_0}^u \sigma^2(s, u) \cdot b(u, T) ds du$$

$$= f(t_0, T) - f(t_0, t) \cdot b(t, T) + \int_t^T \int_{t_0}^u \sigma(s, T) \cdot \sigma(s, u) ds du$$

and, as $t_0 \le s \le u$ and $t \le u \le T$ is equivalent to $t_0 \le s \le T$ and $\max\{t, s\} \le u \le T$

$$\int_t^T \theta(t_0, u) \cdot b(u, T) du = f(t_0, T) - f(t_0, t) \cdot b(t, T)$$

$$+ \int_{t_0}^T \sigma(s, T) \int_{\max\{t,s\}}^T \sigma(s, u) du ds$$

which is, using (4.21) with $t^* = \max\{t, s\}$

$$= f(t_0, T) - f(t_0, t) \cdot b(t, T)$$

$$+ \int_{t_0}^T \sigma^2(s) \cdot b(s, T) \cdot \left(\begin{array}{c} \frac{b(s, \max\{t,s\})}{a(\max\{t,s\})} - \frac{b(s,T)}{a(T)} \\[2mm] - \int_{\max\{t,s\}}^T \frac{a_t(u)}{a^2(u)} \cdot b(s, u) du \end{array} \right) ds$$

or equivalently

$$f(t_0, T) = f(t_0, t) \cdot b(t, T) + \int_t^T \theta(t_0, u) \cdot b(u, T) du$$

$$- \int_{t_0}^T \sigma^2(s) \cdot b(s, T) \cdot \left(\begin{array}{c} \frac{b(s, \max\{t,s\})}{a(\max\{t,s\})} - \frac{b(s,T)}{a(T)} \\[2mm] - \int_{\max\{t,s\}}^T \frac{a_t(u)}{a^2(u)} \cdot b(s, u) du \end{array} \right) ds.$$

Note that there is no market price of risk involved in this equation. Setting $t = t_0$ as a special case we know that $\max\{t_0, s\} = s$ as $s \ge t_0$ as well as $f(t_0, t_0) = r(t_0)$ and conclude with

$$\int_{t_0}^{T} \theta(t_0, u) \cdot b(u, T) du = f(t_0, T) - r(t_0) \cdot b(t_0, T)$$

$$+ \int_{t_0}^{T} \sigma^2(s) \cdot b(s, T) \cdot \left(\frac{1}{a(s)} - \frac{b(s, T)}{a(T)} - \int_{s}^{T} \frac{a_t(u)}{a^2(u)} \cdot b(s, u) du \right) ds .$$

□

It is interesting to note that the short rate in a Gaussian model with exponential volatility structure has a very special structure. We can see this by rewriting equation (4.19):

$$\begin{aligned} dr(t) &= \left(\theta(t_0, t) + \sigma(t) \cdot \gamma(t) - a(t) \cdot r(t) \right) dt + \sigma(t) dW(t) \\ &= a(t) \cdot \left(\frac{\theta(t_0, t) + \sigma(t) \cdot \gamma(t)}{a(t)} - r(t) \right) dt + \sigma(t) dW(t). \end{aligned}$$

This shows us that the short rate at time t reverts to the expression $(\theta(t_0, t) + \sigma(t) \cdot \gamma(t)) / a(t)$ at a reversion rate of $a(t)$, i.e. if the short rate is too high (greater than $(\theta(t_0, t) + \sigma(t) \cdot \gamma(t)) / a(t)$) it is pulled down since the drift $\theta(t_0, t) + \sigma(t) \cdot \gamma(t) - a(t) \cdot r(t)$ gets negative. On the other hand, the short rate is pulled up if it is too low (smaller than the expression $(\theta(t_0, t) + \sigma(t) \cdot \gamma(t)) / a(t)$) since, in this case, the drift gets positive. This behaviour, briefly called *mean reversion*, can be observed in the market and may be an indication that these models describe quite well the general real-life short-rate movements.

Remark. *As a special case we get for constant $a(t) \equiv a$ that $a_t(u) \equiv 0$ and*

$$b(t, T) = e^{-a \cdot (T-t)},$$

and thus

$$v(t, T) = \int_{t}^{T} \sigma(t, u) du = \frac{\sigma(t)}{a} \cdot \left(1 - e^{-a(T-t)} \right) ,$$

as well as

$$\begin{aligned} f(t_0, T) &= r(t_0) \cdot e^{-a \cdot (T-t_0)} + \int_{t_0}^{T} \theta(t_0, u) \cdot e^{-a \cdot (T-u)} du \\ &\quad - \int_{t_0}^{T} \frac{\sigma^2(s)}{a} \cdot e^{-a \cdot (T-s)} \cdot \left(1 - e^{-a \cdot (T-s)} \right) ds \\ &= r(t_0) \cdot e^{-a \cdot (T-t_0)} + \int_{t_0}^{T} \theta(t_0, u) \cdot e^{-a \cdot (T-u)} du \\ &\quad - \int_{t_0}^{T} \sigma(s, T) \cdot v(s, T) ds. \quad \Box \end{aligned}$$

We will examine a special Gaussian one-factor model with exponential volatility structure which is quite often used in practice in the next section. Because of its mathematical tractability it has become very popular for the pricing of interest-rate derivatives. It also gives a better description of future interest-rate movements than the Black or Black-Scholes model.

4.5.3   The Hull-White Model

Hull and White [HW90] suggested a special Gaussian one-factor model with exponential volatility structure for describing interest-rate movements. As we will see, we can explicitly derive Green's function for this model and therefore will use it as a benchmark model for the evaluation of derivative prices under this technique. To describe the model and embed it into our Heath-Jarrow-Morton framework we set $\sigma(t) \equiv \sigma$ and $a(t) \equiv a$, independent of t, T, i.e. $\sigma(t, T) = \sigma \cdot e^{-a \cdot (T-t)}$ and $v(t, T) = \frac{\sigma}{a} \cdot \left(1 - e^{-a(T-t)}\right)$. As already noted in the previous section this will imply a mean-reverting behaviour of the short rate at time t to $(\theta(t_0, t) + \sigma \cdot \gamma(t))/a$ at a constant reversion rate of a.

Let us start now with a more detailed analysis of the stochastic differential equations for the forward short rates and the short rate of this model. To do this we first examine the implications of the specific parameter choice once they are applied to the arbitrage-free price system of the Heath-Jarrow-Morton framework of Section 4.4.3. We thus have to determine the following integral:

$$\int_{t_0}^{t} \sigma^2(s, T)\, ds = \sigma^2 \int_{t_0}^{t} e^{-2a \cdot (T-s)}\, ds = \frac{\sigma^2}{2a} \left(e^{-2a \cdot (T-t)} - e^{-2a \cdot (T-t_0)}\right)$$

and hence,

$$s_r^2(t_0, t) = \int_{t_0}^{t} \sigma^2(s, t)\, ds = \frac{\sigma^2}{2a} \left(1 - e^{-2a \cdot (t-t_0)}\right).$$

Furthermore, since $v_t(t, T) = -\sigma(t, T)$,

$$\int_{t_0}^{t^*} \sigma(s, T) \cdot v(s, T)\, ds = -\frac{1}{2} \cdot v^2(s, T)\, |_{t_0}^{t^*}$$

$$= \frac{1}{2} \cdot \left(v^2(t_0, T) - v^2(t^*, T)\right)$$

for all $t_0 \leq t^* \leq T$. Especially, setting $t^* = T$, we get

$$\int_{t_0}^{T} \sigma(s, T) \cdot v(s, T)\, ds = \frac{1}{2} \cdot v^2(t_0, T).$$

Hence, applying the HJM arbitrage-free price system and Lemma 4.22, we get the following *arbitrage-free price system for the Hull and White model*:

$$
\begin{aligned}
d_t f(t,T) &= \left[\sigma \cdot e^{-a(T-t)} \cdot v\,(t,T) + \sigma \cdot e^{-a(T-t)} \cdot \gamma(t) \right] dt \\[2mm]
&\quad + \sigma \cdot e^{-a(T-t)} dW(t)\,, \\[2mm]
f(t,T) &= f(t_0,T) + \tfrac{1}{2} \cdot \left(v^2\,(t_0,T) - v^2\,(t,T) \right) \\[2mm]
&\quad + \sigma \cdot e^{-aT} \int_{t_0}^{t} \gamma(s) \cdot e^{as} ds \\[2mm]
&\quad + \sigma \cdot e^{-aT} \int_{t_0}^{t} e^{as} dW(s)\,, \\[2mm]
r(t) &= f(t_0,t) + \tfrac{1}{2} \cdot v^2\,(t_0,t) + \sigma \cdot e^{-at} \int_{t_0}^{t} \gamma(s) \cdot e^{as} ds \\[2mm]
&\quad + \sigma \cdot e^{-at} \int_{t_0}^{t} e^{as} dW(s)\,, \\[2mm]
dr(t) &= \left(\theta(t_0,t) + \sigma \cdot \gamma(t) - a \cdot r(t) \right) dt + \sigma dW(t) \\[2mm]
&\quad \text{with} \\[2mm]
&\quad \theta(t_0,t) = f_T(t_0,t) + a \cdot f(t_0,t) + s_r^2\,(t_0,t)\,.
\end{aligned}
$$

Furthermore, we have the following relation between $f(t_0,T)$ and $r(t_0)$:

$$
f(t_0,T) = r(t_0) \cdot e^{-a(T-t_0)} + \int_{t_0}^{T} \theta(t_0,u) \cdot e^{-a(T-u)} du - \frac{1}{2} \cdot v^2\,(t_0,T)\,. \quad (4.22)
$$

Using Lemma 4.19 we find the following corollary for the distribution of the short rate and the forward rates.

Corollary 4.23 (Hull-White Forward-Rate Distributions) *Let the market price of risk be a deterministic function. Then for all* (t,T), $t_0 \le t \le T \le T^*$, *the forward rate* $f(t,T)$ *is normally distributed (under Q) with expected value*

$$
\begin{aligned}
E_Q[f(t,T)|\mathcal{F}_{t_0}] &= f(t_0,T) + \frac{1}{2} \cdot \left(v^2\,(t_0,T) - v^2\,(t,T) \right) \\[2mm]
&\quad + \sigma \cdot e^{-aT} \int_{t_0}^{t} \gamma(s) \cdot e^{as} ds
\end{aligned}
$$

and variance

$$s_f^2(t_0, t, T) := Var_Q[f(t,T)|\mathcal{F}_{t_0}] = \frac{\sigma^2}{2a}\left(e^{-2a\cdot(T-t)} - e^{-2a\cdot(T-t_0)}\right).$$

Furthermore, for all $t \in [t_0, T^*]$, the short rate $r(t)$ is normally distributed (under Q) with expected value

$$E_Q[r(t)|\mathcal{F}_{t_0}] = f(t_0, t) + \frac{1}{2}\cdot v^2(t_0, t) + \sigma\cdot e^{-at}\int_{t_0}^t \gamma(s)\cdot e^{as}ds$$

and variance

$$s_r^2(t_0, t) := Var_Q[r(t)|\mathcal{F}_{t_0}] = \frac{\sigma^2}{2a}\left(1 - e^{-2a\cdot(t-t_0)}\right).$$

Setting $\gamma \equiv 0$ we get the corresponding properties under the (martingale) measure $\widetilde{Q}$ which are also true for those models having a stochastic market price of risk under Q.

We will now focus on the pricing of discount bonds. As we know from Lemma 4.18, the Hull-White model is a model with affine term structure. Especially, given the short-rate process

$$dr(t) = (\theta(t_0, t) - a\cdot r(t))dt + \sigma d\widetilde{W}(t)$$

under $\widetilde{Q}$, we have to solve the following system of PDEs to find a close form solution for the discount bond prices:

$$\widetilde{A}_t(t,T) - \theta(t_0,t)\cdot B(t,T) + \tfrac{1}{2}\cdot\sigma^2\cdot B^2(t,T) = 0, \quad \widetilde{A}(T,T)=0,$$
$$1 + B_t(t,T) - a\cdot B(t,T) = 0, \quad B(T,T)=0,$$

$$(4.23)$$

for all $(r,t) \in \mathbb{R}\times[t_0,T]$. The solution to the second ODE is given by

$$B(t,T) = \frac{1}{a}\cdot\left(1 - e^{-a\cdot(T-t)}\right) \text{ for all } (r,t)\in\mathbb{R}\times[t_0,T].$$

Hence, the solution to the first PDE is

$$
\begin{aligned}
-\widetilde{A}(t,T) &= \int_t^T \theta(t_0,u)\cdot B(u,T) - \frac{1}{2}\cdot\sigma^2\cdot B^2(u,T)\,du \\
&= \frac{1}{a}\int_t^T \theta(t_0,u)\cdot\left(1 - e^{-a\cdot(T-u)}\right)du \\
&\quad - \frac{\sigma^2}{2a^2}\int_t^T\left(1 - e^{-a\cdot(T-u)}\right)^2 du \\
&= \frac{1}{a}(I_1 - I_2) - I_3
\end{aligned}
$$

with

$$I_1 = \int_t^T \theta\,(t_0, u)\,du$$

$$= \int_t^T f_T(t_0, u) + a \cdot f(t_0, u) + s_r^2\,(t_0, u)\,du$$

$$= f(t_0, T) - f(t_0, t) + a \cdot \int_t^T f(t_0, u)du + \frac{\sigma^2}{2a}\int_t^T 1 - e^{-2a(u-t_0)}du$$

$$= f(t_0, T) - f(t_0, t) + a \cdot (\ln P\,(t_0, t) - \ln P\,(t_0, T))$$

$$+ \frac{\sigma^2}{2a}\int_t^T 1 - e^{-2a(u-t_0)}du$$

$$= f(t_0, T) - f(t_0, t) + a \cdot \ln\left(\frac{P(t_0,t)}{P(t_0,T)}\right) + \frac{\sigma^2}{2a} \cdot \left[u + \frac{1}{2a} \cdot e^{-2a(u-t_0)}\big|_t^T\right]$$

$$= f(t_0, T) - f(t_0, t) + a \cdot \ln\left(\frac{P(t_0,t)}{P(t_0,T)}\right)$$

$$+ \frac{\sigma^2}{2a} \cdot \left[T - t + \frac{1}{2a} \cdot \left(e^{-2a(T-t_0)} - e^{-2a(t-t_0)}\right)\right]$$

$$= f(t_0, T) - f(t_0, t) + a \cdot \ln\left(\frac{P(t_0,t)}{P(t_0,T)}\right)$$

$$+ \frac{\sigma^2}{2a} \cdot (T - t) + \frac{\sigma^2}{4a^2} \cdot \left(e^{-2a(T-t_0)} - e^{-2a(t-t_0)}\right)$$

and, using equation (4.22),

$$I_2 = \int_t^T \theta\,(t_0, u) \cdot e^{-a\cdot(T-u)}du$$

$$= \int_{t_0}^T \theta\,(t_0, u) \cdot e^{-a\cdot(T-u)}du - \int_{t_0}^t \theta\,(t_0, u) \cdot e^{-a\cdot(T-u)}du$$

$$= \int_{t_0}^T \theta\,(t_0, u) \cdot e^{-a\cdot(T-u)}du - e^{-a\cdot(T-t)}\int_{t_0}^t \theta\,(t_0, u) \cdot e^{-a\cdot(t-u)}du$$

$$= f(t_0, T) - r(t_0) \cdot e^{-a\cdot(T-t_0)} + \frac{\sigma^2}{2a^2} \cdot \left(1 - e^{-a(T-t_0)}\right)^2$$

$$- e^{-a\cdot(T-t)} \cdot \left[f(t_0, t) - r(t_0) \cdot e^{-a\cdot(t-t_0)} + \frac{\sigma^2}{2a^2} \cdot \left(1 - e^{-a(t-t_0)}\right)^2\right]$$

$$= f(t_0, T) - e^{-a\cdot(T-t)} \cdot f(t_0, t)$$

$$+ \frac{\sigma^2}{2a^2} \cdot \left[\left(1 - e^{-a(T-t_0)}\right)^2 - e^{-a\cdot(T-t)} \cdot \left(1 - e^{-a(t-t_0)}\right)^2\right]$$

$$= f(t_0, T) - e^{-a\cdot(T-t)} \cdot f(t_0, t)$$

$$+ \frac{\sigma^2}{2a^2} \cdot \left[1 + e^{-2a\cdot(T-t_0)} - e^{-a\cdot(T-t)} - e^{-a\cdot(T-t_0)} \cdot e^{-a\cdot(t-t_0)}\right].$$

Furthermore,

$$I_3 = \frac{\sigma^2}{2a^2} \int_t^T \left(1 - e^{-a \cdot (T-u)}\right)^2 du$$

$$= \frac{\sigma^2}{2a^2} \cdot \int_t^T 1 - 2 \cdot e^{-a \cdot (T-u)} + e^{-2a \cdot (T-u)} du$$

$$= \frac{\sigma^2}{2a^2} \cdot \left[u - \frac{2}{a} \cdot e^{-a \cdot (T-u)} + \frac{1}{2a} \cdot e^{-2a \cdot (T-u)} \Big|_t^T\right]$$

$$= \frac{\sigma^2}{2a^2} \cdot \left[T - t - \frac{2}{a} \cdot \left(1 - e^{-a \cdot (T-t)}\right) + \frac{1}{2a} \cdot \left(1 - e^{-2a \cdot (T-t)}\right)\right]$$

$$= \frac{\sigma^2}{2a^2} \cdot (T-t) + \frac{\sigma^2}{4a^3} \cdot \left[\left(1 - e^{-2a \cdot (T-t)}\right) - 4 \cdot \left(1 - e^{-a \cdot (T-t)}\right)\right]$$

$$= \frac{\sigma^2}{2a^2} \cdot (T-t) + \frac{\sigma^2}{4a^3} \cdot \left[4 \cdot e^{-a \cdot (T-t)} - e^{-2a \cdot (T-t)} - 3\right]$$

Inserting $I_1, I_2,$ and I_3 we get

$$-\widetilde{A}(t,T) = \tfrac{1}{a} \cdot (I_1 - I_2) - I_3$$

$$= \ln\left(\frac{P(t_0,t)}{P(t_0,T)}\right) - B(t,T) \cdot f(t_0,t) + \frac{\sigma^2}{4a^3} \cdot \left[e^{-2a(T-t_0)} - e^{-2a(t-t_0)}\right]$$

$$- \frac{\sigma^2}{2a^3} \cdot \left[1 + e^{-2a \cdot (T-t_0)} - e^{-a \cdot (T-t)} - e^{-a \cdot (T-t_0)} \cdot e^{-a \cdot (t-t_0)}\right]$$

$$- \frac{\sigma^2}{4a^3} \cdot \left[4 \cdot e^{-a \cdot (T-t)} - e^{-2a \cdot (T-t)} - 3\right]$$

$$= \ln\left(\frac{P(t_0,t)}{P(t_0,T)}\right) - B(t,T) \cdot f(t_0,t)$$

$$+ \frac{\sigma^2}{4a^3} \cdot \left[\left(1 - e^{-a \cdot (T-t)}\right)^2 - \left(e^{-a \cdot (t-t_0)} - e^{-a \cdot (T-t_0)}\right)^2\right]$$

$$= \ln\left(\frac{P(t_0,t)}{P(t_0,T)}\right) - B(t,T) \cdot f(t_0,t)$$

$$+ \frac{\sigma^2}{4a^3} \cdot \left[\left(1 - e^{-a \cdot (T-t)}\right)^2 - e^{-2a \cdot (t-t_0)} \cdot \left(1 - e^{-a \cdot (T-t)}\right)^2\right]$$

$$= \ln\left(\frac{P(t_0,t)}{P(t_0,T)}\right) - B(t,T) \cdot f(t_0,t) + \frac{1}{2} \cdot B^2(t,T) \cdot s_r^2(t_0,t).$$

This result is summarized in the following lemma.

Lemma 4.24 (Hull-White Zero-Coupon Bond Prices)
The zero-coupon bond prices in the Hull-White model are given by

$$P(t,T) = P(r,t,T) = A(t,T) \cdot e^{-B(t,T) \cdot r} \quad \text{for all } (r,t) \in \mathbb{R} \times [t_0,T]$$

with

$$B(t,T) = \frac{1}{a} \cdot \left(1 - e^{-a \cdot (T-t)}\right)$$

and

$$\ln A\left(t,T\right) = \ln\left(\frac{P\left(t_0,T\right)}{P\left(t_0,t\right)}\right) + B\left(t,T\right)\cdot f(t_0,t) - \frac{1}{2}\cdot s_P^2\left(t_0,t,T\right),$$

where

$$s_P\left(t_0,t,T\right) := B\left(t,T\right)\cdot s_r\left(t_0,t\right) \ \ with \ \ s_r\left(t_0,t\right) := \sqrt{\frac{\sigma^2}{2a}\cdot\left(1-e^{-2a\cdot(t-t_0)}\right)}.$$

Lemma 4.24 can be used to derive the distribution of the Hull-White zero-coupon bonds. This is slightly more elegant than using Lemma 4.20.

Lemma 4.25 (Hull-White Discount Bond Price Distributions)
Let the market price of risk be a deterministic function. Then for all (t,T), $t_0 \leq t \leq T \leq T^*$, the zero-coupon bond prices are lognormally distributed (under Q), i.e. $\ln P\left(t,T\right)$ is normally distributed (under Q) with expected value

$$\begin{aligned}
E_Q[\ln P\left(t,T\right)|\mathcal{F}_{t_0}] &= \ln\left(\frac{P\left(t_0,T\right)}{P\left(t_0,t\right)}\right) \\
&\quad -\frac{1}{2}\cdot\left(s_P^2\left(t_0,t,T\right) + B\left(t,T\right)\cdot v^2\left(t_0,t\right)\right) \\
&\quad -B\left(t,T\right)\cdot\sigma\cdot e^{-at}\int_{t_0}^{t}\gamma(s)\cdot e^{as}ds
\end{aligned}$$

and variance

$$s_P^2\left(t_0,t,T\right) := Var_Q\left[\ln P\left(t,T\right)|\mathcal{F}_{t_0}\right] = B^2\left(t,T\right)\cdot s_r^2\left(t_0,t\right).$$

Furthermore, for all (t,T), $t_0 \leq t \leq T < T^*$, the zero rates are normally distributed (under Q) with expected value

$$\begin{aligned}
E_Q[R\left(t,T\right)|\mathcal{F}_{t_0}] &= -\frac{1}{T-t}\cdot E_Q[\ln P\left(t,T\right)|\mathcal{F}_{t_0}] \\
&= \frac{1}{T-t}\cdot\ln\left(\frac{P\left(t_0,t\right)}{P\left(t_0,T\right)}\right) \\
&\quad +\frac{1}{2}\cdot\left(\frac{s_P^2\left(t_0,t,T\right) + B\left(t,T\right)\cdot v^2\left(t_0,t\right)}{T-t}\right) \\
&\quad +\frac{B\left(t,T\right)\cdot\sigma}{T-t}\cdot e^{-at}\int_{t_0}^{t}\gamma(s)\cdot e^{as}ds
\end{aligned}$$

and variance

$$s_R^2\left(t_0,t,T\right) := \left(\frac{1}{T-t}\right)^2\cdot Var_Q\left[\ln P\left(t,T\right)|\mathcal{F}_{t_0}\right] = \left(\frac{B\left(t,T\right)\cdot s_r\left(t_0,t\right)}{T-t}\right)^2.$$

Setting $\gamma \equiv 0$ we get the corresponding properties under the (martingale) measure $\widetilde{Q}$ which are also true for those models having a stochastic market price of risk under Q.

Proof. Because of Lemma 4.24 we know that

$$E_Q[\ln P(t,T)|\mathcal{F}_{t_0}] = \ln A(t,T) - B(t,T) \cdot E_Q[r(t)|\mathcal{F}_{t_0}]$$

$$= \ln\left(\frac{P(t_0,T)}{P(t_0,t)}\right) + B(t,T) \cdot f(t_0,t) - \frac{1}{2} \cdot s_P^2(t_0,t,T)$$

$$- B(t,T) \cdot \left(f(t_0,t) + \frac{1}{2} \cdot v^2(t_0,t) + \sigma \cdot e^{-at} \int_{t_0}^t \gamma(s) \cdot e^{as} ds\right)$$

$$= \ln\left(\frac{P(t_0,T)}{P(t_0,t)}\right) - \frac{1}{2} \cdot \left(s_P^2(t_0,t,T) + B(t,T) \cdot v^2(t_0,t)\right)$$

$$- B(t,T) \cdot \sigma \cdot e^{-at} \int_{t_0}^t \gamma(s) \cdot e^{as} ds.$$

Similarly,

$$Var_Q[\ln P(t,T)|\mathcal{F}_{t_0}] = B^2(t,T) \cdot Var_Q[r(t)|\mathcal{F}_{t_0}]$$
$$= B^2(t,T) \cdot s_r^2(t_0,t).$$

The second part of the statement is straightforward.   □

As we have seen in Chapter 2.6 there is a very interesting method to evaluate the prices of derivatives based on Green's function $G : \mathbb{R} \times [0,T] \times \mathbb{R} \times [0,T] \to \mathbb{R}$. We will now show that we can explicitly derive Green's function for the Hull-White model. To do this we restate that Green's function is the so-called fundamental solution of the Cauchy problem which has, within the Hull-White model, the following properties:

a) For any $(r',t') \in \mathbb{R} \times [0,T]$ the function $G_{r',t'}(r,t) = G(r,t,r',t')$ is continuously differentiable in t, twice continuously differentiable in r, and solves the partial differential equation

$$\mathcal{D}G_{r',t'}(r,t) = r \cdot G_{r',t'}(r,t) \quad \text{for all } (r,t) \in \mathbb{R} \times [0,T]$$

with

$$\mathcal{D}G_{r',t'}(r,t) = \frac{\partial}{\partial t} G_{r',t'}(r,t) + (\theta(t) - a \cdot r) \cdot \frac{\partial}{\partial r} G_{r',t'}(r,t)$$

$$+ \frac{1}{2} \cdot \sigma^2 \cdot \frac{\partial^2}{\partial r^2} G_{r',t'}(r,t).$$

b) For any $(r,t) \in \mathbb{R} \times [0,T]$ the function $G_{r,t}(r',t') = G(r,t,r',t')$ is continuously differentiable in t', twice continuously differentiable in r', and solves the *Fokker-Planck* or *forward Kolmogorov equation*

$$\mathcal{D}^*G_{r,t}(r',t') = r' \cdot G_{r,t}(r',t') \quad \text{for all } (r',t') \in \mathbb{R} \times [t,T]$$

with

$$\mathcal{D}^*G_{r,t}(r',t') = -\frac{\partial}{\partial t'} G_{r,t}(x',t') - (\theta(t') - a \cdot r') \cdot \frac{\partial}{\partial r'} G_{r,t}(r',t')$$

$$+ \frac{1}{2} \cdot \sigma^2 \cdot \frac{\partial^2}{\partial r'^2} G_{x,t}(x',t').$$

c) Under technical conditions on $D = D(r,T)$ the solution of the Cauchy problem is given by

$$V_D(r,t) = \int_{-\infty}^{\infty} G(r,t,r',T) \cdot D(r',T)\, dr'.$$

A sufficient set of technical and terminal conditions may be found in Friedman [Fri64] and [Fri75] or Proske [Pro99]. The solution for the Hull-White model is given in Mayer [May98] and stated in the following lemma.

Lemma 4.26 (Green's Function for the Hull-White Model)
Green's function for the Hull White model is given by

$$G(r,t,r',T) = P(r,t,T) \cdot \frac{1}{s_r(t,t') \cdot \sqrt{2\pi}} \cdot e^{-\frac{1}{2 s_r^2(t,t')} \cdot (r'-f(t,t'))^2}$$

for all $(r,t,r',t') \in \mathbb{R} \times [t_0,T] \times \mathbb{R} \times [t_0,T]$. Furthermore, the price at time $t \in [t_0,T]$ of any (European) contingent claim $D = D(r(T),T)$ with maturity T, depending on the short rate $r' := r(T)$ at time T, is given by the Feynman-Kac representation

$$V_D(r,t) = \int_{-\infty}^{\infty} G(r,t,r',T) \cdot D(r',T)\, dr' \quad \text{for all } (r,t) \in \mathbb{R} \times [t_0,T].$$

We will make use of the Feynman-Kac representation in the following chapter to derive the prices for a selection of interest-rate derivatives which will also be examined with respect to their interest-rate risk in Chapter 6. The following lemma is a first application of the Feynman-Kac representation giving us a relation between the zero-coupon bond prices of different maturities at time t.

Lemma 4.27 *For all $t_0 \leq t < T < T' \leq T^*$ and for all $r \in \mathbb{R}$ we have*

$$\frac{P(r,t,T')}{P(r,t,T)} = A(T,T') \cdot e^{-B(T,T') \cdot f(t,T) + \frac{1}{2} \cdot s_P^2(t,T,T')}.$$

Proof. Let $V_D(r,t) := P(r,t,T')$ with $t_0 \leq t < T' \leq T^*$ and $r \in \mathbb{R}$. Furthermore, let $D(r,T) := V_D(r,T) = P(r,T,T')$. Then, using the Feynman-Kac representation, we get for all $(r,t) \in \mathbb{R} \times [t_0,T]$

$$
\begin{aligned}
P(r,t,T') &= V_D(r,t) = \int_{-\infty}^{\infty} G(r,t,r',T) \cdot D(r',T)\, dr' \\
&= P(r,t,T) \cdot \int_{-\infty}^{\infty} \frac{P(r',T,T')}{s_r(t,T) \cdot \sqrt{2\pi}} \cdot e^{-\frac{1}{2 s_r^2(t,T)} \cdot (r'-f(t,T))^2}\, dr'
\end{aligned}
$$

or equivalently,

$$
\begin{aligned}
\frac{P\left(r,t,T'\right)}{P\left(r,t,T\right)} &= A\left(T,T'\right) \int_{-\infty}^{\infty} \frac{e^{-\frac{1}{2s_r^2(t,T)}\cdot\left(r'-f(t,T)\right)^2}}{s_r\left(t,T\right)\cdot\sqrt{2\pi}} \cdot e^{-B(T,T')\cdot r'}\, dr' \\
&= A\left(T,T'\right) \cdot e^{-B\left(T,T'\right)\cdot f(t,T)+\frac{1}{2}\cdot B^2\left(T,T'\right)\cdot s_r^2(t,T)} \\
&\quad \cdot \int_{-\infty}^{\infty} \frac{e^{-\frac{1}{2s_r^2(t,T)}\cdot\left(r'-\left[f(t,T)-B(T,T')\cdot s_r^2(t,T)\right]\right)^2}}{s_r\left(t,T\right)\cdot\sqrt{2\pi}}\, dr' \\
&= A\left(T,T'\right) \cdot e^{-B\left(T,T'\right)\cdot f(t,T)+\frac{1}{2}\cdot s_P^2\left(t,T,T'\right)}.
\end{aligned}
$$

$\square$

4.6   Multi-Factor Models

The one-factor models of the previous Section 4.5 were characterized by only one source of uncertainty. For the short-rate models, for example, we considered the short-rate to act as a driving factor or state variable of our stochastic system. A more general approach consequently involves more than one state variable. The general multi-factor model assumes that there is a set of state variables or factors $F = (F_1, ..., F_m)$, $m \in I\!N$ and a $m-$dimensional Wiener process W such that each factor is an Itô process and the unique strong solution of the differential equation

$$
dF_i = \mu^F\left(F_i, t\right) dt + \sigma_i^F\left(F_i, t\right) dW\left(t\right) = \mu^F\left(F_i, t\right) dt + \sum_{j=1}^{m} \sigma_{ij}^F\left(F_i, t\right) dW_j\left(t\right)
$$

for all $t \in [t_0, T^*]$ and for all $i = 1, ..., m$. For more details on the existence and uniqueness of stochastic differential equations see Section 2.6. It should be noted that for ease of exposition we put the number of state variables or factors equal to the dimension of the Wiener process, i.e. the sources of uncertainty. It may quite well be possible that the first number is greater than the second. Furthermore, it is assumed that a selection of zero rates $R\left(t,T\right)$, $T \in \mathcal{T}_n := \{T_1, ..., T_n\}$ with $t_0 \leq T_1 \leq T_2 \leq \cdots \leq T_n \leq T^*$ can be explained by

$$
R\left(t,T\right) = g_T\left(F\left(t\right)\right), \quad \text{for all } t \in [t_0, T], T \in \mathcal{T}_n,
$$

and for some functions $g_T : I\!R^m \to I\!R$. Among others, such models are studied by Duffie and Singleton [DS94b]. If the functions g_T are affine, the changes of the zero rates are linear functions of the state variables or factors. Such models are called *linear multi-factor models*. If the number of factors is much less than the number of zero rates explained by a multi-factor model, the complexity of a portfolio risk analysis can be reduced dramatically (see Section 6.4 for more details). The case of a one-factor short-rate model is covered by setting

$$r\left(t\right) = R\left(t,t\right) = g_0\left(\mathsf{F}\left(t\right)\right) \text{ for } t \in [t_0, T^*]$$

for some function $g_0 : {I\!\!R}^m \to {I\!\!R}$. Popular choices are $g_0\left(\mathsf{F}\left(t\right)\right) = \frac{1}{2} \cdot \|\mathsf{F}\left(t\right)\|^2$ or $g_0\left(\mathsf{F}\left(t\right)\right) = a_0'\mathsf{F}\left(t\right)$ for some $a_0 \in {I\!\!R}^m$. If F is a Gaussian Markov process, the first case is called a squared-Gauss-Markov process while in the second case r is a Gaussian process. In most two-factor models one of the state variables is the short rate while the other may be some economic variable. Brennan and Schwartz [BS82], for example, use a two-factor model of the short rate and a long-term interest rate, also called consol rate. Longstaff and Schwartz [LS92] developed a two-factor model for the short rate and the volatility which, in their model, is considered to be stochastic. Hull and White [HW94a] invented a two-factor model for the short rate and a stochastic drift factor. Other multi-factor models are studied, e.g., by Jamshidian [Jam91] and [Jam96] or Duffie and Kan [DK94] and [DK96]. A detailed analysis of multi-factor models may be rather lengthy and is beyond the scope of this book. Beside the original papers, the interested reader may also refer to Rebonato [Reb96] for an overview of one- or two-factor interest-rate models.

4.7   LIBOR Market Models

Up to now we discussed the general Heath-Jarrow-Morton framework as well as different models of the short and forward rates. Especially the one-factor short-rate models are mathematically particularly tractable and very often lead to closed-form solutions for the prices of contingent claims. Nevertheless, these rates usually do not exist in practice and the calibration of the models is not always really satisfying. Brace, Gątarek, and Musiela [BGM97] and Miltersen, Sandmann, and Sondermann [MSS97] introduced a new class of models called LIBOR market models. These models describe the evolution of real market rates and lead to pricing formulas which have the Black-Scholes form and are therefore very familiar to traders and easy to calibrate. Their name goes back to the *London InterBank Offered Rate*, or briefly *LIBOR*, which represents the rate of interest earned on Eurodollars deposited by one bank with another bank. Beside the *EURIBOR*, the *Euro InterBank Offered Rate* introduced in 1999, the LIBOR is the most commonly used index rate for the so-called floating rate notes (see Section 5.4.1 for more details). Due to the name of the models we will use the LIBOR as a representative of the floating index or market rate within this section. A *Floating Rate Note* or briefly *FRN* is a certificate of deposit issued by a bank or company raising funds to finance its business activities in which the interest or coupon to be paid changes periodically. There are FRNs that reset their coupon daily, weekly, monthly (1-month LIBOR), quarterly (3-month LIBOR), or semi-annually (6-month LIBOR). Typically FRNs have maturities from 18 months to five years. Because the LIBOR is a floating interest rate, let us denote this short-term zero rate at time t for the period $[T_1, T_2]$ by $R_L(t, T_1, T_2)$. It is important to note that the LIBOR is, by market convention, a linear interest rate, i.e. a one dollar investment deposited at time T_1 will have a final value of $1 + R_L(t, T_1, T_2) \cdot (T_2 - T_1)$ at time T_2 if we enter into this (forward) agreement at time t. Note that the interest to be paid is fixed at time T_1, the beginning of the period, but interest is paid at time T_2, the end of the period. We consider the interest-rate market to be arbitrage-free and complete with equivalent martingale measure $\widetilde{Q}$. Since the zero-coupon bonds of different maturities are the primary traded assets in our interest-rate market, the assumption of Theorem 3.31b) is satisfied. Furthermore, we get the following implicit definition for the LIBOR:

$$P(t, T_2) \cdot [1 + R_L(t, T_1, T_2) \cdot (T_2 - T_1)] = P(t, T_1),$$

for $t_0 \leq t \leq T_1 \leq T_2 \leq T^*$, where $P(t, T)$ is the price at time t of a zero-coupon bond with maturity T, $t_0 \leq t \leq T \leq T^*$, and T^* denotes the maximum time horizon. Solving for R_L, we get for all $t_0 \leq t \leq T_1 \leq T_2 \leq T^*$:

$$R_L(t, T_1, T_2) = \frac{1}{T_2 - T_1} \cdot \left(\frac{P(t, T_1)}{P(t, T_2)} - 1 \right) = \frac{P^F(t, T_1, T_2) - 1}{T_2 - T_1}, \quad (4.24)$$

where $P^F(t, T_1, T_2)$ denotes the forward price at time t for the forward maturity time T_2 of a zero-coupon bond with maturity time T_1 as it was already defined in Section 4.4.4. The time T_1 is called the maturity time of the forward LIBOR, the time from T_1 to T_2 is called the *tenor*, and the expression $1/(T_2 - T_1)$ is called the *accrual factor* or *daycount fraction*. If $t = T_1$, $R_L(T_1, T_2) := R_L(T_1, T_1, T_2)$ is called the *spot LIBOR* at time T_1. To ensure that equation (4.24) is well defined and that we construct a model with positive LIBORs, our first *assumption* is that the initial term structure of the zero-coupon bond prices is strictly positive and strictly decreasing, i.e.

$$P(t_0, T) > 0 \text{ and } P(t_0, T_1) > P(t_0, T_2) \qquad \text{(LM0)}$$

for all $T, T_1, T_2 \in [t_0, T^*]$ with $T_1 < T_2$.

4.7.1   The Discrete-Tenor Model

In most markets only a finite number of forward LIBORs of one specific tenor ΔT, usually 3 or 6 months, are actively traded. We therefore assume that there are n forward LIBORs with maturity times[9] T_i,
$i = 1, ..., n$, $t_0 = T_0$, and $T_{n+1} = T^*$. Using $P(t, T_i)$ as a numéraire we
already know by the change of numéraire Theorem 3.31 that the forward
prices $\left(P^F(t, T_i, T_{i+1})\right)_{t \in [t_0, T_i]}$ and hence the forward LIBORs
$(R_L(t, T_i, T_{i+1}))_{t \in [t_0, T_i]}$ are martingales under the T_{i+1}−forward measure
$Q^{T_{i+1}}$, $i = 1, ..., n$. Hence, quite naturally, we make the *assumption* that
the forward LIBOR processes $(R_L(t, T_i, T_{i+1}))_{t \in [t_0, T_i]}$ follow the stochastic
differential equation

$$dR_L(t, T_i, T_{i+1}) = R_L(t, T_i, T_{i+1}) \cdot \sigma_i(t) \, dW^{T_{i+1}}(t), \qquad \text{(LM1)}$$

$i = 1, ..., n$, where $W^{T_{i+1}}$ is a m−dimensional Wiener process under $Q^{T_{i+1}}$
and the forward LIBOR volatility process $(\sigma_i(t))_{t \in [t_0, T_i]}$ is assumed to be
strictly positive and bounded for all $i = 1, ..., n$. The solution to $(LM1)$ is
given by

$$R_L(t, T_i, T_{i+1}) = R_L(t_0, T_i, T_{i+1}) \cdot e^{\int_{t_0}^{t} \sigma_i(s) dW^{T_{i+1}}(s) - \frac{1}{2} \int_{t_0}^{t} \|\sigma_i(s)\|^2 ds},$$

$i = 1, ..., n$. Hence, because of assumption $(LM0)$, $R_L(t, T_i, T_{i+1}) > 0$ for
all $t \in [t_0, T_i]$, $i = 1, ..., n$. We will show a little later how the $Q^{T_{i+1}}$−Wiener
processes can be successively constructed for all $i = 1, ..., n$, starting with
the $Q^{T_{n+1}}$−Wiener process $W^{T_{n+1}}$. Under the additional *assumption* that
the forward LIBOR volatility process

$$(\sigma_i(t))_{t \in [t_0, T_i]} \text{ is deterministic} \qquad \text{(LM2)}$$

for all $i = 1, ..., n$, we know by Theorem 2.43 that the forward LIBORs
$R_L(t, T_i, T_{i+1})$ are lognormally distributed under the T_{i+1}−forward measure $Q^{T_{i+1}}$ where the variance of $\ln R_L(t, T_i, T_{i+1})$ under the T_{i+1}−forward
measure $Q^{T_{i+1}}$ is given by

$$Var_{Q^{T_{i+1}}} [\ln R_L(t, T_i, T_{i+1}) \,|\, \mathcal{F}_{t_0}] := s_i^2(t_0, t) = \int_{t_0}^{t} \|\sigma_i(s)\|^2 \, ds, \ i = 1, ..., n.$$

For a pricing *example* let us consider the following payment

$$\begin{aligned} D(T_{i+1}) &= L \cdot (T_{i+1} - T_i) \cdot \max\{R_L(T_i, T_i, T_{i+1}) - R_X, 0\} \\ &= L \cdot (T_{i+1} - T_i) \cdot \max\{R_L(T_i, T_{i+1}) - R_X, 0\} \end{aligned}$$

---

[9] Note, that the length $T_{i+1} - T_i$, $i = 1, ..., n$, necessary for calculating the accrual
factor or daycount fraction is evaluated using a specific algorithm called the *daycount
convention* (see Section 5.1 for more details). Unfortunately, the daycount convention
may differ between different markets and may also be not exactly equal to the tenor ΔT
which makes the calculation of real world prices sometimes a little tedious.

to be made at time T_{i+1} and depending on the forward LIBOR $R_L(t, T_i, T_{i+1})$ at time T_i, the so-called cap rate R_X, and the notional amount L. This is the typical terminal payment of a caplet which we will discuss in detail within Section 5.6.2. This pay-off compensates the owner of the caplet for a loan payment which he wanted to be capped at a rate of R_X if the LIBOR exceeds the cap rate R_X at time T_i for the period $[T_i, T_{+1}]$, $i = 1, ..., n$. If we choose the zero-coupon bond price process $(P(t, T_{i+1}))_{t \in [t_0, T_{i+1}]}$ as a numéraire we know by the change of numéraire formula of Corollary 3.32 that the price

$$Caplet^{LM}(t, T_i, R_X) := Caplet^{LM}(t, T_i, T_{i+1}, R_X)$$

of the caplet is given by

$$\frac{Caplet^{LM}(t, T_i, R_X)}{P(t, T_{i+1}) \cdot L \cdot (T_{i+1} - T_i)} = E_{Q^{T_{i+1}}} \left[\max\{R_L(T_i, T_i, T_{i+1}) - R_X, 0\} | \mathcal{F}_t \right].$$

Hence, under the assumptions $(LM1)$ and $(LM2)$ the price of the caplet at time t_0 is given by

$$\frac{Caplet^{LM}(t_0, T_i, R_X)}{P(t_0, T_{i+1}) \cdot L \cdot (T_{i+1} - T_i)} = R_L(t_0, T_i, T_{i+1}) \cdot N(d_1) - R_X \cdot N(d_2) \quad (4.25)$$

with

$$d_1 = \frac{\ln\left(\frac{R_L(t_0, T_i, T_{i+1})}{R_X}\right) + \frac{1}{2} \cdot s_i^2(t_0, T_i)}{s_i(t_0, T_i)} \quad \text{and} \quad d_2 = d_1 - s_i(t_0, T_i).$$

Correspondingly, the price of a floorlet defined by the final payment

$$D(T_{i+1}) = L \cdot (T_{i+1} - T_i) \cdot \max\{R_X - R_L(T_i, T_i, T_{i+1}), 0\}$$

is given by

$$\frac{Floorlet^{LM}(t_0, T_i, R_X)}{P(t_0, T_{i+1}) \cdot L \cdot (T_{i+1} - T_i)} = R_X \cdot N(-d_2) - R_L(t_0, T_i, T_{i+1}) \cdot N(-d_1).$$
$$(4.26)$$

As we will see in Section 5.6.2 caplets and floorlets are the basic elements for pricing caps and floors which are nothing other than a portfolio of caplets or floorlets. In the market, traders usually quote the prices of caplets (caps) and floorlets (floors) in terms of Black volatilities σ_i^{Black}, i.e. the volatilities entered into the Black model used to price the caplets and floorlets. The Black model will be discussed in more detail in Section 5.8.3. Comparing the Black caplet and floorlet formulas (see equations (5.28) and (5.29)) to equations (4.25) and (4.26) we see that we can *calibrate* the LIBOR market model to the market prices by simply ensuring that the caplet and floorlet volatilities of the LIBOR market model equal the implied Black caplet and floorlet volatilities quoted in the market, i.e.

$$s_i\left(t_0, T_i\right) = \sigma_i^{Black} \cdot \sqrt{T_i - t_0}, \, i = 1, ..., n.$$

Examples for the calibration of the LIBOR market model to real market data can be found, e.g., in Hull and White [HW00], Brace, Gątarek, and Musiela [BGM97], or Pedersen and Schumacher [PS00].

The drawback of using a LIBOR market model is that the dynamics of the different forward LIBORs $R_L\left(t, T_i, T_{i+1}\right)$ are specified by the different probability measures $Q^{T_{i+1}}$, $i = 1, ..., n$, making each forward LIBOR a martingale with respect to its own probability measure. For pricing derivatives which only depend on one forward LIBOR such as caplets or floorlets, this is not a problem. But if we want to evaluate contingent claims with more complicated pay-offs we need to model the joint behaviour of different forward LIBORs under a single measure. As we will see, the terminal measure $Q^{T_{n+1}}$ is a good choice for this single measure. To show this we first consider the Radon-Nikodým derivative $L_i\left(T_i\right) := dQ^{T_i}/dQ^{T_{i+1}}$ for a change from the measure Q^{T_i} to the measure $Q^{T_{i+1}}$, $i = 1, ..., n$. Using the change of numéraire formula (Corollary 3.32) we know that the Radon-Nikodým derivative is defined by

$$
\begin{aligned}
L_i\left(t\right) &= \frac{dQ^{T_i}}{dQ^{T_{i+1}}}|\mathcal{F}_t = \frac{P\left(t_0, T_{i+1}\right)}{P\left(t_0, T_i\right)} \cdot \frac{P\left(t, T_i\right)}{P\left(t, T_{i+1}\right)} \\
&= \frac{P\left(t_0, T_{i+1}\right)}{P\left(t_0, T_i\right)} \cdot \left[1 + R_L\left(t, T_i, T_{i+1}\right) \cdot \left(T_{i+1} - T_i\right)\right]
\end{aligned}
$$

for all $t \in \left[t_0, T_i\right]$, $i = 1, ..., n$, where we used equation (4.24) to derive the last statement. Together with assumption $(LM1)$ we get for all $t \in \left[t_0, T_i\right]$

$$
\begin{aligned}
dL_i\left(t\right) &= \frac{P\left(t_0, T_{i+1}\right)}{P\left(t_0, T_i\right)} \cdot \left(T_{i+1} - T_i\right) dR_L\left(t, T_i, T_{i+1}\right) \\
&= \frac{P\left(t_0, T_{i+1}\right)}{P\left(t_0, T_i\right)} \cdot \left(T_{i+1} - T_i\right) \cdot R_L\left(t, T_i, T_{i+1}\right) \cdot \sigma_i\left(t\right) dW^{T_{i+1}}\left(t\right) \\
&= \frac{R_L\left(t, T_i, T_{i+1}\right) \cdot \left(T_{i+1} - T_i\right)}{1 + R_L\left(t, T_i, T_{i+1}\right) \cdot \left(T_{i+1} - T_i\right)} \cdot \sigma_i\left(t\right) \cdot L_i\left(t\right) dW^{T_{i+1}}\left(t\right) \\
&= L_i\left(t\right) \cdot \gamma_i\left(t\right) dW^{T_{i+1}}\left(t\right), \quad\quad\quad\quad (4.27)
\end{aligned}
$$

with

$$\gamma_i\left(t\right) := \gamma\left(t, T_i, T_{i+1}\right) := \frac{R_L\left(t, T_i, T_{i+1}\right) \cdot \left(T_{i+1} - T_i\right)}{1 + R_L\left(t, T_i, T_{i+1}\right) \cdot \left(T_{i+1} - T_i\right)} \cdot \sigma_i\left(t\right),$$

$i = 1, ..., n$, where $W^{T_{i+1}}$ is a $m-$dimensional Wiener process under $Q^{T_{i+1}}$. Hence,

$$L_i\left(t\right) = e^{\int_{t_0}^t \gamma_i(s) dW^{T_{i+1}}(s) - \frac{1}{2} \int_{t_0}^t \|\gamma_i(s)\|^2 ds}, \, t \in \left[t_0, T_i\right]. \quad\quad (4.28)$$

Since $R_L(t, T_i, T_{i+1}) > 0$ under the T_{i+1}–forward measure $Q^{T_{i+1}}$, we know that

$$\frac{R_L(t, T_i, T_{i+1}) \cdot (T_{i+1} - T_i)}{1 + R_L(t, T_i, T_{i+1}) \cdot (T_{i+1} - T_i)} \in (0, 1] \text{ under } Q^{T_{i+1}}.$$

From $(LM1)$ it follows that $(\gamma_i(t))_{t \in [t_0, T_i]}$ is bounded and therefore holds the Novikov condition. Applying Girsanov's theorem (Theorem 2.41) we can now define the Q^{T_i}–Wiener process W^{T_i} using the $Q^{T_{i+1}}$–Wiener process $W^{T_{i+1}}$ by

$$W^{T_i}(t) = W^{T_{i+1}}(t) - \int_{t_0}^t \gamma_i(s)\, ds \text{ for all } t \in [t_0, T_i], \tag{4.29}$$

$i = 1, ..., n$. Applying equation (4.29) for $i = n - 1$ we get for all $t \in [t_0, T_i]$ from assumption $(LM1)$:

$$
\begin{aligned}
\frac{dR_L(t, T_{n-1}, T_n)}{R_L(t, T_{n-1}, T_n)} &= \sigma_{n-1}(t)\, dW^{T_n}(t) \\
&= \sigma_{n-1}(t) \left[dW^{T_{n+1}}(t) - \gamma_n(t)\, dt \right] \\
&= -\gamma_n(t) \cdot \sigma_{n-1}(t)\, dt + \sigma_{n-1}(t)\, dW^{T_{n+1}}(t).
\end{aligned}
$$

Iteratively repeating this procedure, we receive

$$\frac{dR_L(t, T_i, T_{i+1})}{R_L(t, T_i, T_{i+1})} = -\sum_{k=i+1}^n \gamma_k(t) \cdot \sigma_i(t)\, dt + \sigma_i(t)\, dW^{T_{n+1}}(t) \tag{4.30}$$

for all $t \in [t_0, T_i]$, $i = 1, ..., n$. Since $\sum_{k=i+1}^n \gamma_k(t) \cdot \sigma_i(t) > 0$ for all $i = 1, ..., n-1$, only the forward LIBOR $(R_L(t, T_n, T_{n+1}))_{t \in [t_0, T_n]}$ is a martingale with respect to the terminal measure $Q^{T_{n+1}}$. It should be noted that we simultaneously constructed a family of forward prices $(P^F(t, T_i, T_{i+1}))_{t \in [t_0, T_i]}$ via the stochastic differential equation

$$
\begin{aligned}
dP^F(t, T_i, T_{i+1}) &= (T_{i+1} - T_i)\, dR_L(t, T_i, T_{i+1}) \\
&= (P^F(t, T_i, T_{i+1}) - 1) \\
&\quad \cdot \left(-\sum_{k=i+1}^n \gamma_k(t) \cdot \sigma_i(t)\, dt + \sigma_i(t)\, dW^{T_{n+1}}(t) \right).
\end{aligned}
$$

Equation (4.30) can now be used for a *Monte Carlo simulation* of the forward LIBORs under the terminal measure $Q^{T_{n+1}}$. We therefore have to simulate the LIBORs $R_L(t, T_i, T_{i+1})$, given $R_L(t_0, T_i, T_{i+1})$, at the grid points $t = T_1, ..., T_i$ for all $i = 1, ..., n$. We do this by using a discretization of equation (4.30). To be more precise, for each $l \in \{0, ..., i-1\}$ we simulate all values

$$R_L\left(T_{l+1}, T_i, , T_{i+1}\right) - R_L\left(T_l, T_i, T_{i+1}\right)$$
$$= -\sum_{k=i+1}^{n} \gamma_k\left(T_l\right) \cdot R_L\left(T_l, T_i, T_{i+1}\right) \cdot \sigma_i\left(T_l\right) \cdot \left(T_{l+1} - T_l\right)$$
$$+ R_L\left(T_l, T_i, T_{i+1}\right) \cdot \sigma_i\left(T_l\right) \cdot \left(W^{T_{n+1}}\left(T_{l+1}\right) - W^{T_{n+1}}\left(T_l\right)\right),$$

$i = 1, ..., n$, where the simulation of the Wiener process is drawn from the equation

$$W^{T_{n+1}}\left(T_{l+1}\right) = W^{T_{n+1}}\left(T_l\right) + \varepsilon_l \cdot \sqrt{T_{l+1} - T_l}$$

with independent standard normally distributed random variables ε_l, $l = 1, ..., i - 1$. Given the simulated LIBORs we get the simulation of the zero-coupon bond prices using equation (4.24) by

$$P\left(T_l, T_{i+1}\right) = \prod_{k=l}^{i} \frac{P\left(T_l, T_{k+1}\right)}{P\left(T_l, T_k\right)} = \prod_{k=l}^{i} \frac{1}{1 + R_L\left(T_l, T_k, T_{k+1}\right) \cdot \left(T_{k+1} - T_k\right)}$$

for $l = 1, ..., i$ and $i = 1, ..., n$. These simulations can now be used to evaluate contingent claims depending on an arbitrary set of forward LIBORs. For example, by the change of numéraire formula (Corollary 3.32), we know that the price at time $t \in [t_0, T_{i+1}]$ of a contingent claim $D = D\left(T_{i+1}\right)$ with maturity or pay-off at time T_{i+1} is given by

$$
\begin{aligned}
V_D\left(t\right) &= P_0\left(t\right) \cdot E_{\widetilde{Q}}\left[\widetilde{D}\left(T_{i+1}\right) | \mathcal{F}_t\right] \\
&= P\left(t, T_{i+1}\right) \cdot E_{Q^{T_{i+1}}}\left[D\left(T_{i+1}\right) | \mathcal{F}_t\right] \\
&= P\left(t, T_{n+1}\right) \cdot E_{Q^{T_{n+1}}}\left[\frac{D\left(T_{i+1}\right)}{P\left(T_{i+1}, T_{n+1}\right)} | \mathcal{F}_t\right].
\end{aligned}
$$

We therefore get the price of the contingent claim at time $t = t_0$ by simply taking the arithmetic average of the simulated values for $D\left(T_{i+1}\right) / P\left(T_{i+1}, T_{n+1}\right)$ and multiply the result with $P\left(t_0, T_{n+1}\right)$. For more details and examples the interested reader may refer to Pelsser [Pel00].

4.7.2 The Continuous-Tenor Model

In this section we will extend the discrete-tenor to a continuous-tenor model, i.e. a LIBOR market model where all LIBORs $R_L\left(t, T_\Delta, T\right)$, $t_0 \leq t \leq T_\Delta \leq T \leq T_{n+1}$ with $T_{n+1} = T^*$ and $T_\Delta = T - \Delta T$ are specified. Especially, we will explicitly construct the cash account process $\left(P_0\left(t\right)\right)_{t \in [t_0, T_{n+1}]}$ and the corresponding measure $Q^{P_0\left(T_{n+1}\right)}$, equivalent to Q, such that the discounted zero-coupon bond prices are $Q^{P_0\left(T_{n+1}\right)} -$

martingales for all $T \in [t_0, T_{n+1}]$, i.e. $Q^{P_0(T_{n+1})} = \widetilde{Q}$. We hereby assume that we are already given a discrete-tenor model as specified in Section 4.7.1. It is therefore sufficient to consistently fill the gaps between the grid points $t_0 = T_0, T_1, ..., T_{n+1}$ where we *assume*, according to assumptions $(LM1)$ and $(LM2)$, that each of the forward LIBOR processes $(R_L(t, T_\Delta, T))_{t \in [t_0, T_\Delta]}$, $T \leq T_{n+1}$, follows the stochastic differential equation

$$dR_L(t, T_\Delta, T) = R_L(t, T_\Delta, T) \cdot \sigma_{T_\Delta}(t) \, dW^T(t), \qquad \text{(LMC)}$$

under the T−forward measure Q^T where W^T is a m−dimensional Wiener process under Q^T and the forward LIBOR volatility process $(\sigma_{T_\Delta}(t))_{t \in [t_0, T_\Delta]}$ is assumed to be deterministic, strictly positive and bounded for all $T \in [t_0, T_{n+1}]$.

Note that there are no forward LIBORs $R_L(t, T_\Delta, T)$ for $T_\Delta > T_n$. We therefore start with the time interval $(T_n, T_{n+1}]$ and make the following *assumption*:

For $\alpha(T) \in [0, 1]$ and for all $t_0 \leq t \leq T_n \leq T \leq T_{n+1}$ we set

$$R_L(t, T_n, T) \cdot (T - T_n) = R_L(t, T_n, T_{n+1}) \cdot \alpha(T) \cdot (T_{n+1} - T_n) \quad \text{(LM3)}$$

or equivalently,

$$R_L(t, T, T_{n+1}) \cdot (T_{n+1} - T) = R_L(t, T_n, T_{n+1}) \cdot [1 - \alpha(T)] \cdot (T_{n+1} - T_n).$$

Hence, $\alpha(T_n) = 0$ and $\alpha(T_{n+1}) = 1$. Furthermore, $(R_L(t, T_n, T))_{t \in [t_0, T_n]}$ as well as $(R_L(t, T, T_{n+1}))_{t \in [t_0, T]}$ are $Q^{T_{n+1}}$−martingales under assumption $(LM3)$ with

$$
\begin{aligned}
dR_L(t, T_n, T) &= \alpha(T) \cdot \frac{T_{n+1} - T_n}{T - T_n} dR_L(t, T_n, T_{n+1}) \\
&= \alpha(T) \cdot \frac{T_{n+1} - T_n}{T - T_n} \cdot R_L(t, T_n, T_{n+1}) \cdot \sigma_n(t) \, dW^{T_{n+1}}(t) \\
&= R_L(t, T_n, T) \cdot \sigma_n(t) \, dW^{T_{n+1}}(t)
\end{aligned}
$$

and $\alpha(T)$ can be extracted from the initial zero rate curve by

$$R_L(t_0, T, T_{n+1}) \cdot (T_{n+1} - T) = R_L(t, T_n, T_{n+1}) \cdot [1 - \alpha(T)] \cdot (T_{n+1} - T_n)$$

or equivalently[10], using equation (4.24) and assumption $(LM0)$,

$$\alpha\left(T\right) = \frac{P\left(t_0, T_n\right) - P\left(t_0, T\right)}{P\left(t_0, T_n\right) - P\left(t_0, T_{n+1}\right)} \in [0, 1].$$

In our discrete time setting we can interpret the value $P_0\left(t\right)$ of the cash account at time $t \in [t_0, T_{n+1}]$ as the amount of money accumulated from time t_0 up to time t by rolling over a series of zero-coupon bonds with the shortest maturities available. We can therefore evaluate $P_0\left(t\right)$ for $t \in \{t_0, T_1, ..., T_{n+1}\}$ via the forward zero-coupon bond prices $P^F\left(T_i, T_i, T_{i+1}\right)$, $i = 0, 1, ..., n$, by

$$P_0\left(T_i\right) = \prod_{j=0}^{i-1} P^F\left(T_j, T_j, T_{j+1}\right), \ i = 0, 1, ..., n+1.$$

Because of equation (4.24), we know that

$$\begin{aligned} P_0\left(T_{i+1}\right) &= P_0\left(T_i\right) \cdot P^F\left(T_i, T_i, T_{i+1}\right) \\ &= P_0\left(T_i\right) \cdot [1 + R_L\left(T_i, T_i, T_{i+1}\right) \cdot \left(T_{i+1} - T_i\right)] \\ &> P_0\left(T_i\right) \end{aligned}$$

under the $T_{i+1}-$forward measure $Q^{T_{i+1}}$ since $R_L\left(T_i, T_i, T_{i+1}\right) > 0$ under $Q^{T_{i+1}}$, $i = 0, ..., n$. Quite naturally, we define

$$P_0\left(T\right) := P_0\left(T_n\right) \cdot [1 + R_L\left(T_n, T_n, T\right) \cdot \left(T - T_n\right)]$$

for each $T \in [T_n, T_{n+1}]$.

Remark. *Because of*

$$P_0\left(T_{n+1}\right) = P_0\left(T_n\right) \cdot [1 + R_L\left(T_n, T_n, T_{n+1}\right) \cdot \left(T_{n+1} - T_n\right)],$$

we get, using assumption $(LM3)$,

$$\begin{aligned} P_0\left(T\right) &= P_0\left(T_n\right) \cdot [1 + R_L\left(T_n, T_n, T\right) \cdot \left(T - T_n\right)] \\ &= P_0\left(T_n\right) \cdot [1 + R_L\left(T_n, T_n, T_{n+1}\right) \cdot \alpha\left(T\right) \cdot \left(T_{n+1} - T_n\right)] \quad (4.31) \\ &= P_0\left(T_n\right) \cdot \left(\begin{array}{c} 1 - \alpha\left(T\right) \\ +\alpha\left(T\right) \cdot [1 + R_L\left(T_n, T_n, T_{n+1}\right) \cdot \left(T_{n+1} - T_n\right)] \end{array} \right) \\ &= \left(1 - \alpha\left(T\right)\right) \cdot P_0\left(T_n\right) + \alpha\left(T\right) \cdot P_0\left(T_{n+1}\right). \end{aligned}$$

---

[10] Note, that the easiest way to ensure assumption $(LM3)$ is to choose

$$R_L\left(t_0, T, T_{n+1}\right) := R_L\left(t_0, T_n, T_{n+1}\right)$$

which leads us to

$$\alpha\left(T\right) = \frac{T - T_n}{T_{n+1} - T_n}.$$

It can be easily seen that this equation is equivalent to

$$\ln P_0\left(T\right) = \left(1 - \alpha^*\left(T\right)\right)\cdot \ln P_0\left(T_n\right) + \alpha^*\left(T\right)\cdot \ln P_0\left(T_{n+1}\right)$$

with

$$\alpha^*\left(T\right) = \frac{\ln\left(1 + \alpha\left(T\right)\cdot\left[\frac{P_0(T_{n+1})}{P_0(T_n)} - 1\right]\right)}{\ln\left(\frac{P_0(T_{n+1})}{P_0(T_n)}\right)} \in [0,1]$$

or

$$\alpha\left(T\right) = \frac{\left(\frac{P_0(T_{n+1})}{P_0(T_n)}\right)^{\alpha^*(T)} - 1}{\left[\frac{P_0(T_{n+1})}{P_0(T_n)} - 1\right]} \in [0,1]\,,$$

which is the assumption used in Musiela and Rutkowski [MR97], p. 348. □

Applying Corollary 3.32 we are now able to calculate the Radon-Nikodým derivative of the $T-$forward measure Q^T with respect to the $T_{n+1}-$forward measure $Q^{T_{n+1}}$ via

$$\frac{dQ^T}{dQ^{T_{n+1}}}\Big|_{\mathcal{F}_t} = \frac{P\left(t,T\right)\cdot P\left(t_0,T_{n+1}\right)}{P\left(t_0,T\right)\cdot P\left(t,T_{n+1}\right)} = \frac{P\left(t_0,T_{n+1}\right)}{P\left(t_0,T\right)}\cdot P^F\left(t,T,T_{n+1}\right)$$

for all $t_0 \leq t \leq T$ by

$$
\begin{aligned}
\frac{dQ^T}{dQ^{T_{n+1}}} &= \frac{dQ^T}{dQ^{T_{n+1}}}\Big|_{\mathcal{F}_T} = \frac{P\left(t_0,T_{n+1}\right)}{P\left(t_0,T\right)\cdot P\left(T,T_{n+1}\right)}\\
&= \frac{P\left(t_0,T_{n+1}\right)}{P\left(t_0,T\right)}\cdot P^F\left(T,T,T_{n+1}\right).
\end{aligned}
$$

Furthermore, since

$$
\begin{aligned}
P^F\left(t,T,T_{n+1}\right) &= 1 + R_L\left(t,T,T_{n+1}\right)\cdot\left(T_{n+1} - T\right)\\
&= 1 + R_L\left(t,T_n,T_{n+1}\right)\cdot\left(1 - \alpha\left(T\right)\right)\cdot\left(T_{n+1} - T_n\right)
\end{aligned}
$$

it follows, using the notation $L_T\left(t\right) := \frac{dQ^T}{dQ^{T_{n+1}}}\Big|_{\mathcal{F}_t}$,

$$
\begin{aligned}
dL_T\left(t\right) &= \frac{P\left(t_0,T_{n+1}\right)}{P\left(t_0,T\right)}dP^F\left(t,T,T_{n+1}\right)\\
&= \frac{P\left(t_0,T_{n+1}\right)}{P\left(t_0,T\right)}\cdot\left(1 - \alpha\left(T\right)\right)\cdot\left(T_{n+1} - T_n\right)dR_L\left(t,T_n,T_{n+1}\right)\\
&= \frac{R_L\left(t,T,T_{n+1}\right)\cdot\left(T_{n+1} - T\right)}{1 + R_L\left(t,T,T_{n+1}\right)\cdot\left(T_{n+1} - T\right)}\cdot\sigma_n\left(t\right)\cdot L_T\left(t\right)dW^{T_{n+1}}\left(t\right)\\
&= \gamma\left(t,T,T_{n+1}\right)\cdot L_T\left(t\right)dW^{T_{n+1}}\left(t\right)
\end{aligned}
$$

with

$$\gamma(t, T, T_{n+1}) = \frac{R_L(t, T, T_{n+1}) \cdot (T_{n+1} - T)}{1 + R_L(t, T, T_{n+1}) \cdot (T_{n+1} - T)} \cdot \sigma_n(t)$$

$$= \frac{R_L(t, T_n, T_{n+1}) \cdot (1 - \alpha(T)) \cdot (T_{n+1} - T_n)}{1 + R_L(t, T_n, T_{n+1}) \cdot (1 - \alpha(T)) \cdot (T_{n+1} - T_n)} \cdot \sigma_n(t)$$

for all $t_0 \le t \le T$, where $W^{T_{n+1}}$ is a m-dimensional Wiener process under $Q^{T_{n+1}}$. Hence,

$$L_T(t) = e^{\int_{t_0}^{t} \gamma(s, T, T_{n+1}) dW^{T_{n+1}}(s) - \frac{1}{2} \int_{t_0}^{t} \|\gamma(s, T, T_{n+1})\|^2 ds}, \ t \in [t_0, T].$$

Since $R_L(t, T_n, T_{n+1}) > 0$ under the T_{n+1}-forward measure $Q^{T_{n+1}}$ and $\alpha(T) \in (0, 1]$, we know that

$$\frac{R_L(t, T_n, T_{n+1}) \cdot (1 - \alpha(T)) \cdot (T_{i+1} - T_i)}{1 + R_L(t, T_i, T_{i+1}) \cdot (1 - \alpha(T)) \cdot (T_{i+1} - T_i)} \in (0, 1] \text{ under } Q^{T_{n+1}}.$$

From $(LM1)$ it follows that $(\gamma(t, T, T_{n+1}))_{t \in [t_0, T]}$ is bounded and therefore satisfies the Novikov condition. Applying Girsanov's theorem (Theorem 2.41) we can now define the Q^T-Wiener process W^T using the $Q^{T_{n+1}}$-Wiener process $W^{T_{n+1}}$ by

$$W^T(t) = W^{T_{n+1}}(t) - \int_{t_0}^{t} \gamma(s, T, T_{n+1}) \, ds \text{ for all } t \in [t_0, T].$$

Hence, using assumption (LMC), we get for all $t \in [t_0, T]$

$$\frac{dR_L(t, T_\Delta, T)}{R_L(t, T_\Delta, T)} = \sigma_{T_\Delta}(t) \, dW_T(t)$$

$$= -\sigma_{T_\Delta}(t) \cdot \gamma(t, T, T_{n+1}) \, dt + \sigma_{T_\Delta}(t) \, dW^{T_{n+1}}(t).$$

If we let $\widehat{T}_{n+1} := T$ play the role of T_{n+1} in Section 4.7.1, setting $\widehat{T}_i := \widehat{T}_{i+1} - \Delta T$,

$$\gamma_{\widehat{T}_i}(t) := \gamma\left(t, \widehat{T}_i, \widehat{T}_{i+1}\right) := \frac{R_L\left(t, \widehat{T}_i, \widehat{T}_{i+1}\right) \cdot \left(\widehat{T}_{i+1} - \widehat{T}_i\right)}{1 + R_L\left(t, \widehat{T}_i, \widehat{T}_{i+1}\right) \cdot \left(\widehat{T}_{i+1} - \widehat{T}_i\right)} \cdot \sigma_{\widehat{T}_i}(t),$$

and

$$W^{\widehat{T}_i}(t) = W^{\widehat{T}_{i+1}}(t) - \int_{t_0}^{t} \gamma_{\widehat{T}_i}(s) \, ds$$

for all $t \in [t_0, T]$, $i = 1, ..., n$, we get, according to equation (4.30),

$$\frac{dR_L\left(t, \widehat{T}_i, \widehat{T}_{i+1}\right)}{R_L\left(t, \widehat{T}_i, \widehat{T}_{i+1}\right)} = -\sum_{k=i+1}^{n} \gamma_{\widehat{T}_k}(t) \cdot \sigma_{\widehat{T}_i}(t) \, dt + \sigma_{\widehat{T}_i}(t) \, dW^T(t)$$

$$= -\sum_{k=i+1}^{n+1} \gamma_{\widehat{T}_k}(t) \cdot \sigma_{\widehat{T}_i}(t) \, dt + \sigma_{\widehat{T}_i}(t) \, dW^{T_{n+1}}(t)$$

for all $t \in [t_0, T]$, $i = 1, ..., n$, with

$$\gamma_{\widehat{T}_{n+1}}(t) := \gamma\left(t, \widehat{T}_{n+1}, T_{n+1}\right) = \gamma\left(t, T, T_{n+1}\right).$$

This equation can now be used for a Monte Carlo simulation of all forward LIBORs under the terminal measure $Q^{T_{n+1}}$ as described in Section 4.7.1 and hence for pricing contingent claims depending on arbitrary LIBORs $R_L(t, T)$, $t_0 \leq t \leq T \leq T_{n+1}$. We can also complete our definition of the cash account for all $\widehat{T}_i \in (T_{i-1}, T_i)$, by setting

$$P_0\left(\widehat{T}_i\right) := \frac{P_0\left(\widehat{T}_{i+1}\right)}{1 + R_L\left(\widehat{T}_i, \widehat{T}_i, \widehat{T}_{i+1}\right) \cdot \left(\widehat{T}_{i+1} - \widehat{T}_i\right)}, \, i = 1, ..., n.$$

Corresponding to our change of numéraire example we get the following lemma.

Lemma 4.28 *Using the definitions from above, the equivalent martingale measure $\widetilde{Q}$ on $\left(\Omega, \mathcal{F}_{T_{n+1}}\right)$ is defined by*

$$\frac{d\widetilde{Q}}{dQ^{T_{n+1}}} = P_0\left(T_{n+1}\right) \cdot P\left(t_0, T_{n+1}\right),$$

i.e. its Radon-Nikodým derivative with respect to the probability measure $Q^{T_{n+1}}$. Furthermore,

$$P(t, T) = P_0(t) \cdot E_{\widetilde{Q}}\left[\frac{1}{P_0(T)} | \mathcal{F}_t\right]$$

for all $t_0 \leq t \leq T \leq T_{n+1}$.

Proof. Let the measure $Q^{P_0(T_{n+1})}$ on $\left(\Omega, \mathcal{F}_{T_{n+1}}\right)$ be defined by its Radon-Nikodým derivative with respect to the probability measure $Q^{T_{n+1}}$

$$\frac{dQ^{P_0(T_{n+1})}}{dQ^{T_{n+1}}}|_{\mathcal{F}_t} = \frac{P_0(t) \cdot P(t_0, T_{n+1})}{P_0(t_0) \cdot P(t, T_{n+1})} = \frac{P_0(t) \cdot P(t_0, T_{n+1})}{P(t, T_{n+1})}$$

for all $t_0 \leq t \leq T_{n+1}$, i.e.

$$\frac{dQ^{P_0(T_{n+1})}}{dQ^{T_{n+1}}} = \frac{dQ^{P_0(T_{n+1})}}{dQ^{T_{n+1}}}|_{\mathcal{F}_{T_{n+1}}} = P_0\left(T_{n+1}\right) \cdot P\left(t_0, T_{n+1}\right).$$

Then $Q^{P_0(T_{n+1})}$ is equivalent to Q since $\frac{dQ^{P_0(T_{n+1})}}{dQ^{T_{n+1}}} > 0$ and $Q^{T_{n+1}}$ is equivalent to $\widetilde{Q}$ and therefore to Q. Furthermore, because $(R_L(t, T, T_{n+1}))_{t \in [t_0, T]}$

is a $Q^{T_{n+1}}$−martingale, we get

$$
\begin{aligned}
P\left(t,T\right) &= P\left(t,T_{n+1}\right)\cdot P^{F}\left(t,T,T_{n+1}\right)\\
&= P\left(t,T_{n+1}\right)\cdot\left[1+R_{L}\left(t,T,T_{n+1}\right)\cdot\left(T_{n+1}-T\right)\right]\\
&= P\left(t,T_{n+1}\right)\cdot E_{Q^{T_{n+1}}}\left[1+R_{L}\left(T,T,T_{n+1}\right)\cdot\left(T_{n+1}-T\right)|\mathcal{F}_{t}\right]\\
&= P\left(t,T_{n+1}\right)\cdot E_{Q^{T_{n+1}}}\left[\frac{1}{P\left(T,T_{n+1}\right)}|\mathcal{F}_{t}\right]\\
&= P_{0}\left(t\right)\cdot E_{Q^{P_{0}\left(T_{n+1}\right)}}\left[\frac{1}{P_{0}\left(T\right)}|\mathcal{F}_{t}\right]
\end{aligned}
$$

where the last equation follows from the change of numéraire Theorem 3.31. Using Lemma 2.17a) we also know that $\left(\widetilde{P}\left(t,T\right)\right)_{t\in[t_{0},T]}$ with

$$
\widetilde{P}\left(t,T\right)=\frac{P\left(t,T\right)}{P_{0}\left(t\right)}=E_{Q^{P_{0}\left(T_{n+1}\right)}}\left[\frac{1}{P_{0}\left(T\right)}|\mathcal{F}_{t}\right]
$$

is a $Q^{P_{0}\left(T_{n+1}\right)}$−martingale for all $T\in[t_{0},T_{n+1}]$. Hence,

$$
Q^{P_{0}\left(T_{n+1}\right)}\in I\!M=I\!M\left(Q\right),
$$

i.e. $Q^{P_{0}\left(T_{n+1}\right)}=\widetilde{Q}$, since $|I\!M|=1$.   □

4.8   Credit Risk Models

It should be noted that the stochastic interest-rate models discussed in this book do not cover all kinds of risk faced by a fixcd-income portfolio trader. We basically concentrated on market risk and were not concered about the possibility of one party in a financial contract going into default or changing its rating. The risk of price-changes with respect to such events is called credit or default risk. The modelling of default risk has become a popular area of research in recent years. Even if an adequate presentation of the main results in pricing defaultable (zero-) coupon bonds or credit derivatives is beyond the scope of this book we try to give a brief overview of some of the main results with respect to this topic. There are basically two approaches for modelling default risk in zero-coupon and coupon bonds. The structural approach goes back to Merton [Mer74], who showed, under rather restrictive assumptions, that the value of a firm's assets can be described by an Itô process and that a defautable (zero-) coupon bond can be priced as the difference between the today's value of the company's assets minus the value of a call option on the company's assets with a maturity time equal to the maturity of the (zero-) coupon bond and an exercise price equal to the notional of the defaultable (zero-) coupon bond.

Under the many articles dealing with generalizations of Merton's model the reader may refer to Black and Cox [Bla76], Geske [Ges77], Ho and Singer [HS84], Shimko, Tejima, and Van Deventer [STD93], Kim, Ramaswamy, and Sundaresan [KRS93], or Longstaff and Schwartz [LS95].

The models of the second approach are usually called reduced-form or intensity-based models and do not explicitly consider the relation between default and asset value. They rather model default by the stopping time of some given hazard-rate process and thus specify the default process exogenously. As a consequence, reduced-form models do not rely on the observability of the firm's assets. For more information on these models the reader may refer to Jarrow and Turnbull [JT95], Lando [Lan94], [Lan96], and [Lan98], Schönbucher [Sch96], or Duffie and Singleton [DS94b].

While the structural models only have limited success in explaining the behaviour of credit spreads the drawback of the reduced-form models is the missing link between the firm's value and corporate default. New models which overcome many of the weaknesses of the previous two approaches but still combine their advantages can be found in Cathcart and El-Jahel [CEJ98] or in Schmid and Zagst [SZ00]. The latter allow for a Hull-White or a generalized CIR process to describe the short rate. The quality of the firm is modelled by a CIR type of uncertainty or signaling process which also has an influence on the explicitly modelled mean-reverting short-rate spread. They show how the model parameters can be estimated by market data, derive a closed-form solution for defaultable zero-coupon bonds and show how credit derivatives may be priced using a tree-based method.

5

Interest-Rate Derivatives

There is a great variety of financial derivatives built on our primary traded assets and with additional features such as optionality or agreements with respect to future points in time. This chapter is dedicated to describing and evaluating some of these products, from both a mathematical and practical point of view. It will also be shown how the different pricing models and techniques of the previous sections can be applied to the pricing of interest-rate derivatives.

We start with a brief discussion on how the financial market defines the time between two specific dates in Section 5.1. Probably the simplest financial instrument derived from the zero-coupon bonds is a portfolio of zero-coupon bonds which is, under special assumptions on the notional of the zero-coupon bonds, called a coupon bond and described in Section 5.2. Coupon and zero-coupon bonds are the underlying instruments for the forward agreements and futures discussed in Sections 5.3 and 5.4. Zero bonds are also the main building block for another family of interest-rate instruments, the interest-rate swaps, which are presented in Section 5.5. To describe all the various types of swaps would be beyond the scope of this book. The interested reader may also refer to Fabozzi, Fabozzi, and Pollack [FFP91], p. 1155-1242. An interesting overview of financial futures is given, e.g. in Fitzgerald, Lubochinski, and Thomas [FLT93]. The pricing of the previous instruments only needs basic arbitrage arguments or simple expectations with respect to the martingale or forward measure.

The first and also easiest optioned financial instrument is an option on a zero-coupon bond which is priced in Section 5.6.1. We do this by showing an application of Green's function methods to deriving an option price for the Hull-White model. It is shown in Sections 5.6.2 and 5.6.3 that these zero-coupon bond options build the platform for the pricing of caps, floors and coupon bond options. In Section 5.6.4 we show how options on interest-rate swaps can be priced. This is done by applying the change of numéraire theorem in the special form of the newly developed swap market models.

All the previous interest-rate options are considered to be market standard. Contingent claims with payoffs more complicated than that of standard (European) interest-rate call or put options are called exotic interest-rate options. An overview of some of these products is given in Section 5.7. Here, again, we will apply the Hull-White model to derive closed-form solutions for the option prices.

Each model we use for pricing interest-rate derivatives has to be fitted to market data. Section 5.8 gives an overview of different sources of interest-rate information expressed by yield or zero-rate curves, market prices or quoted volatilities. The latter are always quoted with respect to a benchmark model which is, most of the time, a version of the Black model. We will show in Sections 5.8.2 and 5.8.3 how this model is used in the market to price specific products such as caps and floors, options on coupon bonds, or swaptions. A practical case study on how Black volatility information can be transformed to an implied volatility curve for the Hull-White model is shown in Section 5.8.4.

5.1   Daycount Conventions

Daycount conventions are typically used when interest rates are quoted. They define the way in which interest is accrued over time. A daycount convention is expressed as a fraction X/Y. If we have to calculate the time between two dates t and T, $t_0 \leq t \leq T \leq T^*$, according to a daycount convention X/Y, X defines how the number of days between the two dates has to be calculated and Y is the total number of days in the reference period (one year). For our convenience we denote the time between t and T, calculated according to the daycount convention X/Y by $(T-t)_{X/Y}$, i.e.

$$(T-t)_{X/Y} := \frac{\text{number of days between } t \text{ and } T \text{ according to convention}}{\text{number of days in the reference period}}.$$

If $R(t,T)$ denotes the interest rate for the period from t to T, quoted according to the daycount convention X/Y, the (accrued) interest earned in this period is given by $R(t,T) \cdot (T-t)_{X/Y}$. Unfortunately there appear to be different daycount conventions in different markets and even for different financial instruments. This tells us that the number of years between two dates may be different depending on the daycount convention we use. It is therefore important to realize for which daycount convention a specific interest rate is quoted. Standards for daycount conventions have been published, e.g., by the *International Swaps and Derivatives Association (ISDA)* or the *International Securities Market Association (ISMA)*. Let us, to get a first impression, have a look on the most important daycount conventions according to the ISDA standards[1].

The *act/act* method uses the actual number of days between the two dates t and T as well as for the reference period. In normal years the actual number of days between t and T will be divided by 365, in leap-years we divide by 366 days. If part of the period from t to T falls in a leap-year, these days will be divided by 366 and the remaining days will be divided by 365 before we finally add both results.

The *act/360* method uses the actual number of days between the two dates t and T and assumes 360 days for the reference period (per year).

The *30E/360* method assumes 30 days per month (actual number of days if the month is covered by less than 30 days) and 360 days per year. If T is February 28 or 29, we only count for 28 or 29 days in this month.

The *30/360* method equals the *30E/360* method with one exception: Suppose there is more than 30 days between t and T. If T falls on the 31st of a month but t was not the 30th or 31st of a month, then 31 days will be used for the last month.

As already mentioned, applications of the different methods can differ between the various markets and are due to changes over time. At present[2], the *act/act* method is used, e.g., for Eurobonds and U.S. Treasury bonds. The *act/360* method is used, e.g., for money market instruments such as the U.S. Treasury bills or the floating side of a swap. The *30E/360* method is usually used, e.g., for the fixed side of a swap. The *30/360* method is used, e.g., for U.S. corporate and municipal bonds. It may also be used for the fixed side of a swap. If we do not assign a specific daycount convention to any difference in time we implicitly assume that we are calculating according to the daycount convention *act/act*.

5.2   Coupon Bonds

While a zero-coupon or discount bond pays back a fixed notional amount or face value at maturity with no payments in between, a coupon-bearing bond or briefly coupon bond is characterized by the fact that the holder of the

---

[1] Because the differences between the ISDA and the ISMA standards are very little, we will not distinguish between the two methods in the sequel.

[2] See also Fabozzi, Fabozzi, and Pollack [FFP91], p. 92-94 or Hull [Hul00], p. 98-99 and p. 127-128 for more details.

coupon bond receives some periodic payment during the life of the coupon bond called the coupon. Coupon bonds issued in the European markets typically have coupon payments once per year. Coupon bonds issued in the United States may have semi-annual payments, i.e. half of the coupon is paid twice per year. Let us assume that the holder of a coupon bond gets payments

$$C\left(T_i\right) \text{ at time } T_i, \ i = 1, ..., n \text{ with } t_0 \leq T_1 < T_2 < \cdots < T_n = T_B \leq T^*,$$

and T_B denoting the maturity time of the coupon bond. For $i = 1, ..., n-1$ these payments are coupon payments, for $i = n$ this payment is the coupon payment plus the notional amount or face value of the coupon bond which, at maturity, is payed back to the investor or holder of the coupon bond. Hence, a coupon bond is nothing other than a portfolio of zero-coupon bonds with maturity T_i and a notional amount of $C\left(T_i\right)$, $i = 1, ..., n$, or, in other words, a portfolio of $C\left(T_i\right)$ zero-coupon bonds with maturity T_i, $i = 1, ..., n$, and a face value of 1. So the price $Bond\left(t, T_B, C\right)$ of a coupon bond $\left(T_B, C\right)$ with coupon payments $C = \left(C\left(T_1\right), ..., C\left(T_n\right)\right)$ and maturity T_B at time $t \in [t_0, T_1]$ is given by

$$Bond\left(t, T_B, C\right) = \sum_{i=1}^{n} C\left(T_i\right) \cdot P\left(t, T_i\right). \tag{5.1}$$

If we concentrate on short-rate models, we will denote this by

$$Bond\left(r, t, T_B, C\right) = \sum_{i=1}^{n} C\left(T_i\right) \cdot P\left(r, t, T_i\right). \tag{5.2}$$

The price $Bond\left(t, T_B, C\right)$ is sometimes called the *dirty price* of a coupon bond and it is interesting to note that this is usually not the price quoted at the exchange. Suppose that the last coupon payment took place at time t_0. Then the *accrued interest* $AI\left(t_0, t, C\right)$ at time $t \in [t_0, T_1]$ is defined by

$$AI\left(t_0, t, C\right) := C\left(T_1\right) \cdot \frac{\left(t - t_0\right)_{DC(B)}}{\left(T_1 - t_0\right)_{DC(B)}}$$

with $DC\left(B\right)$ denoting the daycount convention used for the coupon bond[3]. The quoted price or *clean price* $Bond_{clean}\left(t_0, t, T_B, C\right)$ of the coupon bond at time t is defined by

$$Bond_{clean}\left(t_0, t, T_B, C\right) := Bond\left(t, T_B, C\right) - AI\left(t_0, t, C\right).$$

---

[3] Usually, the daycount convention used for Eurobonds or U.S. Treasury bonds is $DC\left(B\right) = act/act$. However, other conventions are possible. For more details see Section 5.1.

This means that whenever we buy a coupon bond quoted at a clean price of $Bond_{clean}(t_0, t, T_B, C)$, the cash price we have to pay is

$$Bond(t, T_B, C) = Bond_{clean}(t_0, t, T_B, C) + AI(t_0, t, C),$$

i.e. we have to compensate the seller of the coupon bond for the interest accrued since the last coupon date because we and not the seller will receive the coupon with the next coupon payment.

Sometimes coupon bond prices are calculated using a constant zero rate for all maturities. This zero rate can be extracted either from the zero-rate curve or from a given coupon bond price $Bond(t, T_B, C)$ by solving the equation

$$Bond(t, T_B, C) = \sum_{i=1}^{n} C(T_i) \cdot (1+y)^{-(T_i-t)_{DC(B)}}$$

for $y \geq 0$. The solution $y(t, T_B, C)$ is usually called the *yield-to-maturity* of the coupon bond (T_B, C) at time $t \in [t_0, T_1]$. Note that the same daycount convention is used for calculating the yield-to-maturity and the accrued interest. A transformation to other daycount conventions is straightforward.

Example. On October 20, 2000, we would like to evaluate the 6.5% German government bond maturing at $T_B = 10/14/2005$ with annual coupon payments. The market data for the zero rates and the corresponding zero-coupon bond prices or discount factors is given as follows:

time	zero rate (in %)	discount factor (in %)
10/14/01	4.955	95.24
10/14/02	4.996	90.56
10/14/03	5.038	86.04
10/14/04	5.046	81.79
10/14/05	5.054	77.68

Hereby, the zero rates are continuous rates in act/act quotation. Using equation (5.1), the dirty price of the zero-coupon bond at time $t = 10/20/2000$

and with respect to a notional of 100 Euro can therefore be calculated as[4]

$$
\begin{aligned}
Bond\,(t, T_B, C) &= \sum_{i=1}^{n} C\,(T_i) \cdot P\,(t_0, T_i) \\
&= 100 \cdot P\,(t_0, T_5) + 6.5 \cdot \sum_{i=1}^{5} P\,(t_0, T_i) \\
&= 100 \cdot 0.7768 + 6.5 \cdot 4.3131 \\
&= 105.715.
\end{aligned}
$$

The accrued interest for the past 6 days since the last coupon payment at time $t_0 = 10/14/2000$ is given by

$$
\begin{aligned}
AI\,(t_0, t, C) &= C\,(T_1) \cdot \frac{(t - t_0)_{act/act}}{(T_1 - t_0)_{act/act}} \\
&= 6.5 \cdot \frac{6}{365} = 0.107.
\end{aligned}
$$

We thus get a clean price at time t of

$$
Bond_{clean}\,(t_0, t, T_B, C) = Bond\,(t, T_B, C) - AI\,(t_0, t, C) = 105.608.
$$

The corresponding yield-to-maturity which solves the equation

$$
105.715 = \sum_{i=1}^{n} C\,(T_i) \cdot (1 + y)^{-(T_i - t)_{act/act}}
$$

is given by $y\,(t, T_B, C) = 5.193\%$. ☐

5.3   Forward Agreements on Coupon Bonds

A forward contract is an agreement to buy or sell a financial instrument at a certain future time T, called the maturity time, and for a certain price X, called the delivery price. Forward contracts are usually set up directly between two financial institutions or a financial institution and one of its clients and thus are very often non-standardized and traded over the counter (OTC), i.e. not on an exchange. The party who agrees to buy the underlying financial instrument is said to have a long position, the party who agrees to sell the financial instrument is said to have a short position.

---

[4] Because we used a specific (averaged) zero rate curve, market prices (dirty) usually differ from the dirty prices derived from the zero rate curve. Indeed, the market price (dirty) for the considered bond on October 20, 2000 has been 106.0418.

The delivery price is chosen at time $t = T_0 \in [t_0, T]$ when both parties enter into the contract so that the value $V_{Forward}(T_0, T, X)$ of the forward contract to both parties is zero, i.e. it costs nothing to enter into a forward contract either holding a short or a long position. Chosen that way, the delivery price is called the *forward price* of the underlying financial instrument, denoted by $Forward(T_0, T)$. Thus, $V_{Forward}(T_0, T, Forward(t_0, T)) = 0$. The forward contract is settled at maturity when the holder of the short position has to deliver the underlying financial instrument and the holder of the long position has to pay a cash amount given by the delivery price. Thus, even if the value of the forward contract is zero at the beginning $t = T_0$ it may become positive or negative as the price of the underlying financial instrument changes over time. Hence, it is very important to distinguish between the forward price and the value of the forward contract. For an evaluation let $I(t, T) = \sum_{i=1}^{k} C(T_i) \cdot P(t, T_i)$ with $t_0 \leq T_0 \leq t \leq T_1 < \cdots < T_k \leq T$, $k \in I\!N$ denote the present value at time t of the non-stochastic income or cash flow stream $C = (C(T_1), ..., C(T_k))$ to be received from the underlying financial instrument during the life of the forward contract. Furthermore, let us denote the price of the underlying financial instrument at time t, $t_0 \leq T_0 \leq t \leq T \leq T^*$, by $S(t)$ and let us assume that there is no difference in the continuous zero rate for borrowing and lending. We consider the following two portfolios set up at time t:

A: One long position in the forward contract with delivery price X and time to maturity T plus a zero-coupon bond with a notional amount of X, also with time to maturity T.

B: One unit of the underlying financial instrument plus borrowings of an amount of $I(t, T)$ structured in k zero-coupon bonds with a time to maturity of T_i and a notional of $C(T_i)$, $i = 1, ..., k$.

The income from the interest payments of the underlying financial instrument in portfolio B can be used to repay the borrowings while the sale of the maturing zero-coupon bond of portfolio A gives us exactly the amount of money X to buy the underlying via the forward contract. This leads us to the following value table for the portfolios A and B at time t and time T:

$Portfolio/Time$	t	T
A	$V_{Forward}(t, T, X) + X \cdot P(t, T)$	$S(T)$
B	$S(t) - I(t, T)$	$S(T)$

Having the same terminal value and under the assumption of no arbitrage possibilities the two portfolios must have the same value at time t, i.e.

$$S(t) - I(t, T) = V_{Forward}(t, T, X) + X \cdot P(t, T)$$

or equivalently, for the value of the forward contract at time t,

$$V_{Forward}(t, T, X) = S(t) - I(t, T) - X \cdot P(t, T).$$

At time $t = T_0$, when the forward is set up, we insert the forward price $Forward(T_0, T)$ for X to get

$$
\begin{aligned}
S(T_0) - I(T_0, T) &= V_{Forward}(T_0, T, Forward(T_0, T)) \\
&\quad + Forward(T_0, T) \cdot P(T_0, T) \\
&= 0 + Forward(T_0, T) \cdot P(T_0, T),
\end{aligned}
$$

or equivalently

$$
Forward(T_0, T) = (S(T_0) - I(T_0, T)) \cdot P^{-1}(T_0, T).
$$

Basically, S may be the price of any financial instrument with a non-stochastic income stream. Nevertheless, we will use this equation especially for coupon bonds. To do this, let us consider a coupon bond with payments

$C(T_i)$ at time T_i, $i = 1, ..., n$, $t_0 \leq T_0 \leq T_1 < T_2 < \cdots < T_n = T_B \leq T^*$.

Furthermore, let $k := \max\{i \in \{1, ..., n\} : T_i \leq T\}$ and

$$
I(T_0, T) = \sum_{i=1}^{k} C(T_i) \cdot P(T_0, T_i)
$$

be the present value of the coupon payments up to the maturity time T of the forward contract. Then the forward price of that coupon bond for the maturity time T at time T_0 is given by

$$
Forward(T_0, T, T_B, C) = (Bond(T_0, T_B, C) - I(T_0, T)) \cdot P^{-1}(T_0, T).
$$
(5.3)

Note that the forward price $Forward(T_0, T, T_B, C)$, such as the coupon bond price $Bond(T_0, T_B, C)$, is a dirty price which includes the accrued interest for the time from the last coupon payment date $T_k \in [T_0, T]$ to time T. Also note that the forward price $Forward(T_0, T, T_B)$ at time T_0 for the maturity time T of a zero-coupon bond which matures at time T_B is given by

$$
Forward(T_0, T, T_B) = \frac{P(T_0, T_B)}{P(T_0, T)} = P(T_0, T, T_B).
$$

Example. On October 20, 2000, i.e. $t = T_0 = 10/20/2000$, we would like to price a forward agreement on the 6.5% German government bond maturing at $T_B = 10/14/2005$ which we already discussed in the example of Section 5.2. The forward agreement is for June 20, 2002, i.e. the maturity time of the forward rate agreement is $T = 06/20/2002$. The additional market data for the zero rates (quoted in act/act) and the corresponding discount factors is given as follows

time	zero rate (in %)	discount factor (in %)
10/14/01	4.955	95.24
06/20/02	4.983	92.03

Furthermore, the dirty price of the coupon bond at time $t = T_0$ was calcultatd to $Bond\,(t, T_B, C) = 105.715$. The only coupon payment between times t and T takes place at time $T_1 = 10/14/2001$. The present value of this payment at time t is

$$I\,(t, T) = C\,(T_1) \cdot P\,(t, T_1) = 6.5 \cdot 0.9524 = 6.191$$

Using equation (5.3), we get a forward price at time $t = T_0$ of

$$
\begin{aligned}
Forward\,(t, T, T_B, C) &= (Bond\,(t, T_B, C) - I\,(t, T)) \cdot P^{-1}\,(t, T) \\
&= (105.715 - 6.191) \cdot \frac{1}{0.9203} \\
&= 108.143.
\end{aligned}
$$

$\square$

5.4   Interest-Rate Futures

A financial futures contract is an agreement between a trader who could act as a buyer or seller of a futures contract and the futures exchange or its clearing house. If the trader acts as a buyer he agrees to take, if he acts as a seller, he agrees to make delivery of a certain amount of a financial instrument at a price fixed when the contract is set up and at a designated time T, called the maturity or delivery date of the futures contract. Standard delivery months are March, June, September and December. Examples for a financial instrument could be a stock, an index, a currency or a coupon bond. For some futures contracts the settlement at maturity is in cash rather than physical delivery. The buyer of a futures contract is said to have a long position, the seller is said to have a short position in the futures contract. Briefly spoken, the trader is either long or short the futures. When the trader takes a position in a futures contract he must deposit a minimum amount of money per contract as specified by the exchange on a so-called *margin account*. As the price for the futures changes, the value of the traders position will also change resulting in a market gain or loss at the end (close) of or even during each trading day. This process is referred to as *marking to market*. Should the market value of the traders position fall below a lower bound determined by the exchange the trader has to provide an additional amount of money. This process is known as *margin call*. If the traders position increases he may withdraw money from his account. Hence, a futures position may quite well involve some cash flows prior to the delivery date. Describing the margin procedure and other (optional) specifications of the different futures contracts would be beyond the scope of this book. The interested reader may refer to, e.g., Fabozzi, Fabozzi, and Pollack [FFP91], Hull [Hul00], Fitzgerald, Lubochinski, and Thomas [FLT93] or Steiner, Meyer, and Luttermann [SML94] for more details on this topic.

The futures contracts on interest-rate instruments or briefly interest-rate futures contracts may be divided into futures on short-term instruments and futures on intermediate- or long-term instruments. Members of the first group are the Eurodollar and Treasury bill futures traded on the International Money Market (IMM) of the Chicago Mercantile Exchange (CME) or the EURIBOR futures traded on the EUREX, the 1998 merger of the Deutsche Terminbörse (DTB) and the Swiss Options and Financial Futures Exchange (SOFFEX). These instruments, briefly called short-term interest-rate futures, will be discussed in Section 5.4.1. Members of the second group are the Treasury bond and Treasury note futures traded on the Chicago Board of Trade (CBOT) or the (Euro-) Bobl and (Euro-) Bund futures traded on the EUREX. Another important futures exchange is the London International Financial Futures and Options Exchange (LIFFE). Since the underlyings of all these futures are coupon bonds, we simply refer to them as coupon-bond futures contracts. A detailed presentation is given in Section 5.4.2.

5.4.1 Short-Term Interest-Rate Futures

Futures contracts based on a short-term index rate, i.e. an interest rate or some sort of interest on a reference instrument, are called *short-term interest-rate futures*. If the index rate is dependent on a short-term Treasury bill these futures are called Teasury bill futures. If the index rate is an inter bank offered interest rate such as the LIBOR or EURIBOR, the corresponding financial futures are called Eurodollar or EURIBOR futures. The quotes for Eurodollar or EURIBOR futures are often used to derive a zero-rate curve. The corresponding zero rates are called Eurodollar or EURIBOR strip rates. The strip procedure usually ignores the difference between expected futures rates and forward rates. We will show, how we can overcome this inaccuracy using a specific interest-rate model.

Treasury Bills and Treasury Bill Futures

Treasury bills are, such as all U.S. Treasury securities, backed by the U.S. government and are therefore considered to be free of credit risk. Since the U.S. Treasury market is one of the most active and most liquid markets in the world, interest rates on Treasury securities are very popular interest-rate benchmarks in international capital markets. While the Treasury bonds, with original time to maturity of more than ten years, and Treasury notes, with original time to maturity between two and ten years, belong to the category of coupon bonds, the Treasury bills are zero-coupon bonds with original time to maturity of one year or less. Quotes on Treasury bills are on a bank discount basis not on a price basis. Suppose that $P(t,T)$ is the price at time t of a Treasury bill, i.e. a zero-coupon bond, with maturity time T and relative to a notional of $L = 1$. Then the quote on a Treasury bill, also called the *bank discount rate* or *bankers' discount yield*, $R_{BD}(t,T)$ at time t for the maturity time T, is defined by

$$R_{BD}(t,T) = \frac{1 - P(t,T)}{(T-t)_{DC(TB)}},$$

where the index $DC(TB)$ indicates the daycount convention for the Treasury bills and is defined to be $DC(TB) = act/360$. Equivalently, given a bank discount rate at time t for the maturity time T of $R_{BD}(t,T)$, the price of the corresponding Treasury bill can be easily calculated as

$$P(t,T) = 1 - R_{BD}(t,T) \cdot (T-t)_{DC(TB)}.$$

If, for example, the quote at time t for a Treasury bill with 90 days to maturity time T is $R_{BD}(t,T) = 6\%$, the price of the Treasury bill would be

$$P(t,T) = 1 - 0.06 \cdot \frac{90}{360} = 98.5\%.$$

The corresponding zero-coupon rate $R_L(t,T)$, also called the *Treasury bill yield*, is expressed as a linear interest rate and given by

$$R_L(t,T) = \frac{1 - P(t,T)}{P(t,T) \cdot (T-t)_{DC(TB)}} = \frac{R_{BD}(t,T)}{P(t,T)}.$$

It represents the annualized zero-coupon rate earned by the owner of the Treasury bill if he buys the Treasury bill at time t and sells it at maturity time T. Hence,

$$P(t,T) = \frac{1}{1 + R_L(t,T) \cdot (T-t)_{DC(TB)}}.$$

To give an example, the above bank discount rate of $R_{BD}(t,T) = 6\%$ quoted for a Treasury bill with 90 days to maturity corresponds to a Treasury bill yield of

$$R_L(t,T) = \frac{R_{BD}(t,T)}{P(t,T)} = \frac{6\%}{98.5\%} = 6.09\%.$$

We hereby assumed that the daycount convention for the bank discount rate and the Treasury bill yield are the same. However, other definitions may be possible leading to straightforward adjustments of the previous equations.

Now let $R_{BD}(t, T_i, T_{i+1})$ denote the *forward bank discount rate* and $P(t, T_i, T_{i+1})$ the forward price of the Treasury bill at time t for the time interval $[T_i, T_{i+1}]$, $i \in \mathbb{N}$. Then $R_{BD}(T_i, T_i, T_{i+1}) = R_{BD}(T_i, T_{i+1})$ and $P(T_i, T_i, T_{i+1}) = P(T_i, T_{i+1})$. The *Treasury bill futures contract* was the first contract on a short-term debt instrument at the International Money Market (IMM) of the Chicago Mercantile Exchange (CME). The (fictitious) financial instrument underlying the Treasury bill future is a Treasury bill with exactly 90 days to maturity at the expiration day of the future and the contract is on a notional amount of $L = 1$ *Mio. US\$*. Hence, if T_i is the maturity time of an actually traded Treasury bill futures contract, the maturity time of the underlying Treasury bill is $T_{i+1} = T_i + 90$ days and the index $i \in \mathbb{N}$ runs through all actually traded futures contracts. For example, the underlying financial instrument of a Treasury bill futures contract which matures in 120 days from now is a Treasury bill with $120 + 90 = 210$ days time to maturity. The maturity months for the Treasury bill futures contracts are March, June, September, and December. The last trading day is the business day preceding the first day of the delivery month on which a 13-week Treasury bill is issued and a one-year Treasury bill has 13 weeks remaining to maturity. The first delivery date is the first business day following the last trading day and delivery must usually take place within a three-day period and with actually traded Treasury bills. At the delivery date, the time to maturity of a Treasury bill which is due to a physical delivery may slightly differ from the 90 days of the underlying (fictitious) Treasury bill of the futures contract. However, because all Treasury bills of the same maturity, no matter whether they are new issues or older ones with the same remaining life, are equivalently traded in the money market, there is basically only one deliverable financial instrument which can be delivered from the seller of the Treasury bill futures contract. In return, the buyer of the Treasury bill futures contract has to pay the seller a cash amount depending on the futures price. The *Treasury bill futures quote* $F_{BD}(t, T_i) := F_{BD}(t, T_i, T_{i+1})$ at time $t \in [t_0, T_i]$ for the maturity time T_i of the future and $T_{i+1} = T_i + 90$ days is defined as a price quote equal to 1 minus the implied futures Treasury bill quote or implied futures bank discount rate $R_{BD}^F(t, T_i, T_{i+1})$, i.e.

$$F_{BD}(t, T_i) = 1 - R_{BD}^F(t, T_i, T_{i+1}),$$

where the actual futures quote is expressed in per cent. The corresponding *cash futures price* $F(t, T_i) := F(t, T_i, T_{i+1})$ which is used for the daily marking to market is given by

$$
\begin{aligned}
F(t, T_i) &= 1 - [1 - F_{BD}(t, T_i)] \cdot (T_{i+1} - T_i)_{DC(TB)} \\
&= 1 - R_{BD}^F(t, T_i, T_{i+1}) \cdot (T_{i+1} - T_i)_{DC(TB)}
\end{aligned}
$$

for a notional amount of 1. The *invoice* or *market futures price* for the daily marking to market is based on the notional amount of the futures contract an therefore equal to

$$L \cdot F\left(t, T_i\right) = 1,000,000 \cdot F\left(t, T_i\right).$$

For *example*, a Treasury bill futures quote for a given futures maturity time T_i of $F_{BD}\left(t, T_i\right) = 93\% = 0.93$ tells us that the corresponding Treasury bill is traded in the futures market at an implied future bank discount rate of $R_{BD}^F\left(t, T_i, T_{i+1}\right) = 7\%$ with $T_{i+1} = T_i + 90$ days. Hence, the cash futures price is equal to

$$
\begin{aligned}
F\left(t, T_i\right) &= 1 - R_{BD}^F\left(t, T_i, T_{i+1}\right) \cdot \left(T_{i+1} - T_i\right)_{DC(TB)} \\
&= 100\% - 7\% \cdot \frac{90}{360} = 98.25\%
\end{aligned}
$$

giving an invoice price for the daily marking to market of $L \cdot 98.25\% = 982,500 \ US\$$. Consequently, a change of $1bp$ (1 basis point) or 0.01% in the futures quote leads to a change in the invoice price of

$$L \cdot 0.01\% \cdot 0.25 = 25 \ US\$.$$

At maturity time T_i, the seller of the futures contract has to physically deliver a Treasury bill as pointed out above. On the other hand, the buyer of the Treasury bill futures contract has to pay the seller a cash amount equal to the invoice price of

$$
\begin{aligned}
L \cdot F\left(T_i, T_i\right) &= L \cdot \left[1 - R_{BD}\left(T_i, T_{i+1}\right) \cdot \left(T_{i+1} - T_i\right)_{DC(TB)}\right] \\
&= L \cdot P\left(T_i, T_{i+1}\right).
\end{aligned}
$$

Depending on the Treasury bill which is delivered, the length $\left(T_{i+1} - T_i\right)_{DC(TB)}$ of the time interval from the maturity time T_i of the future to the maturity time T_{i+1} of the Treasury bill may slightly differ from 0.25. If we are given an arbitrage-free, complete interest-rate market with equivalent martingale measure $\widetilde{Q}$, the cash futures price at time $t \in [t_0, T_i]$ is defined by[5]

$$F\left(t, T_i\right) = E_{\widetilde{Q}}\left[P\left(T_i, T_{i+1}\right) | \mathcal{F}_t\right],$$

---

[5] For a motivation within the generalized Black-Scholes model see Section 3.5. Using this definition, the expected cash flow at time $t' + \Delta t \in [t, T_i]$ with respect to the last settlement date t' as seen from time t is

$$
\begin{aligned}
E_{\widetilde{Q}}\left[F\left(t' + \Delta t, T_i\right) - F\left(t', T_i\right) | \mathcal{F}_t\right] &= E_{\widetilde{Q}}\left[E_{\widetilde{Q}}\left[P\left(T_i, T_{i+1}\right) | \mathcal{F}_{t' + \Delta t}\right] | \mathcal{F}_t\right] \\
&\quad - E_{\widetilde{Q}}\left[E_{\widetilde{Q}}\left[P\left(T_i, T_{i+1}\right) | \mathcal{F}_{t'}\right] | \mathcal{F}_t\right] \\
&= E_{\widetilde{Q}}\left[P\left(T_i, T_{i+1}\right) | \mathcal{F}_t\right] - E_{\widetilde{Q}}\left[P\left(T_i, T_{i+1}\right) | \mathcal{F}_t\right] \\
&= 0
\end{aligned}
$$

because of Lemma 2.6. Following the same arguments as in Section 3.5, this is the reason why it costs nothing to enter into a futures contract.

which leads us to an *implied futures Treasury bill quote* of

$$R_{BD}^F(t, T_i, T_{i+1}) = \frac{1 - F(t, T_i)}{(T_{i+1} - T_i)_{DC(TB)}} = 4 \cdot \left(1 - E_{\tilde{Q}}[P(T_i, T_{i+1}) | \mathcal{F}_t]\right).$$

Equivalently, the quoted Treasury bill futures price is given by

$$
\begin{aligned}
F_{BD}(t, T_i) &= 1 - R_{BD}^F(t, T_i, T_{i+1}) = 1 - \left[\frac{1 - F(t, T_i)}{(T_{i+1} - T_i)_{DC(TB)}}\right] \\
&= 1 - 4 \cdot \left(1 - E_{\tilde{Q}}[P(T_i, T_{i+1}) | \mathcal{F}_t]\right).
\end{aligned}
$$

If the difference between futures and forward contracts is ignored we can calculate the cash futures price as well as the quoted Treasury bill futures price directly from the zero-rate curve at time $t \in [t_0, T_i]$. To do this, let the forward price $P(t, T_i, T_{i+1})$ at time t of a zero-coupon bond with maturity time T_{i+1} calculated for a forward maturity time T_i be extracted from the discount curve at time t by

$$P(t, T_i, T_{i+1}) = \frac{P(t, T_{i+1})}{P(t, T_i)}.$$

Under the given assumption, the cash futures price is equal to the forward price, i.e.

$$F(t, T_i) = P(t, T_i, T_{i+1}).$$

In this case, the implied futures Treasury bill quote is

$$R_{BD}^F(t, T_i, T_{i+1}) = 4 \cdot (1 - P(t, T_i, T_{i+1}))$$

and the Treasury bill future would be quoted at

$$F_{BD}(t, T_i) = 1 - R_{BD}^F(t, T_i, T_{i+1}).$$

To give an *example*, if the discount curve at time t tells us that $P(t, T_i) = 98.5\%$ and $P(t, T_{i+1}) = 97\%$, the forward price $P(t, T_i, T_{i+1})$ is equal to 98.48%, the implied futures Treasury bill quote is

$$R_{BD}^F(t, T_i, T_{i+1}) = 4 \cdot (1 - P(t, T_i, T_{i+1})) = 6.09\%,$$

and the Treasury bill futures quote is

$$F_{BD}(t, T_i) = 1 - R_{BD}^F(t, T_i, T_{i+1}) = 93.91\%.$$

Eurodollar and EURIBOR Futures

Eurodollars are deposits of U.S. dollars in institutions outside the United States. The rate of interest earned on Eurodollars deposited by one bank with another is called the Eurodollar interest rate. It is also known as the *London InterBank Offered Rate* or briefly *LIBOR*. Since 1999 the floating index rate of the Euro money market is expressed by the *Euro InterBank Offered Rate* or briefly *EURIBOR*. Due to the increasing need of short-term Eurodollar lending and borrowing, the IMM introduced the Eurodollar Time Deposit futures contract or briefly *Eurodollar futures* contract in 1981 which is tied to the LIBOR. In 1999 the European Banking Federation (FBE) sponsored the introduction of the *EURIBOR futures* contract which is tied to the EURIBOR and traded at the EUREX as well as at the LIFFE. Both contracts are designed to protect its owner from fluctuations in the 3-month LIBOR or 3-month EURIBOR for a 90-day period. The contract size or notional is $L = 1$ *Mio. US$* for the Eurodollar futures and $L = 1$ *Mio. Euro* for the EURIBOR futures contract. The maturity months for both futures contracts are March, June, September, and December with up to five or even more years to maturity. The daycount convention for both futures contracts is $DC\,(EF) = act/360$. Since there is no underlying cash security, neither futures contract allows for physical delivery. Instead, settlement is made in cash at the last trading day which is the second business day before the third Wednesday of the respective maturity month. The final settlement price for both futures contracts is set equal to 1 minus the 3-month LIBOR or EURIBOR, respectively, where the actual settlement price is expressed in per cent. Let $R_L\,(t, T_i, T_{i+1})$ denote the annualized linear forward index rate, either LIBOR or EURIBOR, for the (3-month) time interval $[T_i, T_{i+1}]$ quoted at time $t \in [t_0, T_i]$ with $T_{i+1} = T_i + 90$ days. Hereby, the index $i \in I\!N$ runs through all actually traded futures contracts. Hence, the final settlement price $F_L\,(T_i, T_i, T_{i+1})$ for the corresponding futures contract with maturity date T_i is

$$F_L\,(T_i, T_i, T_{i+1}) = 1 - R_L\,(T_i, T_i, T_{i+1}),$$

where the actual futures quote is expressed in per cent. Given the discount curve at time T_i we know that

$$
\begin{aligned}
R_L\,(T_i, T_{i+1}) &= R_L\,(T_i, T_i, T_{i+1}) \\
&= \frac{1}{(T_{i+1} - T_i)_{DC(EF)}} \cdot \left(\frac{1}{P\,(T_i, T_{i+1})} - 1 \right) \\
&= 4 \cdot \left(\frac{1}{P\,(T_i, T_{i+1})} - 1 \right),
\end{aligned}
$$

or equivalently,

$$P\,(T_i, T_{i+1}) = \frac{1}{1 + R_L\,(T_i, T_{i+1}) \cdot 0.25}.$$

At time $t \in [t_0, T_i]$ both futures are quoted in terms of a price which is equal to 1 minus the annualized futures index rate $R_L^F(t, T_i, T_{i+1})$ in per cent, i.e. the futures contract with maturity date T_i is quoted as

$$F_L(t, T_i) := F_L(t, T_i, T_{i+1}) = 1 - R_L^F(t, T_i, T_{i+1})$$

in per cent. Hence, as t tends to T_i, $1 - F_L(t, T_i)$ converges to the 3-month LIBOR or EURIBOR. The futures quote corresponds to a *cash price* $F_C(t, T_i)$ which is used for the daily marking to market and defined by

$$
\begin{aligned}
F_C(t, T_i) &= 1 - [1 - F_L(t, T_i)] \cdot (T_{i+1} - T_i)_{DC(EF)} \\
&= 1 - R_L^F(t, T_i, T_{i+1}) \cdot 0.25
\end{aligned}
$$

for a notional amount of 1. As for the Treasuy bill future, the *invoice* or *market price* for the daily marking to market is based on the notional amount of the futures contract and therefore set to

$$L \cdot F_C(t, T_i) = 1,000,000 \cdot F_C(t, T_i).$$

or equivalently

$$L \cdot [1 - (1 - F_L(t, T_i)) \cdot 0.25].$$

Consequently, a change of $1bp$ (1 basis point) or 0.01% in the futures quote leads to a change in the invoice price of

$$L \cdot 0.01\% \cdot 0.25 = 25 \ US\$ \ (Euro).$$

All gains and losses of the futures contract are marked to market on a daily basis by the futures exchange and credited or debited to the owner's cash or margin account. If we are given an arbitrage-free, complete interest-rate market with equivalent martingale measure $\widetilde{Q}$, the futures index rate $R_L^F(t, T_i, T_{i+1})$ at time $t \in [t_0, T_i]$ is defined by

$$R_L^F(t, T_i, T_{i+1}) = E_{\widetilde{Q}}\left[R_L^F(T_i, T_i, T_{i+1}) | \mathcal{F}_t\right] = E_{\widetilde{Q}}\left[R_L(T_i, T_{i+1}) | \mathcal{F}_t\right].$$

The corresponding cash futures price is given by[6]

$$F_C(t, T_i) = 1 - E_{\widetilde{Q}}\left[R_L(T_i, T_{i+1}) | \mathcal{F}_t\right] \cdot 0.25$$

which leads to a futures quote of

$$F_L(t, T_i) = 1 - E_{\widetilde{Q}}\left[R_L(T_i, T_{i+1}) | \mathcal{F}_t\right].$$

---

[6] Using the same arguments as above, the expected cash flows at each future settlement date, as seen from time t, are equal to zero.

If the difference between futures and forward rates is ignored we can calculate the cash futures price as well as the futures quote directly from the zero-rate curve at time $t \in [t_0, T_i]$. To do this let the forward index rate $R_L(t, T_i, T_{i+1})$ at time t for the time interval $[T_i, T_{i+1}]$ be extracted from the discount curve at time t by

$$R_L(t, T_i, T_{i+1}) = 4 \cdot \left(\frac{P(t, T_i)}{P(t, T_{i+1})} - 1 \right).$$

Under the given assumption, the cash futures price is equal to

$$F_C(t, T_i) = 1 - R_L(t, T_i, T_{i+1}) \cdot 0.25$$

and the future would be quoted at

$$F_L(t, T_i) = 1 - R_L(t, T_i, T_{i+1}).$$

To give an *example*, if the discount curve at time t tells us that $P(t, T_i) = 98.5\%$ and $P(t, T_{i+1}) = 97\%$, the forward index rate $R_L(t, T_i, T_{i+1})$ is equal to 6.19%, the cash futures price is $F_C(t, T_i) = 1 - R_L(t, T_i, T_{i+1}) \cdot 0.25 = 98.45\%$, and the futures quote is $F_L(t, T_i) = 1 - R_L(t, T_i, T_{i+1}) = 93.81\%$.

The Eurodollar and EURIBOR Strip Rates

Eurodollar and EURIBOR futures can be used to lock in LIBOR and EU-RIBOR based zero-coupon rates for horizons up to the longest futures maturity time which will be expressed for continuous compounding here. These zero-coupon rates are called the continuous *Eurodollar* and *EURIBOR strip rates* and depend on the futures we use. They are expressed in money market terms on an $DC(EF) = act/360$ daycount convention from which they may be converted to the traders convenience. Let $T_1, ..., T_N$ be the maturity dates of the actually traded index rate futures considered for the futures strip and the index rate may be LIBOR or EURIBOR. The futures quotes are $F_L(t, T_i)$, $i = 1, ..., N$, at time $t \in [t_0, T_1]$ and $R_L(t, T_1)$ is the index rate at time t for the time interval $[t, T_1]$. Let $(T_{i+1} - T_i)_{DC(EF,ltd)}$ denote the days to the last trading day (ltd) of the futures contract maturing at time T_{i+1} as seen from time T_i, $i = 0, ..., N - 1$, with $T_0 := t$ and calculated using the daycount convention $DC(EF)$. Furthermore, let $T \in (T_{N-1}, T_N]$ be the time horizon up to which we want to calculate the strip rates $R(t, T'_j)$, $j = 1, ..., N$, with $T'_j := T_j$, $j = 1, ..., N - 1$, and $T'_N := T$. Then

$$P^{-1}(t, T'_j) = \prod_{i=0}^{j-1} \left(1 + R_L^F(t, T_i, T_{i+1}) \cdot \tau_i \right) = \prod_{i=0}^{j-1} \left[1 + (1 - F_L(t, T_i)) \cdot \tau_i \right]$$

with

$$\tau_i := \begin{cases} (T_{i+1} - T_i)_{DC(EF,ltd)} & , i \in \{0, ..., N-2\} \\ \min\left\{ (T - T_{N-1})_{DC(EF)}, (T_N - T_{N-1})_{DC(EF,ltd)} \right\} & , i = N-1 \end{cases}$$

and

$$1 - F_L\left(t, T_0\right) = R_L^F\left(t, T_0, T_1\right) = R_L^F\left(T_0, T_0, T_1\right) = R_L\left(T_0, T_1\right) = R_L\left(t, T_1\right)$$

is the price at time t of a zero-coupon bond with maturity T_j', $j = 1, ..., N$. The zero-coupon bond can now be transformed to the continuous (or any other) index strip rate via

$$R\left(t, T_j'\right) = -\frac{1}{\left(T_j' - t\right)_{DC(EF)}} \cdot \ln P\left(t, T_j'\right)$$

$$= \frac{1}{\left(T_j' - t\right)_{DC(EF)}} \cdot \sum_{i=0}^{j-1} \ln\left[1 + \left(1 - F_L\left(t, T_i\right)\right) \cdot \tau_i\right],$$

j=1,...,N.

Example. We want to calculate the 180 days EURIBOR strip rate using the following settlement prices of the December 2000 and March 2001 EURIBOR futures at October 19, 2000 (time t) as they were published by the Handelsblatt

Index i	Futures Quote	Implied Futures Rate	Distance τ_i
0 (10/19/00)	–	4.91	60 days
1 (Dec 00)	94.82	5.18	91 days
2 (Mar 01)	94.83	5.17	29 days

where we took the 2-month EURIBOR as a proxy for the 60 day zero rate as it is published on the Reuters page EURIBORRECAP02. The price at time t of a zero-coupon bond with maturity time T in 180 days from t is given by

$$P^{-1}\left(t, T\right) = \prod_{i=0}^{2}\left(1 + R_L^F\left(t, T_i, T_{i+1}\right) \cdot \tau_i\right)$$

$$= \left(1 + \frac{4.91\% \cdot 60}{360}\right) \cdot \left(1 + \frac{5.18\% \cdot 91}{360}\right) \cdot \left(1 + \frac{5.17\% \cdot 29}{360}\right)$$

$$= 102.564\%.$$

Hence, we are talking about a 180 days EURIBOR strip rate of

$$R\left(t, t + 180 \text{ days}\right) = \frac{360}{180} \cdot \ln 1.0256 = 5.063\%.$$

☐

Convexity Adjustments

If we construct a zero-rate curve using the strip rate procedure explained above, we usually ignore the difference between expected futures rates and forward rates. However, since we are dealing with stochastic interest rates it may be inappropriate to use the implied futures rates instead of the forward zero rates. As soon as we can relate the expected or implied futures rate to the forward zero rate the difference between the two would be the correction term which solves our problem. However, the expected futures rate depends on the interest-rate model we use for our calculations. We therefore illustrate the correction principle by applying the Hull-White model of Section 4.5.3. Following Lemma 4.25, we know that the implied futures rate $R_L^F\left(t, T_i, T_{i+1}\right)$ at time t for the time interval $[T_i, T_{i+1}]$ satisfies the equation

$$
\begin{aligned}
1 + R_L^F\left(t, T_i, T_{i+1}\right) \cdot \tau_i &= E_{\widetilde{Q}}[1 + R_L\left(T_i, T_{i+1}\right) \cdot \tau_i | \mathcal{F}_t] \\
&= E_{\widetilde{Q}}[P^{-1}\left(T_i, T_{i+1}\right) | \mathcal{F}_t] \\
&= e^{E_{\widetilde{Q}}[\ln P^{-1}(T_i, T_{i+1}) | \mathcal{F}_t] + \frac{1}{2} \cdot s_P^2(t, T_i, T_{i+1})} \\
&= \frac{P\left(t, T_i\right)}{P\left(t, T_{i+1}\right)} \cdot e^{s_P^2(t, T_i, T_{i+1}) + \frac{1}{2} \cdot B(T_i, T_{i+1}) \cdot v^2(t, T_i)} \\
&= [1 + R_L\left(t, T_i, T_{i+1}\right) \cdot \tau_i] \cdot \\
&\qquad \cdot e^{s_P^2(t, T_i, T_{i+1}) + \frac{1}{2} \cdot B(T_i, T_{i+1}) \cdot v^2(t, T_i)},
\end{aligned}
$$

or equivalently

$$
\begin{aligned}
R_L\left(t, T_i, T_{i+1}\right) \cdot \tau_i &= \left[1 + R_L^F\left(t, T_i, T_{i+1}\right) \cdot \tau_i\right] \cdot \\
&\qquad \cdot e^{-s_P^2(t, T_i, T_{i+1}) - \frac{1}{2} \cdot B(T_i, T_{i+1}) \cdot v^2(t, T_i)} - 1,
\end{aligned}
$$

with

$$
B\left(t, T_i\right) = \frac{1}{a} \cdot \left(1 - e^{-a \cdot (T_i - t) DC(EF)}\right), \quad v\left(t, T_i\right) = \sigma \cdot B\left(t, T_i\right)
$$

and

$$
s_P\left(t, T_i, T_{i+1}\right) = B\left(T_i, T_{i+1}\right) \cdot s_r\left(t, T_i\right)
$$

as well as

$$
s_r\left(t, T_i\right) = \sqrt{\frac{\sigma^2}{2a} \cdot \left(1 - e^{-2a \cdot (T_i - t) DC(EF)}\right)}.
$$

In other words, the error term $\varepsilon_{HW}\left(t, T_i, T_{i+1}\right)$ for the Hull-White model is given by

$$
\varepsilon_{HW}\left(t, T_i, T_{i+1}\right) = R_L^F\left(t, T_i, T_{i+1}\right) - R_L\left(t, T_i, T_{i+1}\right)
$$

$$
= \frac{\left[1 + R_L^F\left(t, T_i, T_{i+1}\right) \cdot \tau_i\right] \cdot \left(1 - e^{-s_P^2(t, T_i, T_{i+1}) - \frac{1}{2} \cdot B(T_i, T_{i+1}) \cdot v^2(t, T_i)}\right)}{\tau_i}
$$

dependent on the Hull-White model parameters a and σ, and the futures rate $R_L^F(t, T_i, T_{i+1})^7$ at time t for the time interval $[T_i, T_{i+1}]$. This means that the implied futures rates must be adjusted by the corresponding error term if we want to calculate the correct zero-rate curve out of the futures quotes. This correction is known as *convexity adjustment*.

Example. We want to calculate the error terms for the futures rates in the futures strip rates example from above. We use the Hull-White model with the parameters $a = 0.1$ and $\sigma = 2\%$. The resulting error terms and forward rates are listed in the following table

Index i	Implied Futures Rate	Error Term	Forward Rate
0 (10/19/00)	4.91	0	–
1 (Dec 00)	5.18	0.002	5.178
2 (Mar 01)	5.17	0.005	5.165

leading to an adjusted 180 days EURIBOR strip rate of

$$R(t, t + 180 \text{ days}) = 5.061\%.$$

$\square$

5.4.2 Coupon Bond Futures

This section is dedicated to futures on medium- and long-term coupon bonds. The probably most successful futures contract is the Treasury bond or briefly *T-bond future* traded at the CBOT. The financial instrument underlying the 30 year T-bond future is a T-bond with 6% semi-annual coupon and more than 15 years time to maturity. Because this is a fictitious instrument, any US government bond with a time to maturity of more than 15 years on the first day of the delivery month and not callable within 15 years from that day may be delivered. The face value of the T-bond futures contract is 100,000 US$. The financial instrument underlying the *(Euro) Bobl future* is a coupon bond issued by the German government or the Treuhandanstalt with 6% annual coupon and 4.5 to 5.5 years time to maturity. Delivery can take place with any German government bond, German government debt obligation, German government Treasury note or exchange traded debt security of the Treuhandanstalt having a time to maturity within the range from 4.5 to 5.5 years at delivery. The financial instrument underlying the *(Euro) Bund future* is a German government bond

---

[7] Note, that we used the same daycount convention for the continuous forward zero rate $R_L(t, T_i, T_{i+1})$ as we did for the futures rate $R_L^F(t, T_i, T_{i+1})$. Changes to other daycount conventions are straightforward.

with 6% annual coupon and 8.5 to 10.5 years time to maturity. Delivery can take place with any German government bond of a time to maturity within the range from 8.5 to 10.5 years. Since 1999, the face value of the Bobl and Bund futures contract is 100, 000 EURO. Before 1999 it was 250, 000 DEM. Because all quotes are made in terms of the fictitious underlying coupon bond we must be able to convert this underlying to any deliverable coupon bond. This is done using the so-called conversion factors which are published by the exchange for all deliverable bonds. Beside some minor adjustments of the time to maturity and coupon payment dates of a deliverable bond its *conversion factor* is equal to the clean price of the bond on the first day of the delivery month under the assumption that the zero rates for all maturities are equal to the coupon of the respective underlying, i.e. 6% for the T-bond, Bobl and the Bund future, and that the notional amount equals 1. For more details and the underlyings of other futures contracts see, e.g. Hull [Hul00] or Fabozzi, Fabozzi, and Pollack [FFP91]. It is important to note that the conversion factor for a given bond and a given delivery month is constant through time and will not be affected by changes in the price of the bond or the corresponding futures contract.

At delivery the seller of the futures contract has the right to choose which one of the deliverable bonds he would like to deliver. The buyer, on the other side, is obliged to pay the seller a cash amount of futures price times the conversion factor of the delivered bond plus the accrued interest of that bond. The option to choose which coupon bond to deliver is called the *delivery option* of the seller. While the EUREX has set the delivery within two days from the tenth day of the delivery month, at the CBOT the seller, within some guidelines, may decide when during the delivery month he will deliver the coupon bond. The seller thus has a *timing option* he could exercise to his advantage. At the CBOT the seller can give a notice to the clearing house of the intention to deliver after the exchange has closed (until 8 p.m. Chicago time). Since the futures settlement price, which is the basis for the invoice, is fixed at 2 p.m. Chicago time and the T-bonds continue trading until 4 p.m. this gives the seller a third option called the *wild card play*. These options in the hand of the seller will decrease the market price of the future. Also, the more options the seller of the futures contract gets the more complicated it will be to price the futures. For the purpose of exposition we will ignore these options for the evaluation of the futures price here.

To get an understanding for the price of the futures let us assume for a minute that there is only one deliverable bond defined by its time to maturity and cash flow vector (T_B, C) with $C = (C(T_1), ..., C(T_n))$, $t_0 \leq T_1 < \cdots < T_n = T_B \leq T^*$, and let us consider the following trading strategy: at time $t \in [t_0, T_1]$, with t_0 assumed to be the last coupon date, we buy a bond at a dirty price of

$$Bond(t, T_B, C) = Bond_{clean}(t_0, t, T_B, C) + AI(t_0, t, C).$$

We finance this amount of money with a loan at a linear interest rate $R_L(t,T)$, i.e. discounting is done using the linear expression $1 + R_L(t,T) \cdot (T-t)_{DC(R_L)}$, until the maturity date T of the future where $DC(R_L)$ denotes the daycount convention used[8] for R_L. At the same time we sell at no cost a futures contract with maturity date T at a futures (clean) price of $F(t,T,T_B,C)$. At maturity this future will supply us with a cash amount of money of

$$F(t,T,T_B,C) \cdot Conv(T_B,C) + AI(t_0,T,C),$$

where $Conv(T_B,C)$ denotes the conversion factor of the bond considered and it is assumed that there will be no further coupon payments between t_0 and T. Hence, at time T, we have to deliver the bond at a clean price of $F(t,T,T_B,C)$ and use the amount of money we get to pay back the loan. Assuming no arbitrage possibilities, this procedure should leave us with nothing in hand. The total gain must therefore be equal to zero, i.e.[9]

$$F(t,T,T_B,C) \cdot Conv(T_B,C) + AI(t_0,T,C)$$
$$-Bond(t,T_B,C) \cdot \left[1 + R_L(t,T) \cdot (T-t)_{DC(R_L)}\right] = 0$$

or equivalently[10]

---

[8] If the linear interest rate is a money market rate as the EURIBOR (LIBOR), it is usually quoted on an *act*/360 basis (see Section 5.1 or Fabozzi, Fabozzi, and Pollack [1991], p. 1201).

[9] If $t = T$ the left side of the equation, multiplied by -1, simplyfies to the expression

$$Bond_{clean}(t,T_B,C) - F(t,T,T_B,C) \cdot Conv(T_B,C)$$

which is also called the *basis of the coupon bond* (T_B,C) at time t and with respect to the future with maturity time T.

[10] If there are coupon payments at times $T_1,...,T_k$ with $k := \max\{i \in \{1,...,n\} : T_i \leq T\}$ and a present value of

$$I(t,T) = \sum_{i=1}^{k} C(T_i) \cdot P(t,T_i),$$

the price of the future would be given by

$$F(t,T,T_B,C) = \frac{[Bond(t,T_B,C) - I(t,T)] \cdot \left[1 + R_L(t,T) \cdot (T-t)_{DC(R_L)}\right]}{Conv(T_B,C)}$$
$$- \frac{AI(T_k,T,C)}{Conv(T_B,C)}.$$

Note, that $Bond(t,T_B,C)$ is the dirty price of the bond (T_B,C) and not the quoted or clean price. As mentioned above, it can be derived from the quoted bond price $Bond_{clean}(t_0,t,T_B,C)$ by

$$Bond(t,T_B,C) = Bond_{clean}(t_0,t,T_B,C) + AI(t_0,t,C).$$

$$F\left(t, T, T_B, C\right) = \frac{Bond\left(t, T_B, C\right) \cdot \left[1 + R_L\left(t, T\right) \cdot \left(T - t\right)_{DC(R_L)}\right]}{Conv\left(T_B, C\right)}$$

$$- \frac{AI\left(t_0, T, C\right)}{Conv\left(T_B, C\right)}. \tag{5.4}$$

Unfortunately, there is not only one but a whole (finite) set of coupon bonds which may be used for delivery. We thus have to find that deliverable coupon bond which excludes any arbitrage opportunity with one of the deliverable bonds using the above trading strategy. If we denote this coupon bond by the time to maturity and cash flow vector (T_B^*, C^*), we must have

$$\begin{aligned} F\left(t, T\right) &= F\left(t, T, T_B^*, C^*\right) \\ &= \min_{\text{deliverable bonds } (T_B, C)} F\left(t, T, T_B, C\right). \end{aligned}$$

It should be noted that, unlike the forward price, the quoted *futures price* $F\left(t, T\right)$ is a clean price which does not include any accrued interest for the time-period $[t_0, t]$. The *invoice* or *market futures price* for the daily marking to market is based on the notional amount of the futures contract and therefore equal to

$$L \cdot F\left(t, T\right) = 100,000 \cdot F\left(t, T\right).$$

All gains and losses of the futures contract are marked to market on a daily basis by the futures exchange and credited or debited to the owner's cash or margin account.

The bond (T_B^*, C^*) is known as the *cheapest-to-deliver bond*. The strategy of buying a bond and selling a future at the same time is closely related to the so-called repurchase agreements or repos. A *repo* is an agreement where the owner of a financial instrument agrees to sell it to a counterparty and buy it back at a slightly higher price later, i.e. the counterparty is providing a loan with the financial instrument as an insurance. The interest earned from that deal is called the *repo rate*. Given the futures price, we can calculate the implied interest rate $R_{L,imp}\left(t, T, T_B, C\right)$ for the bond (T_B, C) by

$$R_{L,imp}\left(t, T, T_B, C\right) = \frac{1}{(T-t)_{DC(R_L)}} \cdot \left[\frac{F(t,T) \cdot Conv(T_B,C) + AI(t_0,T,C)}{Bond(t,T_B,C)} - 1\right]. \tag{5.5}$$

If $1 + R_L\left(t, T\right) \cdot \left(T - t\right) = P^{-1}\left(t, T\right)$, i.e. the interest for a loan equals that of an investment in the bond market, we simply get

$$F\left(t, T, T_B, C\right) = \frac{Forward\left(t, T, T_B, C\right) - AI\left(T_k, T, C\right)}{Conv\left(T_B, C\right)}.$$

This expression is the interest we earn by buying a coupon bond at time t and selling it via the futures at time T. The other way round, the market sells us a coupon bond at time t and agrees to buy it back at time T. Therefore, $R_{L,imp}(t, T, T_B, C)$ is called the *implied repo rate* for the bond (T_B, C) and is also denoted by $\text{Repo}_{imp}(t, T, T_B, C)$. Note that with this definition we get

$$\text{Repo}_{imp}(t, T, T_B, C) - R_L(t, T) = \frac{Conv(T_B, C)}{Bond(t, T_B, C) \cdot (T - t)_{DC(R_L)}} \cdot [F(t, T) - F(t, T, T_B, C)].$$

Hence, the repo rate for a specific bond is always lower than the rate $R_L(t, T)$. Equality holds for the cheapest-to-deliver bond, i.e.

$$\text{Repo}_{imp}(t, T, T_B^*, C^*) = R_L(t, T).$$

Therefore, searching for the chaepest-to-deliver bond is equal to searching for the deliverable bond with the highest repo rate, since this bond gives us the highest possible interest under all coupon bonds we could buy now and deliver via the futures.

Example. To evaluate the December Bobl future we are given the following market data on October 20, 2000, i.e. $t = 10/20/2000$, for the deliverable coupon bonds and a notional of 100 Euro each.

coupon	maturity (T, T_B)	dirty price	conv. factor
5.00 %	08/19/2005	100.5343	0.959996
6.50 %	10/14/2005	106.0418	1.020193
6.00 %	01/05/2006	108.4777	1.000000
6.00 %	02/16/2006	107.8842	0.999756

The maturity of the future is $T = 12/07/2000$ and the linear interest rate $R_L(t, T)$, quoted act/act, is 4.78115% with $(T - t)_{act/act} = 0.131507$. Using equations (5.4) and (5.5) with t_0 denoting the last coupon payment of the respective deliverable coupon bonds before time t, we get

t_0	$(T - t_0)_{act/act}$	$AI(t_0, T, C)$	$F(t, T, T_B, C)$	Repo_{imp}
08/19/00	0.301370	1.506849	103.812478	3.50312%
10/14/00	0.147945	0.961644	103.653816	4.65424%
01/05/00	0.920548	5.523288	*103.636469*	*4.78155%*
02/16/00	0.805479	4.832877	103.754965	3.94614%

Hence, the chepest-to-deliver bond is the 6% coupon bond with maturity 01/05/2006 leading to a price for the December Bobl future of 103.636. The market price for the December Bobl future on the same day was 103.61 and therefore a little lower than our theoretical price, as expected. ▢

5.5   Interest-Rate Swaps

An interest-rate swap is an agreement between two parties, called the counterparties, to regularly exchange interest-rate payments based on a notional or principal amount L which we set to one if nothing else is mentioned. The notional amount doesn't change hands but is only for the purpose of determining the size of the interest payments. The most common type of a swap is the so-called plain vanilla interest-rate swap sometimes also called *fixed-for-floating interest rate swap*. In this one party, the *fixed leg* of the swap, commits to make a set of payments at a predetermined fixed rate of interest, also called the fixed-rate coupon, at the fixed leg payment dates $\widetilde{T}_1, ..., \widetilde{T}_{N_{fix}}$. At the same time the other party, the *floating leg* of the swap, agrees to make a set of floating interest payments usually based on a floating rate index[11] or briefly floating index R_L, also called the variable-rate coupon, at the floating leg payment dates[12] $T_1, ..., T_N$. The exact floating rate payment may include a spread added or subtracted from the floating index known as the *floating spread*. Both, fixed and floating interest, start accruing on the swap's effective date or start date $T_0 = \widetilde{T}_0$ and cease accruing on the swaps maturity date $T_S = \max\left\{ T_N, \widetilde{T}_{N_{fix}} \right\}$. At the start date T_0 the floating index is fixed first as it is at each of the following payment dates $T_1, ..., T_{N-1}$ for the periods $[T_i, T_{i+1}]$, $i = 0, ..., N - 1$, and actual payments from the floating leg are made at the floating leg payment dates $T_1, ..., T_N$. The lifetime of the swap, i.e. $T_S - T_0$, is called its *tenor* and the interest payments may be exchanged many times a year during the tenor of the swap. The trade date t_0 is the date on which the counterparties commit to the swap. If $t_0 < T_0$, the swap starts at some future date T_0 (relative to t_0). These swaps are also called *deferred swaps* or *forward swaps*. In the sequel we show how the swap can be priced at any time $t \in \left[t_0, \min\left\{ T_1, \widetilde{T}_1 \right\} \right]$, i.e. between the trade date and the first payment date of the swap[13]. The swap is named after the fixed leg of the swap. If a party wants to enter a swap agreement where it pays fixed interest it is looking for a *payer swap*, if the party wants to pay on the basis of a floating rate and receive fixed interest payments it is looking for a *receiver swap*. Swap agreements are tailored for the OTC market and can therefore be written with very long maturities usually ranging from one year to over fifteen years. Because swaps include multiperiod interest payments they are

---

[11] Until 1999 this floating rate index usually was the LIBOR. Since 1999 this index is replaced by the EURIBOR.

[12] Usually, the floating leg pays quarterly or semi-annually while the fixed leg pays semi-annually or annually.

[13] Of course, the valuation could be done for any time $t \geq t_0$. We only have to adjust the coupon stream and the accrued interest of each leg according to the coupons which are paid before time t.

typical instruments for hedging multiperiod interest-rate risk. This is one of the reasons why the swap market has grown so rapidly since the 1970s. To determine the value of a swap we consider the fixed leg and the floating leg of the swap separately and assume that the interest-rate market is arbitrage-free and complete with equivalent martingale measure $\widetilde{Q}$.

5.5.1   Floating Leg and Floating Rate Notes

Let $R_L(t, T_i, T_{i+1})$ denote the floating rate index at time $t \in [t_0, T_i]$ for the interest period $[T_i, T_{i+1}]$, $i = 0, ..., N-1$. A floating interest payment at time T_{i+1} is based on the floating index fixing $R_L(T_i, T_i, T_{i+1})$ at time T_i, $i = 0, ..., N-1$, and a constant spread rate $S \in \mathbb{R}$. Using the change of numéraire Theorem 3.31, equation (4.24), and the fact that $(R_L(t, T_i, T_{i+1}))_{t \in [t_0, T_i]}$ is a martingale under the T_{i+1}−forward measure $Q^{T_{i+1}}$ (see Section 4.7.1 for more details), the value $V_{float}(t, T_i, T_{i+1}, S)$ at time $t \in [t_0, T_i]$ of this payment for a notional of $L = 1$ is given by

$$\frac{V_{float}(t, T_i, T_{i+1}, S)}{P(t, T_{i+1})} = E_{Q^{T_{i+1}}}[(R_L(T_i, T_i, T_{i+1}) + S) \cdot$$
$$\cdot (T_{i+1} - T_i)_{DC(FL)} | \mathcal{F}_t]$$
$$= (R_L(t, T_i, T_{i+1}) + S) \cdot (T_{i+1} - T_i)_{DC(FL)}$$
$$= \frac{P(t, T_i)}{P(t, T_{i+1})} - 1 + S \cdot (T_{i+1} - T_i)_{DC(FL)},$$

or equivalently,

$$V_{float}(t, T_i, T_{i+1}, S) = P(t, T_i) - P(t, T_{i+1}) + P(t, T_{i+1}) \cdot S(T_{i+1}),$$

where $S(T_{i+1}) := S \cdot (T_{i+1} - T_i)_{DC(FL)}$ denotes the spread payment at time T_{i+1}, $i = 0, ..., N-1$, and $DC(FL)$ stands for the daycount convention used for the floating leg of the swap[14]. Consequently, the value[15] $V_{float}(t, T_0, T_N, S)$ of the floating leg at time $t \in [t_0, T_0]$ is given by

$$V_{float}(t, T_0, T_N, S) = \sum_{i=0}^{N-1} V_{float}(t, T_i, T_{i+1}, S) \tag{5.6}$$
$$= P(t, T_0) - P(t, T_N) + \sum_{i=0}^{N-1} S(T_{i+1}) \cdot P(t, T_{i+1})$$
$$= P(t, T_0) - P(t, T_N) + S \cdot PVBP_{FL}(t, T_0, T_N),$$

---

[14] Usually, the EURIBOR (LIBOR) as a money market rate is quoted on an *act/360* basis (see Section 5.1 and, e.g., Hull [1997], p. 117 or Fabozzi, Fabozzi, and Pollack [1991], p. 1201).

[15] Note that, by using this notation, we consider the payment dates $T_1, ..., T_{N-1}$ to be fixed as it is for example if the payment dates are equidistant.

with

$$PVBP_{FL}(t, T_0, T_N) := \sum_{i=0}^{N-1} P(t, T_{i+1}) \cdot (T_{i+1} - T_i)_{DC(FL)}$$

denoting the so-called *present value of a basis point (PVBP)* or accrual factor of the floating leg, i.e. the value at time $t \in [t_0, T_1]$ of a periodic fixed-coupon payment of 1 at each of the floating payment dates $T_1, ..., T_N$. If $t \in [T_0, T_1]$ there is no stochastic floating index $R_L(t, T_0, T_1)$, since the rate for the time interval $[T_0, T_1]$ is fixed at time T_0 as $R_L(T_0, T_0, T_1)$ and remains constant in $[T_0, T_1]$. So the value $V_{float}(t, T_0, T_1, S)$ at time $t \in [T_0, T_1]$ of the payment at time T_1 is given by

$$\begin{aligned}
V_{float}(t, T_0, T_1, S) &= P(t, T_1) \cdot [R_L(T_0, T_0, T_1) + S] \cdot (T_1 - T_0)_{DC(FL)} \\
&\quad - P(t, T_1) \cdot \left(\frac{1}{P(T_0, T_1)} - 1\right) + P(t, T_1) \cdot S(T_1) \\
&= \frac{P(t, T_1)}{P(T_0, T_1)} - P(t, T_1) + P(t, T_1) \cdot S(T_1).
\end{aligned}$$

Consequently, the value $V_{float}(t, T_0, T_N, S)$ of the floating leg at time $t \in [T_0, T_1]$ can be derived by

$$V_{float}(t, T_0, T_N, S) = \frac{P(t, T_1)}{P(T_0, T_1)} - P(t, T_N) + \sum_{i=0}^{N-1} S(T_{i+1}) \cdot P(t, T_{i+1}).$$

$$(5.7)$$

Combining equations (5.6) and (5.7), the *value of the floating leg* at time $t \in [t_0, T_1]$ is given by

$$V_{float}(t, T_0, T_N, S) = \widehat{P}(t, T_0, T_1) - P(t, T_N) + S \cdot PVBP_{FL}(t, T_0, T_N),$$

$$(5.8)$$

with

$$\widehat{P}(t, T_0, T_1) := \begin{cases} P(t, T_0) & \text{if } t \in [t_0, T_0] \\ \frac{P(t, T_1)}{P(T_0, T_1)}, & \text{if } t \in [T_0, T_1]. \end{cases}$$

For $t \in [T_0, T_1]$ there is an accrued interest of

$$\begin{aligned}
AI_{float}(T_0, t, S) &= [R_L(T_0, T_0, T_1) + S] \cdot \frac{(t - T_0)_{DC(FL)}}{(T_1 - T_0)_{DC(FL)}} \\
&= \left(\frac{1}{P(T_0, T_1)} - 1 + S\right) \cdot \frac{(t - T_0)_{DC(FL)}}{(T_1 - T_0)_{DC(FL)}}.
\end{aligned}$$

Since there is no accrued interest for $t \in [t_0, T_0]$, the *accrued interest* at time $t \in [t_0, T_1]$ is given by

$$AI_{float}(T_0, t, S) = \left(\frac{1}{P(T_0, T_1)} - 1 + S\right) \cdot \frac{\max\left\{(t - T_0)_{DC(FL)}, 0\right\}}{(T_1 - T_0)_{DC(FL)}}.$$

$$(5.9)$$

For an arbitrary L the correponding values $V_{float}^L (t, T_0, T_N, S)$ and $AI_{float}^L (T_0, t, S)$, $t \in [t_0, T_1]$, can be easily calculated by

$$V_{float}^L (t, T_0, T_N, S) = L \cdot V_{float} (t, T_0, T_N, S)$$

and

$$AI_{float}^L (T_0, t, S) = L \cdot AI_{float} (T_0, t, S).$$

For a Floating Rate Note or briefly FRN with start date T_0, payment dates $T_1, ..., T_N$, and a spread S on the floating index, the notional or principal amount L is paid back at the maturity time T_N of the FRN. Hence, using equation (5.8), the price $FRN (t, T_0, T_N, S)$ of the *Floating Rate Note* at time $t \in [t_0, T_1]$ for $L = 1$ is given by

$$
\begin{aligned}
FRN (t, T_0, T_N, S) &= V_{float} (t, T_0, T_N, S) + P (t, T_N) \\
&= \widehat{P} (t, T_0, T_1) + S \cdot PVBP_{FL} (t, T_0, T_N),
\end{aligned}
$$

with a market value of

$$FRN^L (t, T_0, T_N, S) = L \cdot FRN (t, T_0, T_N, S).$$

According to the coupon-bond systematic $FRN (t, T_0, T_N, S)$ may be called the dirty price of the FRN and is usually not the price quoted at the exchange. The corresponding quoted price or clean price $FRN_{clean} (t, T_0, T_N, S)$ is given by

$$FRN_{clean} (t, T_0, T_N, S) = FRN (t, T_0, T_N, S) - AI_{float} (T_0, t, S)$$

with $t \in [t_0, T_1]$.

5.5.2 Fixed Leg

The fixed-coupon payments are made at times $\widetilde{T}_i$, $i = 1, ..., N_{fix}$. Given a fixed-rate coupon C, the fixed-coupon payment at time $\widetilde{T}_{i+1}$ is equal to $C \left(\widetilde{T}_{i+1} \right) := C \cdot \left(\widetilde{T}_{i+1} - \widetilde{T}_i \right)_{DC(fix)}$ with $DC (fix)$ denoting the daycount convention of the fixed leg[16]. Hence, the present value $V_{fix} \left(t, \widetilde{T}_i, \widetilde{T}_{i+1} \right)$ at time $t \in \left[t_0, \widetilde{T}_i \right]$ of this payment is given by

$$V_{fix} \left(t, \widetilde{T}_i, \widetilde{T}_{i+1}, C \right) = C \left(\widetilde{T}_{i+1} \right) \cdot P \left(t, \widetilde{T}_{i+1} \right),$$

---

[16] Note that this notation is used since the daycount convention of the fixed and floating side is different. Usually, the floating index is quoted in *act/360* notation. Although the market daycount convention for the fixed-rate payments is *30E/360* or sometime *30/360*, other conventions may appear. For example an *act/act* basis is used for the Treasury note rate (see Section 5.1 and, e.g., Hull [1997], p. 117 or Fabozzi, Fabozzi, and Pollack [1991], p. 1196-1205).

$i = 0, ..., N_{fix} - 1$. Consequently, the *value of the fixed leg* at time $t \in \left[t_0, \widetilde{T}_1\right]$, denoted by $V_{fix}\left(t, \widetilde{T}_0, \widetilde{T}_{N_{fix}}, C\right)$, can be calculated by

$$
\begin{aligned}
V_{fix}\left(t, T_0, \widetilde{T}_{N_{fix}}, C\right) &= \sum_{i=0}^{N_{fix}-1} V_{fix}\left(t, \widetilde{T}_i, \widetilde{T}_{i+1}, C\right) \\
&= \sum_{i=0}^{N_{fix}-1} C\left(\widetilde{T}_{i+1}\right) \cdot P\left(t, \widetilde{T}_{i+1}\right) \\
&= C \cdot PVBP\left(t, \widetilde{T}_{N_{fix}}\right),
\end{aligned}
\tag{5.10}
$$

with

$$
PVBP\left(t, T_0, \widetilde{T}_{N_{fix}}\right) := \sum_{i=0}^{N_{fix}-1} P\left(t, \widetilde{T}_{i+1}\right) \cdot \left(\widetilde{T}_{i+1} - \widetilde{T}_i\right)_{DC(fix)},
$$

as above, denoting the so-called *present value of a basis point (PVBP)* or *accrual factor* of the fixed leg, i.e. the value at time $t \in \left[t_0, \widetilde{T}_1\right]$ of a periodic fixed-coupon payment of 1 at each of the payment dates $\widetilde{T}_1, ..., \widetilde{T}_{N_{fix}}$. Note that the value of the fixed leg is closely related to the value at time $t \in \left[t_0, \widetilde{T}_1\right]$ of a coupon bond $\left(\widetilde{T}_{N_{fix}}, C'\right)$ with coupon payments $C' = \left(C\left(\widetilde{T}_1\right), ..., C\left(\widetilde{T}_{N_{fix}-1}\right), C\left(\widetilde{T}_{N_{fix}}\right) + 1\right)$, where we considered that the coupon bond pays back a notional of $L = 1$ at maturity time $\widetilde{T}_{N_{fix}}$ while the fixed leg doesn't. The exact equation is

$$
V_{fix}\left(t, T_0, \widetilde{T}_{N_{fix}}, C\right) = Bond\left(t, \widetilde{T}_{N_{fix}}, C'\right) - P\left(t, \widetilde{T}_{N_{fix}}\right), \ t \in \left[t_0, \widetilde{T}_1\right].
$$

The accrued interest for $t \in \left[t_0, \widetilde{T}_1\right]$ is given by

$$
AI_{fix}\left(T_0, t, C\right) = C\left(\widetilde{T}_1\right) \cdot \frac{\max\left\{(t - T_0)_{DC(fix)}, 0\right\}}{\left(\widetilde{T}_1 - T_0\right)_{DC(fix)}},
\tag{5.11}
$$

where we considered that there is no accrued interest for $t \in [t_0, T_0]$. As above, the correponding values $V_{fix}^L\left(t, T_0, \widetilde{T}_{N_{fix}}, C\right)$ and $AI_{fix}^L\left(T_0, t, C\right)$ for an arbitrary L at time $t \in \left[t_0, \widetilde{T}_1\right]$ can be easily calculated by

$$
V_{fix}^L\left(t, T_0, \widetilde{T}_{N_{fix}}, C\right) = L \cdot V_{fix}\left(t, T_0, \widetilde{T}_{N_{fix}}, C\right)
$$

and

$$
AI_{fix}^L\left(T_0, t, C\right) = L \cdot AI_{fix}\left(T_0, t, C\right).
$$

According to the coupon-bond systematic $V_{fix}\left(t, T_0, \widetilde{T}_{N_{fix}}, C\right)$ may be called the dirty price of the fixed leg. The corresponding quoted price or clean price of the fixed leg for $t \in \left[t_0, \widetilde{T}_1\right]$ is given by

$$
V_{fix}\left(t, T_0, \widetilde{T}_{N_{fix}}, C\right) - AI_{fix}\left(T_0, t, C\right) = Bond_{clean}\left(t, T_0, \widetilde{T}_{N_{fix}}, C'\right)
$$
$$
- P\left(t, \widetilde{T}_{N_{fix}}\right).
$$

5.5.3 Pricing Interest-Rate Swaps

We now combine equations (5.8) and (5.10) of the previous two sections to derive the value $PSwap\left(t, T_0, T_S, C, S\right)$ of a payer swap at time $t \in \left[t_0, \min\left\{T_1, \widetilde{T}_1\right\}\right]$, starting at time T_0 and maturing at time T_S. Again, we suppose that $L = 1$ getting

$$
\begin{aligned}
PSwap\left(t, T_0, T_S, C, S\right) &= V_{float}\left(t, T_0, T_N, S\right) - V_{fix}\left(t, T_0, \widetilde{T}_{N_{fix}}, C\right) \\
&= \widehat{P}\left(t, T_0, T_1\right) - P\left(t, T_N\right) \\
&\quad - C \cdot PVBP\left(t, T_0, \widetilde{T}_{N_{fix}}\right) \\
&\quad + S \cdot PVBP_{FL}\left(t, T_0, T_N\right) \\
&= FRN\left(t, T_0, T_N, S\right) - Bond\left(t, \widetilde{T}_{N_{fix}}, C'\right) \\
&\quad + P\left(t, \widetilde{T}_{N_{fix}}\right) - P\left(t, T_N\right). \quad (5.12)
\end{aligned}
$$

Hence, a coupon bond can be hedged by buying a FRN and selling a payer swap or equivalently, by buying a FRN and a receiver swap[17] where the FRN as well as both legs of the swap have the same maturity time as the bond. The corresponding clean price $PSwap_{clean}\left(t, T_0, T_S, C, S\right)$ can be derived for all $t \in \left[t_0, \min\left\{T_1, \widetilde{T}_1\right\}\right]$ by

$$
\begin{aligned}
PSwap_{clean}\left(t, T_0, T_S, C, S\right) &= PSwap\left(t, T_0, T_S, C, S\right) \\
&\quad - AI_{PSwap}\left(T_0, t, C, S\right) \\
&= FRN_{clean}\left(t, T_0, T_N, S\right) \\
&\quad - Bond_{clean}\left(t, T_0, \widetilde{T}_{N_{fix}}, C'\right) \\
&\quad + P\left(t, \widetilde{T}_{N_{fix}}\right) - P\left(t, T_N\right),
\end{aligned}
$$

---

[17] It should be noted that we hereby assume that the zero bond prices for the bond market and the swap market are the same as well as their movement over time. For many reasons, e.g. different credit risk and liquidity of the markets, this is usually not true in reality.

with

$$AI_{PSwap}(T_0, t, C, S) := AI_{float}(T_0, t, S) - AI_{fix}(T_0, t, C). \tag{5.13}$$

Note that for $t \in [t_0, T_0]$ there is no accrued interest, and so we have

$$PSwap(t, T_0, T_S, C, S) = PSwap_{clean}(t, T_0, T_S, C, S).$$

The value $RSwap(t, T_0, T_S, C, S)$ as well as the clean price or premium of a receiver swap at time $t \in \left[t_0, \min\left\{T_1, \tilde{T}_1\right\}\right]$, starting at time T_0 and maturing at time T_S, can be derived by

$$RSwap(t, T_0, T_S, C, S) = -PSwap(t, T_0, T_S, C, S). \tag{5.14}$$

Swaps are usually quoted for $t \in [t_0, T_0]$, $T_S = T_N = \tilde{T}_{N_{fix}}$, and in the notation of that fixed-coupon rate $y(t, T_0, T_N, S)$ which sets the value of the corresponding swap to zero. Hence, $y(t, T_0, T_N, S)$ is implicitly defined by the equation

$$PSwap(t, T_0, T_N, y(t, T_0, T_N, S), S) = 0. \tag{5.15}$$

This particular fixed-coupon rate is called the (T_0, T_N)−par swap rate at a floating spread of S. Combining equations (5.12) and (5.15) we get for $t \in [t_0, T_0]$

$$
\begin{aligned}
y(t, T_0, T_N, S) &= \frac{P(t, T_0) - P(t, T_N) + S \cdot PVBP_{FL}(t, T_0, T_N)}{PVBP(t, T_0, T_N)} \\
&= y(t, T_0, T_N, 0) + S \cdot \frac{PVBP_{FL}(t, T_0, T_N)}{PVBP(t, T_0, T_N)}.
\end{aligned}
$$

If the payment dates on the fixed and floating leg coincide, we therefore simply get

$$y(t, T_0, T_N, S) = y(t, T_0, T_N) + S.$$

The market convention for standard swaps is to quote the fixed-rate coupon of the swap as an "all-in-cost" rate versus the floating index flat, i.e. with no floating spread added or subtracted from the floating index. To be more precise, the market usually quotes the (T_0, T_N)−par swap rates at a spread of $S = 0$, for a fixed T_0, and a set of standard payment dates of the fixed leg[18]. The corresponding fixed-coupon rates

$$y\left(t, T_0, \tilde{T}_k\right) := y\left(t, T_0, \tilde{T}_k, 0\right)$$

are simply called the $\left(T_0, \tilde{T}_k\right)$−par swap rates.

---

[18] Standard swaps rates are usually quoted for yearly maturities from one to ten years. However, extensions of this range are possible.

5.5.4   The Bootstrap Method

As we have already mentioned, the market quotes on swaps are par swap rates rather than zero rates. An important question therefore is how the zero rates can be extracted from the quoted par swap rates. This transformation is usually done using the so-called bootstrap method. In practice, the zero-rate curve may be completed for short-term maturities using the futures strip rates extracted from the Eurodollar or EURIBOR futures prices as described in Section 5.4.1. Given the zero-coupon bond price $P(t, T_0)$ and the $\left(T_0, \tilde{T}_k\right)$−par swap rates for all $k \in \{1, ..., N_{fix}\}$ with $\left\{\tilde{T}_1, ..., \tilde{T}_{N_{fix}} = T_N\right\} \subseteq \{T_1, ..., T_N\}$ we can derive the (T_0, T_N)−swap discount curve using equation (5.15). Since

$$
\begin{aligned}
P(t, T_0) &= P\left(t, \tilde{T}_k\right) + y\left(t, T_0, \tilde{T}_k\right) \cdot PVBP\left(t, T_0, \tilde{T}_k\right) \\
&= P\left(t, \tilde{T}_k\right) \\
&\quad + y\left(t, T_0, \tilde{T}_k\right) \cdot \sum_{i=0}^{k-1} P\left(t, \tilde{T}_{i+1}\right) \cdot \left(\tilde{T}_{i+1} - \tilde{T}_i\right)_{DC(fix)}
\end{aligned}
$$

we know that

$$
\begin{aligned}
P\left(t, \tilde{T}_k\right) &= \frac{P(t, T_0) - y\left(t, T_0, \tilde{T}_k\right) \cdot \sum_{i=0}^{k-2} P\left(t, \tilde{T}_{i+1}\right) \cdot \left(\tilde{T}_{i+1} - \tilde{T}_i\right)_{DC(fix)}}{1 + y\left(t, T_0, \tilde{T}_k\right) \cdot \left(\tilde{T}_k - \tilde{T}_{k-1}\right)_{DC(fix)}} \\
&= \frac{P(t, T_0) - y\left(t, T_0, \tilde{T}_k\right) \cdot PVBP\left(t, T_0, \tilde{T}_{k-1}\right)}{1 + y\left(t, T_0, \tilde{T}_k\right) \cdot \left(\tilde{T}_k - \tilde{T}_{k-1}\right)_{DC(fix)}}, \quad (5.16)
\end{aligned}
$$

for all $k \in \{1, ..., N_{fix}\}$. Starting with

$$
P\left(t, \tilde{T}_1\right) = \frac{P(t, T_0)}{1 + y\left(t, T_0, \tilde{T}_1\right) \cdot \left(\tilde{T}_1 - T_0\right)_{DC(fix)}}
$$

for $k = 1$ we can now successively derive all zero-coupon bond prices $P\left(t, \tilde{T}_{k+1}\right)$ using $P(t, T_0), ..., P\left(t, \tilde{T}_k\right)$ and equation (5.16) for $k = 2, ..., N_{fix} - 1$. This procedure is usually called the *bootstrap method*. Furthermore, we know by definition, setting $t = T_0$ and $C' := \left(C\left(\tilde{T}_1\right), ..., C\left(\tilde{T}_k\right) + 1\right)$ with

$$
C\left(\tilde{T}_i\right) := y\left(T_0, T_0, \tilde{T}_k\right) \cdot \left(\tilde{T}_i - \tilde{T}_{i-1}\right)_{DC(fix)}, i = 1, ..., k,
$$

that

$$Bond\left(T_0, \widetilde{T}_k, C'\right) := \sum_{i=1}^{k} P\left(T_0, \widetilde{T}_i\right) \cdot C\left(\widetilde{T}_i\right) = 1$$

for all $k \in \{1, ..., N_{fix}\}$. The coupon bonds $\left(\widetilde{T}_k, C'\right)$ are usually called the $\left(T_0, \widetilde{T}_k\right)$ −*par yield bonds*. It is interesting to note that, given the market quote for the (T_0, T_N) −par swap rate $y(t, T_0, T_N)$, the value at time $t \in [t_0, T_0]$ of the correponding payer swap with fixed-coupon rate C can be easliy derived combining equations (5.12) and (5.15) by

$$
\begin{aligned}
PSwap(t, T_0, T_N, C, S) &= (y(t, T_0, T_N, S) - C) \cdot PVBP(t, T_0, T_N) \\
&= (y(t, T_0, T_N) - C) \cdot PVBP(t, T_0, T_N) \\
&\quad + S \cdot PVBP_{FL}(t, T_0, T_N). \qquad (5.17)
\end{aligned}
$$

The correponding formula for the receiver swap is immediate from equation (5.14).

Example. On November 21, 2000 we are given the yearly par swap rates $y(0, 0, k)$ with maturities $k \in \{k = 1, ..., 9\}$. For $t = T_0 = 0$ we are to calculate the yearly discount factors $P(0, k)$ as well as the yearly zero rates $R(0, k)$ and $R_d(0, k)$ for continuous and discrete compounding, i.e.

$$P(0, k) = e^{-R(0,k) \cdot k} = (1 + R_d(0, k))^{-k}, \ k = 0, ..., 9.$$

Following are the results (rates in %) of the bootstrap method:

k	$y(0,0,k)$	$PVBP(0,0,k)$	$P(0,k)$	$R(0,k)$	$R_d(0,k)$
0	–	0.0000	1.0000	–	–
1	5.2700	0.9499	0.9499	5.1358	5.2700
2	5.3300	1.8512	0.9013	5.1943	5.3316
3	5.4040	2.7050	0.8538	5.2679	5.4092
4	5.4800	3.5125	0.8075	5.3450	5.4904
5	5.5600	4.2748	0.7623	5.4279	5.5779
6	5.6325	4.9935	0.7187	5.5044	5.6587
7	5.6950	5.6706	0.6771	5.5715	5.7230
8	5.7425	6.3083	0.6377	5.6228	5.7839
9	5.7875	6.9085	0.6002	5.6728	5.8368

Using equation (5.17) we can now price a 5 year fixed-for-floating payer swap paying annually a fixed-coupon of 5% and receiving EURIBOR semi-annually with a spread of 0 by

$$
\begin{aligned}
PSwap(0, 0, 5, 5\%, 0) &= (y(0, 0, 5) - 5) \cdot PVBP(0, 0, 5) \\
&= (5.5600\% - 5\%) \cdot 4.2748 = 0.0239.
\end{aligned}
$$

□

5.5.5   Other Interest-Rate Swaps

In the previous sections we have concentrated on plain vanilla interest-rate swaps. However, there are many different types of swaps available in the market. We use this section to give a brief overview of the most common types of interest-rate swaps. Instead of a floating and a fixed leg, a swap may also be constructed using two floating legs. In this case the two counterparties exchange variable-rate coupons which may be based, beside LIBOR or EURIBOR, on a commercial paper rate or on the Treasury bill rate to give two examples. These swaps are usually called *floating-for-floating interest-rate swaps* and are mainly used to manage the risk of assets and liabilities when both have an exposure to floating interest rates. If the variable-rate coupons are in a domestic and a foreign currency with both floating rates being applied to the same domestic notional amount, the resulting agreement is known as *differential swap* or briefly *diff swap*.

Another class of interest-rate swaps is designed using a principal or notional amount $L(t)$, $t \in [T_0, T_S]$, which changes its size during the lifetime $[T_0, T_S]$ of the swap. If a risk manager has to consider an amortizing schedule of a loan it might be a good idea to enter into a so-called *amortizing swap* in which the notional amount is reduced over time in a predetermined way. In a *set-up swap*, on the other hand, the notional amount increases over time in a predetermined way. There are also swap agreements where the principal amount is reduced depending on the level of a specific interest rate known as *index amortizing rate swaps* or *indexed principal swaps*. Usually, the lower the interest the greater is the reduction. There are also swaps with variable tenor. In an *extendable swap*, one party has the option to extend the life of the swap. If one party has the right to terminate the swap early, we are talking about a *puttable swap*.

If the fixed-rate coupon is based on an index which doesn't change its time to maturity over time, we talk about constant maturity interest-rate swaps. The most commonly used instrument of this category is an agreement to exchange a specific LIBOR or EURIBOR for a particular swap par rate, for example a 6-month EURIBOR for a 10-year par swap rate every six months for a tenor of five years. These swaps are called *constant maturity swaps* or briefly *CMS swaps*[19]. If a swap is designed to exchange a specific LIBOR or EURIBOR for a particular Treasury rate, we call this agreement a *constant maturity Treasury swap* or briefly *CMT swap*.

---

[19] To be consistent, this swap should be called constant maturity swap swap where the first part of the name, "constant maturity swap", stands for the index rate, i.e. the "constant maturity par swap rate", and the second part indicates the financial agreement, i.e. a "swap".

5.6   Interest-Rate Options

A call (put) option is a contract which gives the owner or buyer of the option the right but not the obligation to buy (sell) an underlying financial instrument from the seller or writer of the option at a specified point in time, called the maturity time or expiration date T, and at a specified price, called the exercise or strike price X. This type of option is called European to distinguish it from the American options for which the expiration date is extended to a specified period of time where the option may be exercised. Within this chapter we concentrate our interest on European options. The buyer of an option is said to be long the option. The seller, also called the writer of the option, on the other hand is said to be short the option. Options may be either over-the-counter, most of the time because they are tailor-made for the clients, or exchange traded and therefore standardized. For the right to exercise the option the buyer has to pay a certain amount of money called the option premium or simply option price. The value of the option at time $t \in [t_0, T^*]$ depends on the value or price $P(t)$ of the underlying financial instrument. At maturity the value of the call option is $\max \{P(t) - X, 0\}$ and the value of the put option is $\max \{X - P(t), 0\}$ where the maximum operator indicates the right to exercise which the owner may let go to waste. Interest-rate options are options with a payoff depending on the level of interest rates either directly or indirectly via some other interest related instrument such as a coupon bond or a swap. In the last two decades the trading volume in interest-rate options increased rapidly with many new products created for the specific needs of the risk or portfolio managers. This chapter is dedicated to give an overview of the definition and pricing of the most popular interest-rate options.

5.6.1   Zero-Coupon Bond Options

One of the simplest interest-rate derivatives is the so-called (European) zero-coupon bond option. The call option gives the holder the right to buy, the put option gives the holder the right to sell the underlying zero-coupon bond by a certain date T at an exercise or strike price X. It is supposed that the zero-coupon bond and the strike price are based on the same notional amount which we will, for ease of simplicity, set to 1. At the beginning of the contract the seller of the option receives an amount of money for selling the option which is the market price of the option. At maturity the holder of a call option will get an amount of money equal to $\max \{P(T, T') - X, 0\}$ where $P(T, T')$ is the price of a zero-coupon bond with time to maturity $T' - T$ at time T. In cases where we base our calculations on a specific short-rate model we will show this by writing $P(r, T, T')$ instead of $P(T, T')$. The holder of the option will exercise the option if $P(T, T') > X$. The call option will not be exercised if $P(T, T') \leq$

X. Correspondingly, the holder of a put option will get an amount of money equal to $\max\{X - P(T,T'),0\}$ at maturity. We will denote the price at time t of a call option with maturity T on a zero-coupon bond with maturity time T' and strike price X by $Call(t,T,T',X)$ or $Call(r,t,T,T',X)$ if we consider a specific short-rate model. Similarly, the correponding put option is denoted by $Put(t,T,T',X)$ or $Put(r,t,T,T',X)$ respectively. The theoretical prices of zero-coupon bond options pretty much depend on the interest-rate model we use. The following lemma gives the prices for these options within the Hull-White model.

Lemma 5.1 (Zero-Coupon Bond Options) *Let $t_0 \leq t \leq T \leq T' \leq T^*$ and let $\mathcal{N}(d)$ denote the value of the cumulative standard normal distribution at $d \in \mathbb{R}$. Then the price at time t of a call option with maturity T and strike price X on a zero-coupon bond with maturity time T' within the Hull-White model is given by*

$$Call(r,t,T,T',X) = P(r,t,T') \cdot \mathcal{N}(d_1) - X \cdot P(r,t,T) \cdot \mathcal{N}(d_2)$$

with

$$d_1 = \frac{\ln\left(\frac{P(r,t,T')}{P(r,t,T)\cdot X}\right) + \frac{1}{2}\cdot s_P^2(t,T,T')}{s_P(t,T,T')}, \quad d_2 = d_1 - s_P(t,T,T'),$$

and $s_P(t,T,T')$ given as in Lemma 4.25. The value of the corresponding put option is given by

$$\begin{aligned} Put(r,t,T,T',X) &= X \cdot P(r,t,T) \cdot \mathcal{N}(-d_2) - P(r,t,T') \cdot \mathcal{N}(-d_1) \\ &= Call(r,t,T,T',X) + X \cdot P(r,t,T) - P(r,t,T'). \end{aligned}$$

The last equation is also called put-call parity.

Proof. Let $D = D(r,T) = \max\{P(r,T,T') - X, 0\}$ with $t_0 \leq t \leq T \leq T' \leq T^*$ and $r \in \mathbb{R}$. Then, using the Feynman-Kac representation, we get for all $(r,t) \in \mathbb{R} \times [t_0,T]$ the price of the call option by

$$\begin{aligned} Call(r,t,T,T',X) &= \int_{-\infty}^{\infty} G(r,t,r',T) \cdot \max\{P(r',T,T') - X,0\}\, dr' \\ &= I_1 - I_2 \end{aligned}$$

with

$$I_1 := \int_{-\infty}^{r^*} G(r,t,r',T) \cdot P(r',T,T')\, dr', \; I_2 := \int_{-\infty}^{r^*} G(r,t,r',T) \cdot X\, dr'$$

and

$$r^* = \frac{1}{B(T,T')} \cdot \ln\left(\frac{A(T,T')}{X}\right) \tag{5.18}$$

$$= \frac{1}{B(T,T')} \cdot \ln\left(\frac{P(r,t,T')}{P(r,t,T) \cdot X}\right) + f(t,T)$$

$$- \frac{1}{2} \cdot B(T,T') \cdot s_r^2(t,T).$$

Hence, using Lemma 4.27,

$$I_1 = \int_{-\infty}^{r^*} G(r,t,r',T) \cdot P(r,T,T')\, dr'$$

$$= P(r,t,T) \cdot A(T,T') \cdot$$

$$\cdot \int_{-\infty}^{r^*} \frac{1}{s_r(t,T) \cdot \sqrt{2\pi}} \cdot e^{-\frac{1}{2s_r^2(t,T)} \cdot (r' - f(t,T))^2} \cdot e^{-r' \cdot B(T,T')}\, dr'$$

$$= P(r,t,T) \cdot \frac{P(r,t,T')}{P(r,t,T)} \cdot e^{B(T,T') \cdot f(t,T) - \frac{1}{2} \cdot B^2(T,T') \cdot s_r^2(t,T,T')}$$

$$\cdot \int_{-\infty}^{r^*} \frac{1}{s_r(t,T) \cdot \sqrt{2\pi}} \cdot e^{-\frac{1}{2s_r^2(t,T)} \cdot (r' - f(t,T))^2} \cdot e^{-r' \cdot B(T,T')}\, dr'$$

$$= P(r,t,T') \cdot$$

$$\cdot \int_{-\infty}^{r^*} \frac{1}{s_r(t,T) \cdot \sqrt{2\pi}} \cdot e^{-\frac{1}{2s_r^2(t,T)} \cdot (r' - [f(t,T) - B(T,T') \cdot s_r^2(t,T)])^2}\, dr'$$

$$= P(r,t,T') \cdot \mathcal{N}(d_1)$$

with (using (5.18))

$$d_1 = \frac{r^* - [f(t,T) - B(T,T') \cdot s_r^2(t,T)]}{s_r(t,T)}$$

$$= \frac{\ln\left(\frac{P(r,t,T')}{P(r,t,T) \cdot X}\right) + \frac{1}{2} \cdot B^2(T,T') \cdot s_r^2(t,T)}{B(T,T') \cdot s_r(t,T)}$$

$$= \frac{\ln\left(\frac{P(r,t,T')}{P(r,t,T) \cdot X}\right) + \frac{1}{2} \cdot s_P^2(t,T,T')}{s_P(t,T,T')}.$$

On the other hand,

$$I_2 = \int_{-\infty}^{r^*} G(r,t,r',T) \cdot X\, dr' = X \cdot \int_{-\infty}^{r^*} G(r,t,r',T)\, dr'$$

$$= X \cdot P(r,t,T) \int_{-\infty}^{r^*} \frac{1}{s_r(t,T) \cdot \sqrt{2\pi}} \cdot e^{-\frac{1}{2s_r^2(t,T)} \cdot (r' - f(t,T))^2}\, dr'$$

$$= X \cdot P(r,t,T) \cdot \mathcal{N}(d_2)$$

with (using (5.18))

$$d_2 = \frac{r^* - f(t,T)}{s_r(t,T)}$$

$$= \frac{\ln\left(\frac{P(r,t,T')}{P(r,t,T)\cdot X}\right) - \frac{1}{2}\cdot s_P^2(t,T,T')}{s_P(t,T,T')}$$

$$= d_1 - s_P(t,T,T').$$

Analogously, we can derive the price of the put option. The put-call parity is straightforward using the well-known equation $\mathcal{N}(-d) = 1 - \mathcal{N}(d)$ for all $d \in I\!R$. $\qquad\qquad\qquad\qquad\qquad\qquad\qquad\qquad\qquad\qquad\square$

5.6.2   Caps and Floors

A very popular interest-rate derivative is the so-called interest-rate cap or briefly cap. It is designed to give the holder of the cap a protection against rising interest rates. To be more precise, a cap guarantees that the interest to be payed on say a floating-rate loan never exceeds a certain predetermined *cap rate* R_X. It does this by ensuring the payment of a certain amount of cash if the agreed floating interest rate R_L exceeds the cap rate. This payment depends on the notional or principal amount of the loan L and the time steps $T_1, ..., T_{K-1}$, $K \in I\!N$, at which the payments are fixed. The resulting time-periods of length $\Delta T_k := (T_{k+1} - T_k)_{DC(R_L)}$, $k = 1, ..., K-1$, are called roll-over periods with $DC(R_L)$ denoting the daycount convention[20] used for the floating interest rate R_L. Hence, at each time step T_k, $k \in \{1, ..., K-1\}$ with T_K denoting the maturity of the cap, the floating rate $R_L(T_k, T_{k+1})$ for the following roll-over period is compared to the cap rate R_X. If the floating rate is lower than the cap rate no payment has to be made, if the floating rate is higher than the cap rate there is a payment of

$$L \cdot \Delta T_k \cdot (R_L(T_k, T_{k+1}) - R_X)$$

to be made at the end of the period, i.e. at time T_{k+1}. Note that the floating rate and the cap rate are supposed to be quoted as linear rates, i.e. discounting is done using the linear expressions $1 + R_L(T_k, T_{k+1}) \cdot \Delta T_k$ or $1 + R_X \cdot \Delta T_k$, respectively. Given the interest rate $R_L(T_k, T_{k+1})$ for the period $[T_k, T_{k+1}]$ the total pay-off at time T_{k+1} is equivalent to a pay-off at time T_k of

$$\frac{L \cdot \Delta T_k}{1 + R_L(T_k, T_{k+1}) \cdot \Delta T_k} \cdot \max\{R_L(T_k, T_{k+1}) - R_X, 0\}.$$

---

[20] As already mentioned, EURIBOR (LIBOR) as a money market rate is usually quoted on an *act/360* basis.

This pay-off can be transformed to

$$L \cdot (1 + R_X \cdot \Delta T_k) \cdot \max \left\{ \frac{1}{1 + R_X \cdot \Delta T_k} - \frac{1}{1 + R_L (T_k, T_{k+1}) \cdot \Delta T_k}, 0 \right\}.$$
$$(5.19)$$

We can easily see that $P(T_k, T_{k+1}) = 1/(1 + R_L (T_k, T_{k+1}) \cdot \Delta T_k)$ is the price of a zero-coupon bond at time T_k maturing at time T_{k+1}. If we set $X := 1/(1 + R_X \cdot \Delta T_k)$, (5.19) transfers to

$$\frac{L}{X} \cdot \max \left\{ X - P(T_k, T_{k+1}), 0 \right\}$$

which is the terminal pay-off of a put option with maturity T_k on a zero-coupon bond with maturity time T_{k+1} and strike price X. The notional amount of that option is L/X. This single period element of a cap is called a *caplet* and can be evaluated within the Hull-White model, using Lemma 5.1, by

$$Caplet (r, t, T_k, T_{k+1}, R_X) = \frac{L}{X} \cdot Put (r, t, T_k, T_{k+1}, X)$$
$$= \frac{L}{X} \cdot (X \cdot P(r, t, T_k) \cdot \mathcal{N}(-d_2) - P(r, t, T_{k+1}) \cdot \mathcal{N}(-d_1))$$
$$= L \cdot (P(r, t, T_k) \cdot \mathcal{N}(-d_2) - (1 + R_X \cdot \Delta T_k) \cdot P(r, t, T_{k+1}) \cdot \mathcal{N}(-d_1)),$$

with

$$d_1 = \frac{\ln \left(\frac{(1 + R_X \cdot \Delta T_k) \cdot P(r, t, T_{k+1})}{P(r, t, T_k)} \right) + \frac{1}{2} \cdot s_P^2 (t, T_k, T_{k+1})}{s_P (t, T_k, T_{k+1})}$$

and

$$d_2 = d_1 - s_P (t, T_k, T_{k+1}).$$

Note that, using this notation, we assumed that our calculations are based on a short-rate model. As we have seen, a *cap* is nothing other than a series of caplets, which directly gives us its price as

$$Cap (r, t, T_K, R_X, \Delta T) = \sum_{k=1}^{K-1} Caplet (r, t, T_k, T_{k+1}, R_X)$$

with $\Delta T = (\Delta T_1, ..., \Delta T_{K-1})$.

An interest-rate floor or briefly floor is defined analogously to an interest-rate cap. The difference is that it sets a lower limit on a floating interest rate, i.e. a payment only has to be made if the floating rate is lower than the cap rate. Using the same notation as above, the payment that has to be made at the end of the roll-over period is given by

$$L \cdot \Delta T_k \cdot \max \left\{ R_X - R_L (T_k, T_{k+1}), 0 \right\}.$$

Applying the same procedure as for the caps this can be transformed into the pay-off

$$\frac{L}{X} \cdot \max \left\{ P\left(T_k, T_{k+1}\right) - X, 0 \right\},$$

which is the terminal pay-off of a zero-coupon bond call option. Hence, a floor is a portfolio of instruments called *floorlets* which are call options on zero-coupon bonds priced, within the Hull-White model, by

$$Floorlet\left(r, t, T_k, T_{k+1}, R_X\right) = \frac{L}{X} \cdot Call\left(r, t, T_k, T_{k+1}, X\right)$$
$$= \frac{L}{X} \cdot \left(P\left(r, t, T_{k+1}\right) \cdot \mathcal{N}\left(d_1\right) - X \cdot P\left(r, t, T_k\right) \cdot \mathcal{N}\left(d_2\right)\right)$$
$$= L \cdot \left(\left(1 + R_X \cdot \Delta T_k\right) \cdot P\left(r, t, T_{k+1}\right) \cdot \mathcal{N}\left(d_1\right) - P\left(r, t, T_k\right) \cdot \mathcal{N}\left(d_2\right)\right)$$

with d_1 and d_2 as defined above. The price of the *floor* is given by

$$Floor\left(r, t, T_K, R_X, \Delta T\right) = \sum_{k=1}^{K-1} Floorlet\left(r, t, T_k, T_{k+1}, R_X\right).$$

For the pricing of caps and floors in the framework of Black's model see Section 5.8. While a cap sets an upper limit and a floor a lower limit on a floating interest rate, a *collar* combines a long position in a cap and a short position in a floor. The price of this instrument is straightforward to calculate as the difference of the corresponding cap and floor prices. A collar is usually designed in a way so that the price of the cap equals the price of the floor leaving the cost for the collar at zero. In this case, the collar is also called *zero-cost collar*.

5.6.3   Coupon-Bond Options

Another important interest-rate derivative is an option on a coupon bond. As we have already seen, the price of a coupon bond with coupon payments $C = \left(C\left(T_1\right), ..., C\left(T_n\right)\right)$ and maturity time T_B, $t_0 \le T_1 < \cdots < T_n = T_B \le T^*$, at time $t \in [t_0, T_1]$ is given by

$$Bond\left(t, T_B, C\right) = \sum_{i=1}^{n} C\left(T_i\right) \cdot P\left(t, T_i\right),$$

or

$$Bond\left(r, t, T_B, C\right) = \sum_{i=1}^{n} C\left(T_i\right) \cdot P\left(r, t, T_i\right)$$

if we consider a specific short-rate model. The call option on a coupon bond gives the holder the right to buy, the put option gives the holder the right to sell the underlying coupon bond at the maturity time T

of the option at an exercise or strike price X. It is assumed that the coupon bond and the strike price are based on the same notional amount which we will, for simplicity, set to 1. At the beginning of the contract the holder of the option has to pay an amount of money for buying the option which is the market price of the coupon-bond option. At maturity the holder of a call option will get an amount of money equal to $\max\{Bond(T, T_B, C) - X, 0\}$ where $Bond(T, T_B, C)$ is the price of the coupon bond (T_B, C) at time T. We will denote the price at time t of a call option with maturity T and strike price X on a coupon bond with maturity time T_B by $Call(t, T, T_B, X, C)$, or $Call(r, t, T, T_B, X, C)$ if we consider a specific short-rate model. Similarly, the correponding put option is denoted by $Put(t, T, T_B, X, C)$ or $Put(r, t, T, T_B, X, C)$. Jamshidian [Jam89] showed that the price of a coupon-bond option can be obtained by evaluating a portfolio of zero-coupon bond options if we use a short-rate model characterized by continuous and strictly decreasing zero-coupon bond price functions $r \longmapsto P(r, T, T_i)$ for all $i \in \{1, ..., n\}$. Let us define the critical short rate r^* implicitly by setting

$$Bond(r^*, T, T_B, C) = X.$$

Since the function $r \longmapsto Bond(r, T, T_B, C)$ is continuous and strictly decreasing we can assume that r^* really exists. Consequently, the holder will exercise the coupon-bond option if the short rate at maturity is lower than this critical short rate, i.e. $r < r^*$ or equivalently

$$Bond(r, T, T_B, C) > Bond(r^*, T, T_B, C) = X.$$

Let the critical exercise prices $X_i(r^*)$, $i = k, ..., n$, be defined by

$$X_i(r^*) := P(r^*, T, T_i),$$

with

$$k := \min\{i \in \{1, ..., n\} : T_i > T\}.$$

Then

$$X = Bond(r^*, T, T_B, C) = \sum_{i=k}^{n} C(T_i) \cdot P(r^*, T, T_i)$$

$$= \sum_{i=k}^{n} C(T_i) \cdot X_i(r^*).$$

Since $r \longmapsto P(r, T, T_i)$ is continuous and strictly decreasing for all $i \in \{k, ..., n\}$, we know that

$$P(r, T, T_i) > X_i(r^*) \quad \text{if and only if} \quad r < r^* \quad \text{for all } i \in \{k, ..., n\}.$$

Hence,

$$\sum_{i=k}^{n} C\left(T_i\right) \cdot \left[P\left(r, T, T_i\right) - X_i\left(r^*\right)\right] > 0 \ \text{ if and only if } \ r < r^*,$$

i.e.

$$\max\left\{\sum_{i=k}^{n} C\left(T_i\right) \cdot \left[P\left(r, T, T_i\right) - X_i\left(r^*\right)\right], 0\right\}$$

$$= \sum_{i=k}^{n} C\left(t_i\right) \cdot \max\left\{P\left(r, T, T_i\right) - X_i\left(r^*\right), 0\right\},$$

and thus

$$\max\left\{Bond\left(r, T, T_B, C\right) - X, 0\right\} = \max\left\{\sum_{i=k}^{n} C\left(T_i\right) \cdot P\left(r, T, T_i\right) - X, 0\right\}$$

$$= \max\left\{\sum_{i=k}^{n} C\left(T_i\right) \cdot \left[P\left(r, T, T_i\right) - X_i\left(r^*\right)\right], 0\right\}$$

$$= \sum_{i=k}^{n} C\left(T_i\right) \cdot \max\left\{P\left(r, T, T_i\right) - X_i\left(r^*\right), 0\right\}.$$

So

$$Call\left(r, t, T, T_B, X, C\right) = \sum_{i=k}^{n} C\left(T_i\right) \cdot Call\left(r, t, T, T_i, X_i\left(r^*\right)\right).$$

Similarly, we can show that

$$Put\left(r, t, T, T_B, X, C\right) = \sum_{i=k}^{n} C\left(T_i\right) \cdot Put\left(r, t, T, T_i, X_i\left(r^*\right)\right)$$

$$= \sum_{i=k}^{n} C\left(T_i\right) \cdot \left[Call\left(r, t, T, T_i, X_i\left(r^*\right)\right) + X_i\left(r^*\right) \cdot P\left(r, t, T\right) - P\left(r, t, T_i\right)\right]$$

$$= Call\left(r, t, T, T_B, X, C\right) + X \cdot P\left(r, t, T\right) - \sum_{i=k}^{n} C\left(T_i\right) \cdot P\left(r, t, T_i\right).$$

We summarize the previous results in the following lemma.

Lemma 5.2 (Coupon Bond Option) *The price at time t of a call option with maturity T and strike price X on a coupon bond (T_B, C) is given by*

$$Call\left(r, t, T, T_B, X, C\right) = \sum_{i=k}^{n} C\left(T_i\right) \cdot Call\left(r, t, T, T_i, X_i\left(r^*\right)\right),$$

where r^* is implicitly given by

$$Bond\,(r^*, T, T_B, C) = X$$

and $Call\,(r, t, T, T_i, X_i\,(r^*))$ is a zero-coupon bond call option with

$$X_i\,(r^*) := P\,(r^*, T, T_i)\,, \quad k := \min\,\{i \in \{1, ..., n\} : T_i > T\}\,.$$

Correspondingly, the price at time t of a put option with maturity T and strike price X on a coupon bond (T_B, C) is given by

$$Put\,(r, t, T, T_B, X, C) = \sum_{i=k}^{n} C\,(T_i) \cdot Put\,(r, t, T, T_i, X_i\,(r^*))$$

$$= Call\,(r, t, T, T_B, X, C) + X \cdot P\,(r, t, T) - \sum_{i=k}^{n} C\,(T_i) \cdot P\,(r, t, T_i)\,.$$

The last equation is called put-call parity for coupon-bond options.

Example. Let us return to the coupon bond and forward agreement example of Sections 5.2 and 5.3. Our actual time is $t_0 = t = 10/20/2000$ and we consider the 6.5% German government bond maturing at $T_B = 10/14/2005$ with annual coupon payments. The forward price (dirty) on this bond for the future time $T = 06/20/2002$ was calculated as 108.141. The parameters of the Hull-White model are given as $a = 0.1$ and $\sigma = 2\%$. The time to expiration of the forward is $T - t = 1.6657$ where we use throughout our example an *act/act* notation. From the discount curve we get the discount factors $P\,(t, T) = 0.920342233$ and $P\,(t, T + 1day) = 0.920214872$, which helps us to calculate the forward rate

$$f\,(t, T) = 365 \cdot \ln\left(\frac{P\,(t, T)}{P\,(t, T + 1day)}\right) = 5.0514\%.$$

We want to calculate the price of an option on our coupon bond with an exercise clean price of 102.00 and maturing at time T using the Hull-White model. It is straightforward to see that the accrued interest at time T is $6.5 \cdot \frac{249}{365} = 4.4342$ because there are 249 days between the last coupon date before time T, i.e. October 14, 2001, and time T. This gives us a dirty exercise price of

$$X = 102.00 + 4.4342 = 106.4342.$$

We can also calculate

$$s_r\,(t, T) = \sqrt{\frac{\sigma^2}{2a} \cdot \left(1 - e^{-2a \cdot (t - t_0)}\right)} = 2.3805\%.$$

To calculate the critical short rate r^* and the corresponding exercise prices

$$X_i\,(r^*) = P\,(r^*, T, T_i) = A\,(T, T_i) \cdot e^{-B(T, T_i) \cdot r^*}$$

we use Lemma 4.24 and the discount factors from the actual discount curve to get the following table.

time	$P(t, T_i)$	$B(T, T_i)$	$s_P(t, T, T_i)$	$A(T, T_i)$
10/14/02	0.9056	0.3128	0.7446%	0.99968
10/14/03	0.8604	1.2347	2.9391%	0.99465
10/14/04	0.8179	2.0688	4.9248%	0.98540
10/14/05	0.7768	2.8236	6.7215%	0.97124

Note that October 14, 2002 is the first coupon payment date after the maturity date T of the option. We can now solve the equation

$$X = \sum_{i=2}^{5} C(T_i) \cdot A(T, T_i) \cdot e^{-B(T,T_i)\cdot r^*}$$

with $T_2 = 10/14/02, ..., T_5 = 10/14/05$ and $C(T_i)$ the corresponding cash flows with $C(T_5) = 106.5$, to get

$$r^* = 5.5993\%.$$

Using Lemma 5.1 we continue with

i	$X_i(r^*)$	$Call(r, t, T, T_i, X_i(r^*))$	$Put(r, t, T, T_i, X_i(r^*))$
2	0.98232	0.00355	0.001973
3	0.92821	0.01344	0.007267
4	0.87762	0.02158	0.011383
5	0.82921	0.02818	0.014531

Applying Lemma 5.2 we get a call price of

$$Call(r, t, T, T_B, X, C) = \sum_{i=2}^{5} C(T_i) \cdot Call(r, t, T, T_i, X_i(r^*)) = 3.2522$$

and a put price of

$$Put(r, t, T, T_B, X, C) = 1.6816.$$

□

We will return to the pricing of coupon-bond options in the framework of Black's model in Section 5.8.

5.6.4 Swaptions

A payer (receiver) swaption is an option on a payer (receiver) swap giving
the owner of the swaption the right to enter at a predetermined time T_0,
the maturity date of the option, into a payer (receiver) swap with swap
start date T_0, equal to the maturity date of the option, swap maturity date
T_S, and at a predetermined "all-in-cost" fixed-rate coupon C relative to the
notional L. As above, the payment dates are denoted by $\widetilde{T}_1, ..., \widetilde{T}_{N_{fix}} = T_S$
for the fixed leg and $T_1, ..., T_N = T_S$ for the floating leg, i.e. the fixed
and the floating leg have the same maturity dates, and the floating spread
S on the variable-rate coupon or the floating index R_L is assumed to be
zero according to the market convention. At time T_0 the owner of a payer
swaption will exercise the option, i.e. decide to enter into the payer swap,
if and only if the value of the payer swap is positive. Using equations (5.12)
and (5.17), the value $PSwaption\,(t, T_0, T_N, C)$ at the option maturity date
$t = T_0$ and for $L = 1$ is given by

$$
\begin{aligned}
PSwaption\,(T_0, T_0, T_N, C) &= \max\left\{PSwap\,(T_0, T_0, T_N, C, 0)\,, 0\right\} \\
&= \max\left\{FRN\,(T_0, T_0, T_N, 0) - Bond\,(T_0, T_N, C')\,, 0\right\} \\
&= \max\left\{1 - Bond\,(T_0, T_N, C')\,, 0\right\} \\
&= \max\left\{(y\,(T_0, T_0, T_N) - C) \cdot PVBP\,(T_0, T_0, T_N)\,, 0\right\}.
\end{aligned}
\tag{5.20}
$$

Hence, the payer (receiver) swaption can be considered as a put (call)
option on the coupon bond (T_N, C') with option maturity time T_0 and
exercise price $X = L$. Considering the last equation in (5.20), the swaption
is exercised at maturity if the fixed-coupon rate C specified in the swaption
contract is lower than the (T_0, T_N) −par swap rate at time T_0. This is how
a swaption is usually priced in the market using the Black model. The
traders consider the par swap rate as the underlying of this derivative
which is supposed to follow a lognormal distribution (see Section 5.8.3 for
more details). In 1998 Jamshidian [Jam98] introduced a new class of models
called *swap market models*. As the LIBOR market models in Section 4.7
these models describe the evolution of real market rates and lead to pricing
formulas which have the Black form and are therefore very familiar to
traders and easy to calibrate. We will now show how these models can be
used for the pricing of payer or receiver swaptions. Since the present value
of a basis point process

$$
PVBP := (PVBP\,(t, T_0, T_N))_{t \in [t_0, T_0]}
$$

is a portfolio of the zero-coupon bonds, i.e. primary traded assets, with
maturities $\widetilde{T}_i$, $i = 1, ..., N_{fix}$, we know that the discounted process

$$
\widetilde{PVBP} := \left(P_0^{-1}(t) \cdot PVBP\,(t, T_0, T_N)\right)_{t \in [t_0, T_0]}
$$

is a $\widetilde{Q}$−martingale. So we can use $PVBP$ as a numéraire with $Q^{\widetilde{T}_1, T_N} :=
Q^{PVBP}$ and we know by the change of numéraire Theorem 3.31 that the

price of the payer swaption at time $t \in [t_0, T_0]$ can be derived by

$$\frac{PSwaption^{SM}(t, T_0, T_N, C)}{L \cdot PVBP(t, T_N)} = E_{Q^{\tilde{T}_1, T_N}}\left[\max\left\{y(T_0, T_0, T_N) - C, 0\right\} | \mathcal{F}_t\right].$$
(5.21)

The swap market model *assumes* that the (T_0, T_N) −par swap rate process $(y(t, T_0, T_N))_{t \in [t_0, T_0]}$ follows the stochastic differential equation

$$dy(t, T_0, T_N) = \sigma_{1,N}(t) \cdot y(t, T_0, T_N) \, dW^{\tilde{T}_1, T_N}(t),$$
(SM)

where $W^{\tilde{T}_1, T_N}$ is a m−dimensional Wiener process under $Q^{\tilde{T}_1, T_N}$ and the volatility process $(\sigma_{1,N}(t))_{t \in [t_0, T_0]}$ is assumed to be strictly positive, bounded and deterministic. Then we know from Theorem 2.43 that the (T_0, T_N) −par swap rate is lognormally distributed under the measure $Q^{\tilde{T}_1, T_N}$, where the variance of $\ln y(t, T_0, T_N)$ under $Q^{\tilde{T}_1, T_N}$ is given by

$$Var_{Q^{\tilde{T}_1, T_N}}\left[\ln y(t, T_0, T_N) | \mathcal{F}_{t_0}\right] := s_{1,N}^2(t_0, t) = \int_{t_0}^{t} \|\sigma_{1,N}(s)\|^2 \, ds.$$

Combining equation (5.21) with assumption (SM) tells us that the value for a payer swaption in the swap market model can be derived by

$$\frac{PSwaption^{SM}(t_0, T_0, T_N, C)}{L \cdot PVBP(t_0, T_0, T_N)} = y(t_0, T_0, T_N) \cdot \mathcal{N}(d_1) - C \cdot \mathcal{N}(d_2), \quad (5.22)$$

with

$$d_1 := \frac{\ln\left(\frac{y(t_0, T_0, T_N)}{C}\right) + \frac{1}{2} \cdot s_{1,N}^2(t_0, T_0)}{s_{1,N}(t_0, T_0)}, \quad d_2 := d_1 - s_{1,N}(t_0, T_0).$$

The corresponding value for a receiver swaption in the swap market model is given by

$$\frac{RSwaption^{SM}(t_0, T_0, T_N, C)}{L \cdot PVBP(t_0, T_0, T_N)} = C \cdot \mathcal{N}(-d_2) - y(t_0, T_0, T_N) \cdot \mathcal{N}(-d_1).$$
(5.23)

In the market, traders usually quote the prices of payer (receiver) swaptions in terms of the corresponding Black volatilities $\sigma_{1,N}^{Black}$, i.e. the volatilities entered into the Black model used to price the swaptions. The Black model will be discussed in more detail in Section 5.8.3. Comparing the Black payer (receiver) swaption formula (see equations (5.32) and (5.33)) to equations (5.22) and (5.23) we see that we can *calibrate* the swap market model to the market prices by simply ensuring that the payer (receiver) swaption volatilities of the swap market model equal the corresponding implied Black swaption volatilities quoted in the market, i.e.[21]

---

[21] Remember that, by convention, we use this notation to indicate an *act/act* daycount convention for the continuous time Black model.

$$s_{1,N}\left(t_0, T_0\right) = \sigma_{1,N}^{Black} \cdot \sqrt{T_0 - t_0}.$$

The drawback of using a swap market model is that it is set up by specifying the dynamics of each $(T_0, T_N)-$par swap rate process $(y\left(t, T_0, T_N\right))_{t \in [t_0, T_0]}$ with respect to the corresponding probability measure $Q^{\tilde{T}_1, T_N}$. For pricing derivatives which only depend on one par swap rate, such as a single payer or receiver swaption, this is not a problem. But if we want to evaluate contingent claims with more complicated payoffs, we need to model the joint behaviour of different par swap rates under a single measure, as was the case in the LIBOR market model. If we want to jointly model the par swap rate processes $(y\left(t, T_n, T_N\right))_{t \in [t_0, T_n]}$ for $T_n \in \left\{T_0, \tilde{T}_1, ..., \tilde{T}_{N_{fix}-1}\right\}$, i.e. a set of par swap rates sharing the same final payment date but with different start dates, the terminal measure Q^{T_N} is a good choice for this single measure. If we want to jointly model a set of par swap rates sharing the same start date but with different final payment dates, i.e. $(y\left(t, T_0, T_n\right))_{t \in [t_0, T_0]}$ for $T_n \in \left\{\tilde{T}_1, ..., \tilde{T}_{N_{fix}}\right\}$, the T_0-forward measure Q^{T_0} can be shown to be a good choice for this single measure. How exactly this is done can be seen, e.g. in Pelsser [Pel00], p. 95-97 or in Musiela and Rutkowski [MR97], p. 410. Neglecting the different daycount conventions of the fixed and floating leg, it can be easily seen that the set of par swap rates $(y\left(t, T_n, T_{n+1}\right))_{t \in [t_0, T_n]}$ for $T_n \in \{T_0, ..., T_{N-1}\}$ leads directly to the LIBOR market model of Section 4.7. In all other cases, however, it can be shown that the LIBOR market model and the swap market model are inconsistent with each other. Another interesting model which leads to closed-form solutions for both cap (floor) and swaption prices, is the *rational lognormal model* developed by Flesaker and Hughston [FH96]. For more details see, e.g. Musiela and Rutkowski [MR97], p. 411-416.

5.7 Interest-Rate Exotics

In the previous sections we discussed interest-rate options which are already considered to be market standard. Contingent claims with payoffs that are more complicated than that of standard (European) interest-rate call and put options are referred to as *exotic interest-rate options* or briefly *interest-rate exotics*. If the value of an interest-rate exotic at maturity depends on the path taken by the underlying financial instrument(s) over time, we call it *path-dependent*. If its value at maturity does not depend on the history of the underlying(s), it is called *path-independent*. Exotic options have first been invented for stock and currency markets. For a more detailed treatment of exotic options in these markets see, e.g. Wilmott [WDH93], p.

148-221, Musiela and Rutkowski [MR97], p. 205-228, Hull [Hul00], p. 458-497, Jarrow and Turnbull [JT00], p. 602-663, or Nelken [Nel96]. It would be beyond the scope of this book to describe all exotic interest-rate options available in the market or discussed in the literature. Especially, we will not go into too many pricing details because the evaluation of interest-rate exotics can become rather lengthy and complicated. Nevertheless, we will try to give a little insight into the pricing of interest-rate exotics and a brief overview of some of the instruments usually available in the over-the-counter market. An interesting elaboration of closed-form solutions for interest-rate exotics using the Hull-White model can be found in Mayer [May98]. He also shows that the conditions of Theorem 2.45 are fullfilled for the Hull-White model, and that the regularity conditions of the Feynman-Kac representation hold for all financial instruments considered here (see also Proske [Pro99]). Other sources for the pricing of specific interest-rate exotics are Turnbull [Tur95] and Musiela and Rutkowski [MR97], p. 401-403.

5.7.1 Digital Options

Probably the easiest type of interest-rate exotics are the so-called digital options. At maturity time $T \in [t_0, T^*]$ and depending on the random variables[22] $V_1(T)$ and $V_2(T)$ on $(\Omega, \mathcal{F}_T)$, the payoff $D_{DC}(T)$ of a digital call option with exercise price X is defined by

$$D_{DC}(T) = D_{DC}(T, X) = \begin{cases} V_2(T) & , \text{if } V_1(T) \geq X \\ 0 & , \text{if } V_1(T) < X. \end{cases}$$

Correspondingly, the payoff $D(T)$ of a digital put option at maturity T and with a strike price X is given by

$$D_{DP}(T) = D_{DP}(T, X) = \begin{cases} V_2(T) & , \text{if } V_1(T) < X \\ 0 & , \text{if } V_1(T) \geq X. \end{cases}$$

A *cash-or-nothing option* is defined by setting $V_2(T) \equiv L \in \mathbb{R}$, where we choose without loss of generality $L = 1$. Such an option pays out a fixed amount of cash equal to 1 if a specific condition is satisfied and zero otherwise. Hereby, the random variable $V_1(T)$ could be the price of a (zero-) coupon bond with maturity time $T' \in [T, T^*]$ at time T or a reference rate of return such as the LIBOR or EURIBOR at time T to give two examples. An *asset-or-nothing option* is given if $V_2(T)$ is chosen to be the value of some financial instrument or asset at maturity time T. Such an asset could

---

[22] Note that we would get a standard call or put option on an underlying financial instrument with final value $V_1(T)$ at maturity time T if $V_2(T) = V_1(T) - X$ or $V_2(T) = X - V_1(T)$.

be, e.g., a (zero-) coupon bond, a swap or even an interest-rate option itself. In the latter case, the resulting asset-or-nothing option is also called a *call-or-nothing* or *put-or-nothing option* as the interest-rate option may be a call or a put. If $V_1(T)$ is considered to be the LIBOR or EURIBOR rate at roll-over time T, $X \geq R_X$ with R_X denoting the cap rate, and if $V_2(T)$ is defined to be the value at time T of the payoff of the corresponding caplet (floorlet), we call this a *digital caplet (floorlet)*. As we build a portfolio of the different digital caplets for the different roll-over periods of a cap we call the resulting financial instrument a *digital cap (floor)*. If, for different exercise prices X_1 and X_2 with $X_1 < X_2$, we consider the final payoff

$$D_{DR}(T) = D_{DC}(T, X_1) - D_{DC}(T, X_2) = \begin{cases} V_2(T) & \text{, if } X_1 \leq V_1(T) \leq X_2 \\ 0 & \text{, else} \end{cases}$$

we defined what is known as a *range digital option*. By construction it can be easily verified that a range digital option is the difference between two digital call options. To give a little insight into the pricing of interest rate exotics we now price a cash-or-nothing as well as an asset-or-nothing option using the Hull-White model of Section 4.5.3.

Lemma 5.3 *Let $T \in [t_0, T^*]$ and $r_0 = r(t_0)$ be the short rate at time t_0. Within the Hull-White model of Section 4.5.3, the price of a cash-or-nothing call option where the cash payment depends on a zero-coupon bond maturing at time $T_1 \in [T, T^*]$ and with an exercise price X is given by*

$$CoNP(r_0, t_0, T, T_1, X) = P(r_0, t_0, T) \cdot N(d_2),$$

with

$$d_2 = \frac{\ln\left(\frac{P(t_0, T_1)}{P(t_0, T) \cdot X}\right) - \frac{1}{2} \cdot s_P^2(t_0, T, T_1)}{s_P(t_0, T, T_1)}$$

and

$$s_P(t_0, T, T_1) := B(T, T_1) \cdot s_r(t_0, T),$$

where

$$s_r(t_0, T) := \sqrt{\frac{\sigma^2}{2a} \cdot \left(1 - e^{-2a \cdot (T - t_0)}\right)}.$$

The price of the corresponding cash-or-nothing put option is

$$CoNP(r_0, t_0, T, T_1, X) = P(r_0, t_0, T) \cdot N(-d_2).$$

Furthermore, the price of an asset-or-nothing call option on a zero-coupon bond maturing at time $T_2 \in [T, T^*]$ where the cash payment depends on a zero-coupon bond with maturity time $T_1 \in [T, T_2]$ and with an exercise price X is given by

$$AoNC(r_0, t_0, T, T_1, T_2, X) = P(r_0, t_0, T_2) \cdot N(-d_1),$$

with

$$d_1 = d_2 + s_P(t_0, T, T_2).$$

The price of the corresponding asset-or-nothing put option is

$$AoNP(r_0, t_0, T, T_1, T_2, X) = P(r_0, t_0, T_2) \cdot N(d_1).$$

Proof. The payoff $D_{CoNC}(T)$ at maturity T of a cash-or-nothing call option where the cash payment depends on a zero-coupon bond maturing at time $T_1 \in [T, T^*]$ and with an exercise price X is given by

$$D_{CoNC}(T) = D_{CoNC}(r, T, X) = \begin{cases} 1 & \text{, if } P(r, T, T_1) \geq X \\ 0 & \text{, if } P(r, T, T_1) < X, \end{cases}$$

where the price of the zero-coupon bond at time $t \in [t_0, T_1]$ is denoted by $P(r, t, T_1)$. Hereby, r represents the dependence of the short rate within the Hull-White model we will use to price the option. The critical value r^* of the short rate is implicitly defined by the equation

$$P(r^*, T, T_1) = X.$$

Using Lemma 4.24, at time t_0, this is equivalent to

$$
\begin{aligned}
r^* &= \frac{1}{B(T, T_1)} \cdot \ln\left(\frac{A(T, T_1)}{X}\right) \\
&= \frac{1}{B(T, T_1)} \cdot \\
&\quad \cdot \left[\ln\left(\frac{P(t_0, T_1)}{P(t_0, T) \cdot X}\right) + B(T, T_1) \cdot f(t_0, T) - \frac{1}{2} \cdot s_P^2(t_0, T, T_1)\right],
\end{aligned}
$$

with

$$s_P(t_0, T, T_1) = B(T, T_1) \cdot s_r(t_0, T)$$

and

$$s_r(t_0, T) = \sqrt{\frac{\sigma^2}{2a} \cdot \left(1 - e^{-2a \cdot (T - t_0)}\right)}.$$

Hence,

$$r^* = \frac{1}{B(T, T_1)} \cdot \ln\left(\frac{P(t_0, T_1)}{P(t_0, T) \cdot X}\right) + f(t_0, T) - \frac{1}{2} \cdot B(T, T_1) \cdot s_r^2(t_0, T).$$

Applying Lemma 4.26, the price $CoNC(r_0, t_0 T, T_1, X)$ of the cash-or-nothing call at time t_0 and with $r_0 = r(t_0)$ is given by

$$CoNC(r_0, t_0, T, T_1, X) = \int_{-\infty}^{\infty} G(r, t_0, r, T) \cdot D_{CoNC}(r, T, X)\, dr,$$

with

$$G\left(r_0, t_0, r, T\right) = P\left(r_0, t_0, T\right) \cdot \frac{1}{s_r\left(t_0, T\right) \cdot \sqrt{2\pi}} \cdot e^{-\frac{1}{2s_r^2\left(t_0, T\right)} \cdot \left(r - f\left(t_0, T\right)\right)^2}$$

for all $r \in I\!R$. Thus,

$$
\begin{aligned}
CoNC\left(r_0, t_0, T, T_1, X\right) &= \int_{-\infty}^{r^*} G\left(r, t_0, r, T\right) dr \\
&= \int_{-\infty}^{r^*} \frac{P\left(r_0, t_0, T\right)}{s_r\left(t_0, T\right) \cdot \sqrt{2\pi}} \cdot e^{-\frac{1}{2s_r^2\left(t_0, T\right)} \cdot \left(r - f\left(t_0, T\right)\right)^2} dr \\
&= P\left(r_0, t_0, T\right) \cdot N\left(\frac{r^* - f\left(t_0, T\right)}{s_r\left(t_0, T\right)}\right) \\
&= P\left(r_0, t_0, T\right) \cdot N\left(d_2\right)
\end{aligned}
$$

with

$$d_2 = \frac{r^* - f\left(t_0, T\right)}{s_r\left(t_0, T\right)} = \frac{\ln\left(\frac{P\left(t_0, T_1\right)}{P\left(t_0, T\right) \cdot X}\right) - \frac{1}{2} \cdot s_P^2\left(t_0, T, T_1\right)}{s_P\left(t_0, T, T_1\right)}.$$

Similarly, the price $CoNP\left(r, t_0, T, T_1, X\right)$ of the cash-or-nothing put at time t_0 and with $r_0 = r\left(t_0\right)$ is given by

$$CoNP\left(r_0, t_0, T, T_1, X\right) = P\left(r_0, t_0, T\right) \cdot N\left(-d_2\right).$$

The payoff $D_{AoN}\left(T\right)$ at maturity T of an asset-or-nothing call option on a zero-coupon bond maturing at time $T_2 \in [T, T^*]$, where the cash payment depends on a zero-coupon bond with maturity time $T_1 \in [T, T_2]$ and with an exercise price X, is given by

$$D_{AoN}\left(T\right) = D_{AoNC}\left(r, T, X\right) = \begin{cases} P\left(r, T, T_2\right) & \text{, if } P\left(r, T, T_1\right) \geq X \\ 0 & \text{, if } P\left(r, T, T_1\right) < X. \end{cases}$$

Another application of Lemma 4.26 tells us that the price $AoNC\left(r_0, t_0, T, T_1, T_2, X\right)$ of the asset-or-nothing call at time t_0 and with $r_0 = r\left(t_0\right)$ is given by

$$
\begin{aligned}
&AoNC\left(r_0, t_0, T, T_1, T_2, X\right) = \int_{-\infty}^{r^*} G\left(r, t_0, r, T\right) \cdot P\left(r, T, T_2\right) dr \\
&= P\left(r_0, t_0, T\right) \cdot A\left(T, T_2\right) \cdot \\
&\quad \cdot \int_{-\infty}^{r^*} \frac{1}{s_r\left(t_0, T\right) \cdot \sqrt{2\pi}} \cdot e^{-B\left(T, T_2\right) \cdot r - \frac{1}{2s_r^2\left(t_0, T\right)} \cdot \left(r - f\left(t_0, T\right)\right)^2} dr \\
&= P\left(r_0, t_0, T\right) \cdot A\left(T, T_2\right) \cdot e^{-f\left(t_0, T\right) \cdot B\left(T, T_2\right) + \frac{1}{2} \cdot s_P^2\left(t_0, T, T_2\right)} \\
&\quad \cdot \int_{-\infty}^{r^*} \frac{1}{s_r\left(t_0, T\right) \cdot \sqrt{2\pi}} \cdot e^{-\frac{1}{2s_r^2\left(t_0, T\right)} \cdot \left(r - \left[f\left(t_0, T\right) - B\left(T, T_2\right) \cdot s_r^2\left(t_0, T\right)\right]\right)^2} dr
\end{aligned}
$$

Using Lemma 4.27, we therefore get

$$\frac{AoNC\,(r_0,t_0,T,T_1,T_2,X)}{P\,(r_0,t_0,T_2)} = N\left(-\frac{r^* - f\,(t_0,T) + B\,(T,T_2)\cdot s_r^2\,(t_0,T)}{s_r\,(t_0,T)}\right)$$

$$= N\,(-d_1),$$

with

$$r^* = \frac{1}{B\,(T,T_1)}\cdot\ln\left(\frac{P\,(t_0,T_1)}{P\,(t_0,T)\cdot X}\right) + f(t_0,T) - \frac{1}{2}\cdot B\,(T,T_1)\cdot s_r^2\,(t_0,T)$$

and

$$d_1 = \frac{r^* - f\,(t_0,T) + B\,(T,T_2)\cdot s_r^2\,(t_0,T)}{s_r\,(t_0,T)}$$

$$= \frac{\ln\left(\frac{P(t_0,T_1)}{P(t_0,T)\cdot X}\right)}{s_P\,(t_0,T,T_1)} - \frac{1}{2}\cdot s_P\,(t_0,T,T_1) + s_P\,(t_0,T,T_2)$$

$$= d_2 + s_P\,(t_0,T,T_2)\,.$$

Similarly, the price $AoNP\,(r_0,t_0,T,T_1,T_2,X)$ of the corresponding asset-or-nothing put option is

$$AoNP\,(r_0,t_0,T,T_1,T_2,X) = P\,(r_0,t_0,T_2)\cdot N\,(d_1)\,.$$

□

As for the caps and floors in Section 5.6.2, the price of a digital cap (digital floor) can be calculated as the sum of digital caplets (digital floorlets) which are digital put (call) options on zero-coupon bonds. Therefore, Lemma 5.3 can be used to price these options within the Hull-White model.

5.7.2 Dual-Strike Caps and Floors

A dual strike cap can be considered to be an interest-rate cap which has a lower cap rate R_{X_1}, an upper cap rate $R_{X_2} \geq R_{X_1}$, and a so-called *trigger rate* b_L. At the beginning of each roll-over period $[T_k, T_{k+1}]$, $k \in \{1, ..., K-1\}$, $t_0 \leq T_1 < \cdots < T_K \leq T^*$, the dual-strike cap is equal to an ordinary cap with cap rate R_{X_1} if the index rate $R_L\,(T_k, T_{k+1})$ is below the trigger rate b_L. If $R_L\,(T_k, T_{k+1})$ is equal or above the trigger rate, the dual-strike cap is equal to an ordinary cap with cap rate R_{X_2}. Hence, the value $D_{DSC}\,(T_k)$ at maturity time T_k, $k \in \{1, ..., K-1\}$, relative to a notional or principal amount L is given by

$$D_{DSC}\,(T_k) = \begin{cases} \frac{L\cdot\Delta T_k\cdot\max\{R_L(T_k,T_{k+1})-R_{X_1},0\}}{1+R_L(T_k,T_{k+1})\cdot\Delta T_k} & \text{, if } R_L\,(T_k,T_{k+1}) < b_L \\[2mm] \frac{L\cdot\Delta T_k\cdot\max\{R_L(T_k,T_{k+1})-R_{X_2},0\}}{1+R_L(T_k,T_{k+1})\cdot\Delta T_k} & \text{, if } R_L\,(T_k,T_{k+1}) \geq b_L. \end{cases}$$

If the dual-strike cap is settled in arrears, the corresponding value $D_{DSCA}(T_k)$ at maturity time T_k, $k \in \{1, ..., K-1\}$, relative to a notional or principal amount L is given by

$$D_{DSCA}(T_k) = \begin{cases} L \cdot \Delta T_k \cdot \max\{R_L(T_k, T_{k+1}) - R_{X_1}, 0\} & \text{, if } R_L(T_k, T_{k+1}) < b_L \\ L \cdot \Delta T_k \cdot \max\{R_L(T_k, T_{k+1}) - R_{X_2}, 0\} & \text{, if } R_L(T_k, T_{k+1}) \geq b_L. \end{cases}$$

For the pricing of a dual-strike cap see, e.g. Musiela and Rutkowski [MR97], p. 401. The corresponding values $D_{DSF}(T_k)$ and $D_{DSFA}(T_k)$ of a dual-strike floor at maturity time T_k, $k \in \{1, ..., K-1\}$ are given by

$$D_{DSF}(T_k) = \begin{cases} \dfrac{L \cdot \Delta T_k \cdot \max\{R_{X_1} - R_L(T_k, T_{k+1}), 0\}}{1 + R_L(T_k, T_{k+1}) \cdot \Delta T_k} & \text{, if } R_L(T_k, T_{k+1}) < b_L \\[2ex] \dfrac{L \cdot \Delta T_k \cdot \max\{R_{X_2} - R_L(T_k, T_{k+1}), 0\}}{1 + R_L(T_k, T_{k+1}) \cdot \Delta T_k} & \text{, if } R_L(T_k, T_{k+1}) \geq b_L \end{cases}$$

and

$$D_{DSCA}(T_k) = \begin{cases} L \cdot \Delta T_k \cdot \max\{R_{X_1} - R_L(T_k, T_{k+1}), 0\} & \text{, if } R_L(T_k, T_{k+1}) < b_L \\ L \cdot \Delta T_k \cdot \max\{R_{X_2} - R_L(T_k, T_{k+1}), 0\} & \text{, if } R_L(T_k, T_{k+1}) \geq b_L \end{cases}$$

if the dual-strike floor is settled in arrears. Hence, the dual strike floor is equal to an ordinary floor with floor rate R_{X_2} if the index rate $R_L(T_k, T_{k+1})$ is equal or above the trigger rate b_L. If $R_L(T_k, T_{k+1})$ is below the barrier rate b_L, the dual-strike floor is equal to an ordinary floor with cap rate R_{X_1}.

5.7.3 Contingent Premium Options

For a contingent premium option, also called *deferred premium option* or *paylater option*, the payment of the option premium is deferred until the maturity time $T \in [t_0, T^*]$ of the option. It has to be paid only if the option is in the money at time T, where we assume that the value of the option at maturity depends on the random variable $V(T)$ on $(\Omega, \mathcal{F}_T)$. Again, $V(T)$ could be the price of a (zero-) coupon bond, a swap or an interest rate at time T. If $CPC(t_0)$ denotes the option premium fixed at time t_0 and to be paid at time T, the value $D_{CPC}(T)$ of a contingent premium call option at maturity time T, relative to a notional or principal amount $L = 1$, and with an exercise price X is given by

$$D_{CPC}(T) = D_{CPC}(T, X) = \begin{cases} V(T) - X - CPC(t_0) & \text{, if } V(T) \geq X \\ 0 & \text{, if } V(T) < X. \end{cases}$$

The corresponding payoff $D_{CPP}(T)$ of a contingent premium put option is given by

$$D_{CPP}(T) = D_{CPP}(T, X) = \begin{cases} 0 & \text{, if } V(T) \geq X \\ X - V(T) - CPP(t_0) & \text{, if } V(T) < X. \end{cases}$$

If we denote the value of an ordinary call option with an exercise price X at maturity time T depending on the random variable $V(T)$ by $D_{Call}(T)$ and the corresponding value of an ordinary put option by $D_{Put}(T)$, we can easily see that

$$D_{CPC}(T) = D_{Call}(T) - CPC(t_0) \cdot \begin{cases} 1 & \text{, if } V(T) \geq X \\ 0 & \text{, if } V(T) < X \end{cases}$$

and

$$D_{CPP}(T) = D_{Put}(T) - CPP(t_0) \cdot \begin{cases} 0 & \text{, if } V(T) \geq X \\ 1 & \text{, if } V(T) < X. \end{cases}$$

Hence, a contingent premium call (put) option is a portfolio of a long position in an ordinary call (put) option and short position in a number of $CPC(t_0)$ $(CPP(t_0))$ cash-or-nothing call (put) options. Let us denote the value at time $t \in [t_0, T]$ of an ordinary call option on an underlying with value $V(T)$ at time T, exercise price X, and maturity time T by $Call(t, T, X)$ and that of the corresponding put option by $Put(t, T, X)$. Furthermore, let the respective values of the contingent premium call and put option be denoted by $CPC(t, T, X)$ and $CPP(t, T, X)$ and that of the cash-or-nothing call and put option by $CoNC(t, T, X)$ and $CoNP(t, T, X)$. Then, by definition of the contingent premium call option, we know that for all $t \in [t_0, T]$,

$$CPC(t, T, X) = Call(t, T, X) - CPC(t_0) \cdot CoNC(t, T, X)$$

and

$$CPP(t, T, X) = Put(t, T, X) - CPP(t_0) \cdot CoNP(t, T, X).$$

Hence, the option premium for the contingent premium call option is given by

$$CPC(t_0) = \frac{Call(t_0, T, X)}{CoNC(t_0, T, X)}$$

and that for the contingent premium put option is

$$CPP(t_0) = \frac{Put(t_0, T, X)}{CoNP(t_0, T, X)}.$$

5.7.4   Other Path-Independent Options

Forward-start options are contracts, agreed to and paid at time $t_0 \in [0, T^*]$, in which the holder, at some future time $T_0 \in [t_0, T]$ and at no additional cost, receives an option with maturity time T and an exercise price equal to the value of the underlying asset at time T_0, i.e. $X = V(T_0)$. We hereby suppose that the value of the underlying asset can be described by the stochastic process $(V(t))_{t \in [t_0, T^*]}$. *Compound options* are options on options. If we are dealing with a call on a put option, sometimes also called *caput*, we would exercise the call option at maturity if the underlying put option is above some previously agreed exercise price. Note that the exercise price and maturity time of the call and the underlying put option may quite well be different. The other types of a compound option are a call on a call, also called *cacall*, a put on a call, known as *pucall*, and a put on a put, called *puput*. Probably the simplest compound option is an option on a zero-coupon bond option. A little more complex is an option on a coupon bond option. There are also options on caps and floors. Options on caps are usually called *captions*, options on floors are known as *floortions*. A *chooser option* gives its owner the right to choose, at maturity, if he would like to get a call or put option on an underlying asset. Because of the right to see the option as one likes it, a chooser option is also called an *as-you-like option*. *Best-of options* are representatives of the so-called *two-colour rainbow options* which are options involving more than one risky asset. Best-of options give the owner the right to get the best of three different possibilities. The first possibility usually is a zero-coupon bond with a given notional, the other two possibilities are functions of interest-rate derivatives such as coupon bonds with different maturities multiplied by some constant weight set for instance to correct for initial price differences. The evaluation of compound and best-of options can be rather tedious and lengthy, depending on the interest-rate model which is used to describe the primary traded assets. For an evaluation of compound and best-of options using the Hull/White-model see, e.g., Mayer [May98], p. 38-61.

5.7.5 Path-Dependent Options

In contrast to the previous options, the value for path-dependent options at maturity depends on the path taken by the underlying financial instrument(s) or asset(s) in the past. Let us suppose that the value of the underlying asset can be described by the stochastic process $(V(t))_{t \in [t_0, T^*]}$. Furthermore, let $T \in [t_0, T^*]$ be the maturity time and X be the exercise price of the option considered. Unlike European options, *American options* can be exercised any time up to the maturity time T, or at least within a specific time interval. If we restrict the possibility of such an early exercise to a certain finite set of dates during the life of the option, we call this type of non-standard option a *Bermuda option*. *Barrier options* are options having a value at maturity which depends on whether or not the underlying stochastic process V crossed a predefined level, called the barrier. *Knock-out options* disappear as soon as V reaches the barrier. If V does so from below (above) the barrier we call this a *up-and-out (down-and-out)* *option*. If the underlying asset is a cap, the resulting knock-out option is called a *knock-out cap*. On the other hand, *knock-in options* become active as soon as V reaches the barrier. If V does so from below (above) the barrier we call this a *up-and-in (down-and-in) option*. An option which makes optimal use of the information available up to time T is the so-called *lookback option*. The call of this type of option pays out the difference between the assets value at maturity and the minimum value of the asset in the period from t_0 to T, i.e. $D_{LBC}(T) = V(T) - \min_{t \in [t_0, T]} V(t)$. The corresponding put pays out the difference between the maximum value of the asset in the period from t_0 to T and the assets value at maturity, i.e. $D_{LBP}(T) = \left(\max_{t \in [t_0, T]} V(t) \right) - V(T)$. If the minimum (maximum) is exchanged by the average of V during at least some part of the life of the option this is called an *average strike call (put) option*. If we stay with the exercise price X and replace $V(T)$ by the average of V during at least some part of the life of the option this is called an *average call (put) option*. Most of the time, especially if we use advanced interest-rate models, there is no closed-form solution available for path-dependent interest-rate exotics. One possible but usually rather time-consuming way to evaluate exotic interest-rate options is the use of (Quasi-) Monte carlo methods (see, e.g. Hull [Hul00], p. 406-415). Another method, which is also used for the evaluation of American options, is the application of interest-rate trees. Hull [Hul00], p. 578-593 describes how the stochastic process of the short rate can be represented by a carefully set-up interest-rate tree. The difference to the well-known stock price trees (see, e.g. Cox, Ross, and Rubinstein [CJR79] or Hull [Hul00], p. 201-213 and 388-406) is that, in an interest-rate tree, the discount rate is not constant but varies from node to node. To represent specific features of the interest-rate process such as the mean reversion, it may be more convenient to use a trinomial instead of a binomial tree, because of the additional degree of freedom offered by the first tree-building method. Hull also shows how the proposed method can be extended to other models. Other sources for the use and implementation of interest-rate trees are, e.g. Black, Derman, and Toy [BDT90], Black and Karasinski [BK91], Hull and White [HW94b], [HW94a], [HW96], Kijima and Nagayama [KN94], [KN96] or Li, Ritchken, and Sankarasubramanian [LRS95]. Tree methods are closely related to the so-called finite-difference methods where, roughly speaking, the partial differential equation of the Cauchy problem (see Definition 2.47) is numerically solved for specific boundary conditions (see, e.g. Wilmott, Dewynne, and Howison [WDH93] or Hull [Hul00], p. 415-424).

5.8   Market Information

The most important market information we need to price financial derivatives consistent to other instruments or underlyings in the market is the zero-rate curve and the volatility curve due to different maturity times. This section gives a brief overview of how we can derive this information out of given market data and market quotes published by a number of brokers via internet or professional data providers. We first deal with the problem of getting adequate zero-rate curve information and the methods which can be applied to transform coupon-bond prices into a consistent zero-rate curve. Second, we discuss the problem of translating volatilities quoted for the Black model into volatilities for a specific interest-rate model such as the Hull-White short-rate model. For the relation of the Black volatilities to the volatilities of the LIBOR and swap market models see Sections 4.7.1 and 5.6.4.

5.8.1   Term Structure

Fortunately, today there are many sources of information on interest-rate data. One example is *The Wall Street Journal* which publishes interest-rate curves each day based on the close of the previous trading day. Another standard source of interest rates on U.S. government securities by maturity is the *Market Roundup* series published by *Salomon Brothers Inc.* which includes information on maturity points ranging from 3 months to 30 years. Information systems and data providers such as *Reuters* or *Bloomberg* publish intraday and historical interest-rate information based on daily, weekly, and monthly market data. Most of the information is available in form of coupon-bond prices and the zero rates have to be evaluated implicitly by mathematical methods. One consequence is that the coupon-bond prices have to be classified by maturity, coupon and liquidity. Rarely traded bonds may not be priced consistently with heavily traded bonds due to older pricing information or other reasons. Therefore, a filter has to be run over the data to ensure the quality of information. *Olsen & Associates* is one company offering filtered market data. A lot of effort has also been put into the derivation of smooth yield curves ensuring that the zero-rate curve doesn't jump from one maturity point to another. Also, there may be a lack of information on many maturity points, which means that we have to derive these points by interpolation. Different methods have been applied to solve these problems. Most of the time these are least-squares methods which differ in the form and the number of terms which are used to fit the model to the market information. Examples are the models of Bradley and Crane [BC73] or Echols and Elliot [EE76] which use the maturity, the logarithm of the maturity, the coupon or the inverse of the maturity as explanatory variables. Other approaches are due to liquidity and tax effects or the smoothness of the forward curves, and make use of spline methods, as in McCulloch [McC71] and [McC75], Litzenberger and Rolfo [LR84], Suits, Mason and Chan [SMC78], or Jordan [Jor84] to name just a few. To explain all these methods would be far beyond the scope of this book. A good overview of this topic may be found in Fabozzi, Fabozzi and Pollack [FFP91], p. 1245-1295.

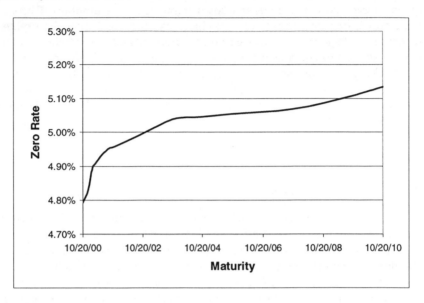

FIGURE 5.1. Continuous zero rate curve of the German government bond market on October 20, 2000

Historical time series can also be used to estimate the parameters of an interest-rate model. Classical methods are the maximum-likelihood method, the general method of moments (GMM) or the Kalman filter (see, e.g. Brown and Dybwig [BD86], Chan, Karolyi, Longstaff, and Sanders [CKLS92], Chen and Scott [CS93] or Gibbons and Ramaswamy [GR93]). Especially the Kalman filter is a good method to estimate the parameters when some of them are time-dependent or if some of the variables or parameters are invisible. However, theses empirical estimations lead to parameters that usually will not fit the market prices of options. So we will describe another method for deriving parameters implied by market prices of interest-rate options in Section 5.8.3. This procedure, by definition, will help us to fit the model parameters to market prices. The idea is comparable to that for deriving implicit volatilities on stock options (see, e.g. Hull [Hul00], p. 255 or Jarrow and Turnbull [JT00], p. 231-233).

5.8.2   Pricing Interest-Rate Options with Black's Model

Especially for the evaluation of futures options, i.e. options on futures for an underlying financial instrument, Black [Bla76] developed a pricing model in a complete Black-Scholes market which we already described in Section 3.5. Today, Black's model is still one of the most popular models for valuing contingent claims under all derivatives traders. Both the lognormal distribution of the underlying price process and the volatility as a measure of uncertainty are assumptions the traders feel quite comfortable with. It is not surprising that Black's model was also adapted to price interest-rate derivatives. To see how this works we consider a *(European) call option on an interest-rate instrument* described by the stochastic process $(S(t))_{t \in [t_0, T^*]}$. Let $T \in [t_0, T^*]$ be the maturity of the option, $F(t, T)$ the futures price of $S(t)$ for the maturity time T, evaluated at time $t \in [t_0, T]$, X the strike price of the option, and σ the volatility of the futures price, which is assumed constant in Black's model. Furthermore, it is assumed that interest rates are deterministic for discounting purposes. Since $F(T, T) = S(T)$, the terminal pay-off of the call option is given by

$$\max\{S(T) - X, 0\} = \max\{F(T, T) - X, 0\}.$$

We could therefore consider the option as a futures call option, and use Lemma 3.30 to derive a call price at time t of

$$Call_F^{Black}(t, T, X) = e^{-\int_t^T r(s)ds} \cdot [F(t, T) \cdot \mathcal{N}(d_1) - X \cdot \mathcal{N}(d_2)], \quad (5.24)$$

with

$$d_1 := \frac{\ln\left(\frac{F(t,T)}{X}\right) + \frac{1}{2} \cdot \sigma^2 \cdot (T - t)}{\sigma \cdot \sqrt{T - t}}, \quad d_2 := d_1 - \sigma \cdot \sqrt{T - t}$$

and $(r(t))_{t \in [t_0, T]}$ denoting the short-rate process, which is, as already mentioned, assumed to be deterministic for discounting purposes. Correspondingly, the price of a (European) put option is given by

$$Put_F^{Black}(t, T, X) = e^{-\int_t^T r(s)ds} \cdot [X \cdot \mathcal{N}(-d_2) - F(t, T) \cdot \mathcal{N}(-d_1)].$$
$$(5.25)$$

It can be easily checked looking at the derivation of the generalized Black option prices in Section 3.5 that we only require two assumptions to be able to price options by the previous formula. The first is that the probability distribution of $S(T)$ at time $t \leq T$ is lognormal with a standard deviation of $\sigma \cdot \sqrt{T - t}$, the second that interest rates are deterministic. As already shown in Section 3.5, futures prices and forward prices are the same under the second assumption. Instead of the futures price we could therefore also use the forward price to value the option contract. We could also relax the assumption that the terminal pay-off is made at the maturity time of the option. Let us suppose that the terminal pay-off is calculated from the value $S(T) = F(T, T)$ at time T but the actual pay-off is delayed until time T', $T \leq T' \leq T^*$. Then the prices for the corresponding (European) call and put options at time t are given by

$$Call_F^{Black}(t, T, T', X) = e^{-\int_t^{T'} r(s)ds} \cdot [F(t,T) \cdot \mathcal{N}(d_1) - X \cdot \mathcal{N}(d_2)],$$
(5.26)

with d_1 and d_2 as above. Correspondingly, the price of a (European) put option is given by

$$Put_F^{Black}(t, T, T', X) = e^{-\int_t^{T'} r(s)ds} \cdot [X \cdot \mathcal{N}(-d_2) - F(t,T) \cdot \mathcal{N}(-d_1)].$$
(5.27)

To summarize, if Black's model of version (5.26) or (5.27) is used for valuing interest-rate derivatives there are three important approximations which are made implicitly:

1. The probability distribution of $S(T)$ at time $t \le T$ has to be lognormal with a standard deviation of $\sigma \cdot \sqrt{T-t}$.

2. Interest rates are supposed to be deterministic for discounting purposes.

3. The forward price of S is equal to the futures price.

   Jamshidian [Jam93] showed that these assumptions have offsetting effects and that Black's model has a stronger theoretical basis than one might expect (see also Hull [Hul00], p. 531-533). But still, assuming that interest rates are stochastic for calculating the resulting terminal pay-off is not consistent with approximation 2 and futures and forward prices are usually not equal when interest rates are stochastic. Nevertheless, many traders use the Black model and prices are often quoted in terms of Black volatilities, as is the case for *caps and floors*.

5.8.3   Black Prices and Volatilities

In Section 5.6.2 we have seen that a cap is a portfolio of *caplets*, where each caplet is due to a special time-period $[T_k, T_{k+1}]$, $k = 1, ..., K-1$, and defined by a total pay-off depending on the value $R_L(T_k, T_{k+1})$ of the floating rate for this period at time T_k. To be more precise, depending on a certain predetermined cap rate R_X, a principal amount L, a maturity time T_K of the cap, and $K-1$ time-periods of size ΔT_k, the terminal pay-off for the period $[T_k, T_{k+1}]$ is defined at time T_k by

$$L \cdot \Delta T_k \cdot \max\{R_L(T_k, T_{k+1}) - R_X, 0\}.$$

However, the pay-off is made at time T_{k+1}. Black's model of equation (5.26) can be used to derive the value of the caplets. We do this with

$R_L\left(t, T_k, T_{k+1}\right)$ denoting the forward rate of the floating rate $R_L\left(T_k, T_{k+1}\right)$ for time T_k, evaluated at time t. Then for $k = 1, ..., K - 1$, we get

$$Caplet_F^{Black}\left(t, T_k, T_{k+1}, R_X\right) = L \cdot \Delta T_k \cdot e^{-\int_t^{T_{k+1}} r(s)ds}$$
$$\cdot \left[R_L\left(t, T_k, T_{k+1}\right) \cdot \mathcal{N}\left(d_1\right) - R_X \cdot \mathcal{N}\left(d_2\right)\right],$$

with

$$d_1 := \frac{\ln\left(\frac{R_L(t, T_k, T_{k+1})}{R_X}\right) + \frac{1}{2} \cdot \sigma^2 \cdot \left(T_k - t\right)}{\sigma \cdot \sqrt{T_k - t}}, \quad d_2 := d_1 - \sigma \cdot \sqrt{T_k - t}.$$

As we have already seen, the *Black cap price* is given as

$$Cap_F^{Black}\left(r, t, T_K, R_X, \Delta T\right) = \sum_{k=1}^{K-1} Caplet_F^{Black}\left(r, t, T_k, T_{k+1}, R_X\right),$$

$$(5.28)$$

with $\Delta T = \left(\Delta T_1, ..., \Delta T_{K-1}\right)$. Correspondingly, for $k = 1, ..., K - 1$, the Black price of the *floorlets* are given by

$$Floorlet_F^{Black}\left(t, T_k, T_{k+1}, R_X\right) = L \cdot \Delta T_k \cdot e^{-\int_t^{T_{k+1}} r(s)ds}$$
$$\cdot \left[R_X \mathcal{N}\left(-d_2\right) - R_L\left(t, T_k, T_{k+1}\right) \mathcal{N}\left(-d_1\right)\right],$$

giving us a *Black price* for the *floor* of

$$Floor_F^{Black}\left(r, t, T_K, R_X, \Delta T\right) = \sum_{k=1}^{K-1} Floorlet_F^{Black}\left(r, t, T_k, T_{k+1}, R_X\right).$$

$$(5.29)$$

To derive a cap price, each caplet must be valued separately, which could be done,

P1. by using a different volatility for each caplet, or

P2. by using the same volatility for all caplets setting up a cap of a specific maturity but varying this volatility depending on the maturity of the cap.

The volatilities derived by procedure *P1* are called *forward forward volatilities*, those derived by procedure *P2* are called *flat* or *forward volatilities*. The volatilities quoted by the brokers are usually flat volatilities for the 3− or 6−month EURIBOR (LIBOR). One of these brokers' pages is the VCAP page from Intercapital Brokers ltd. published on the Reuters information system. Quoted are Black flat volatilities, which can be easily transformed to Black market prices for different cap maturities and cap rates, briefly denoted by

$$Cap_i^{market}, \; i = 1, ..., n.$$

To derive the parameters, e.g., for the Hull-White model, these market prices are then compared to the theoretical cap prices of the Hull-White model of Section 5.6.2 as a function of the parameters a and σ, which we will denote by

$$Cap_i^{HW}(a,\sigma),\, i=1,...,n.$$

The model parameters are then estimated as solution of the optimization problem

$$\left(P_{a,\sigma}^{Cap}\right)\begin{cases} \sum_{i=1}^{n}\left(Cap_i^{HW}(a,\sigma)-Cap_i^{market}\right)^2 \to \min \\ \\ a \geq 0,\, \sigma \geq 0. \end{cases}$$

To solve this problem there are different algorithms available which usually need the partial derivatives of the goal function (see, e.g. Fletcher [Fle90]). Unfortunately, these derivatives may become very complicated in practice. So it is advisable to use an algorithm that makes no explicit use of these partial derivatives. Such an approach is the Downhill-Simplex method published by Nelder and Mead [NM65], which we will use to solve the problem $\left(P_{a,\sigma}^{Cap}\right)$.

Another application of the Black model in the interest-rate market is the pricing of *coupon-bond options*. Let T_B denote the maturity of the coupon bond, T the maturity of the call option on that bond, and

$$C(T_i)\text{ at time }T_i,\, i=1,...,n\text{ with }t_0 \leq T_1 < T_2 < \cdots < T_n = T_B \leq T^*$$

the coupon payments of the bond. Furthermore, let

$$k := \max\{i \in \{1,...,n\} : T_i \leq T\}$$

and

$$I(t_0,T) = \sum_{i=1}^{k} C(T_i) \cdot P(t_0,T_i)$$

be the present value of the coupon payments up to time T, where $P(t_0,T_i)$ denotes the price of a zero-coupon bond with maturity T_i at time t_0. Then, as shown in Section 5.3, the forward price of that coupon bond for the maturity time T at time t_0 is given by

$$Forward(t_0,T) = (Bond(t_0,T_B,C) - I(t_0,T)) \cdot P^{-1}(t_0,T).$$

Note that these prices are dirty and not quoted (see Section 5.2 for an explanation). So the strike price X of the option should be calculated consistently with the dirty price of the coupon bond. Since in most of the exchange-traded bond options the strike price is referred to the quoted or clean price of the coupon bond, it is important in this case to add the coupon bond's accrued interest at the maturity time of the option to the quoted strike price. We take account of this problem by denoting a strike price quoted relative to the clean price of the bond by X_{clean}. Consequently, the notation X stands for a strike price relative to the dirty price of the bond and the relation between X and X_{clean} is given by

$$X = X_{clean} + AI(T_k, T, C).$$

After this modification many traders price the *call option on* that *coupon bond* using equation (5.24) by

$$Call^{Black}(t, T, T_B, X, C) = e^{-\int_t^T r(s)ds} \cdot [Forward(t, T) \cdot \mathcal{N}(d_1) - X \cdot \mathcal{N}(d_2)], \tag{5.30}$$

with

$$d_1 := \frac{\ln\left(\frac{Forward(t,T)}{X}\right) + \frac{1}{2} \cdot \sigma^2 \cdot (T - t)}{\sigma \cdot \sqrt{T - t}}, \quad d_2 := d_1 - \sigma \cdot \sqrt{T - t}.$$

A corresponding pricing formula for a put option can be easily derived using (5.25).

Example. Let us return to the coupon-bond option example of Section 5.6.3. Our actual time is $t_0 = t = 10/20/2000$, and we consider the 6.5% German government bond maturing at $T_B = 10/14/2005$ with annual coupon payments. The forward price (dirty) on this bond for the future time $T = 06/20/2002$ was calculated to be $Forward(t, T) = 108.141$. The time to expiration of the forward is $T - t = 1.6657$ and the discount factor was $P(t, T) = 0.920342233$. We want to calculate the price of an option on this coupon bond with an exercise clean price of $X_{clean} = 102.00$ and maturing at time T using Black's model for coupon-bond options. The dirty exercise price is easily calculated to be $X = 106.4342$ and the Black volatility is given by a quote of $\sigma = 4.6914\%$. Setting

$$e^{-\int_t^T r(s)ds} = P(t, T) = 0.920342233$$

and applying equation (5.30) as well as the corresponding put formula, the call and put prices are given by

$$Call^{Black}(t, T, T_B, X, C) = 3.2519$$

and

$$Put^{Black}(t, T, T_B, X, C) = 1.6813.$$

Figure 5.2 shows the variation of the call and put prices with respect to varying forward prices.   □

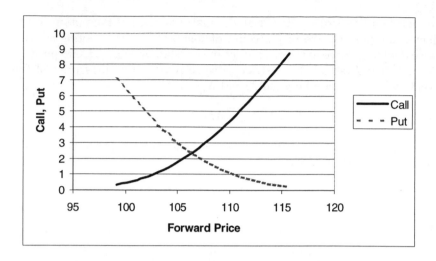

FIGURE 5.2. Variation of the Black call and put prices with forward price

It should be noted that there is a second possibility for how traders can price a call option on a coupon bond using the Black model. Instead of adjusting the strike price they derive what may be called the *clean forward price*, i.e.

$$Forward_{clean}(t,T) := Forward(t,T) - AI(T_k,T,C),$$

and price the call option by

$$Call^{Black}(t,T,T_B,X,C) = e^{-\int_t^T r(s)ds} \cdot \left[\begin{array}{c} Forward_{clean}(t,T) \cdot \mathcal{N}(d_1) \\ -X_{clean} \cdot \mathcal{N}(d_2) \end{array} \right],$$

$$(5.31)$$

with

$$d_1 := \frac{\ln\left(\frac{Forward_{clean}(t,T)}{X_{clean}}\right) + \frac{1}{2} \cdot \sigma^2 \cdot (T-t)}{\sigma \cdot \sqrt{T-t}}, \quad d_2 := d_1 - \sigma \cdot \sqrt{T-t}.$$

Since

$$
\begin{aligned}
\ln\left(\frac{Forward_{clean}(t,T)}{X_{clean}}\right) &= \ln\left(Forward_{clean}(t,T)\right) - \ln\left(X_{clean}\right) \\
&\approx \left(Forward_{clean}(t,T) - 1\right) - \left(X_{clean} - 1\right) \\
&= \left(Forward(t,T) - 1\right) - \left(X - 1\right) \\
&\approx \ln\left(Forward(t,T)\right) - \ln\left(X\right) \\
&= \ln\left(\frac{Forward(t,T)}{X}\right),
\end{aligned}
$$

both methods will lead to approximately the same result as long as the accrued interest is not too large. Unfortunately, bond options are not as heavily exchange-traded as caps and floors. Nevertheless, it is possible to get OTC quotes from many traders usually quoted in terms of Black bond volatilities which are consistent to equation (5.31). The Black bond volatilities can be easily transformed to Black market prices for different option maturities which we will briefly denote by

$$Call_i^{market}, \ i = 1, ..., n.$$

To derive the parameters, e.g., of the Hull-White model these market prices are then compared to the theoretical call prices of the Hull-White model of Section 5.6.3 as a function of the parameters a and σ, which we will denote by

$$Call_i^{HW}(a, \sigma), \ i = 1, ..., n.$$

The model parameters are then estimated as solution of the optimization problem

$$\left(P_{a,\sigma}^{Call}\right) \begin{cases} \displaystyle\sum_{i=1}^{n} \left(Call_i^{HW}(a, \sigma) - Call_i^{market}\right)^2 \to \min \\ a \geq 0, \ \sigma \geq 0. \end{cases}$$

Again, problem $\left(P_{a,\sigma}^{Call}\right)$ can be solved using the Downhill-Simplex method. Of course this method of deriving the parameters of the Hull-White model can also be applied to other derivatives. However, it is quite important to have reliable derivative prices as an input to this procedure. This is most probably the case for cap and floor prices, since these are regularly quoted, for example, from Intercapital Brokers ltd. We will give an example in the following case study. On the other hand, the choice of derivatives used to find the model parameters should be consistent with the contingent claim we would like to price using these parameters. One possibility to ensure this may be to define different baskets of derivatives, characterized for instance by different maturity or strike buckets, and to derive possibly different parameters for each of these classes. The next step would be to assign each contingent claim that has to be evaluated to one of these classes, and to derive its price using the corresponding model parameters.

A third application of the Black model for pricing interest-rate derivatives is the evaluation of *options on payer or receiver swaps*. As we have seen in equation (5.12) a market swaption could be considered as an option on a coupon bond. However, the market usually prices swaptions by considering the par swap rate as the underlying of this derivative, which is supposed to follow a lognormal distribution. To be more precise, let $y(t, T_0, T_N)$ denote the (T_0, T_N)−par swap rates at time $t \in [t_0, T_0]$, and let us assume we are given the problem of pricing a payer swaption with option maturity time T_0, swap maturity date T_S, and a predetermined "all-in-cost" fixed-rate coupon C relative to the notional L as our exercise boundary. As in Section 5.6.4, the payment dates are denoted by $\tilde{T}_1, ..., \tilde{T}_{N_{fix}} = T_S$ for the fixed leg and $T_1, ..., T_N = T_S$ for the floating leg, i.e. the fixed and the floating leg have the same maturity dates, and the floating spread S on the variable-rate coupon or the floating index R_L is assumed to be zero according to the market convention. At time T_0 the owner of a payer swaption will exercise the option, i.e. decide to enter into the payer swap, if and only if the value of the payer swap is positive. Using equation (5.17) this is exactly the case if the predetermined fixed-rate coupon C is lower than the (T_0, T_N)−par swap rate at time $t = T_0$ and the payoff from the swaption is given by a series of cash flows equal to

$$L \cdot \left(\tilde{T}_{i+1} - \tilde{T}_i \right)_{DC(fix)} \cdot \max \{ y(T_0, T_0, T_N) - C, 0 \}$$

made at the swap payment dates $\tilde{T}_{i+1}$, $i = 0, ..., N_{fix} - 1$. To evaluate the value of this payoff, which we simply call a *payer swaplet*, at time $t \in [t_0, T_0]$, we use Black's model of equation (5.26) with $F(t, T_0) := y(t, T_0, T_N)$ denoting the forward (T_0, T_N)−par swap rate for time T_0 at time $t \in [t_0, T_0]$ and the actual pay-off is delayed until time $\tilde{T}_{i+1}$, $T_0 \leq \tilde{T}_{i+1} \leq T_N$, $i = 0, ..., N_{fix} - 1$. Doing so, we get

$$PSwaplet_F^{Black} \left(t, T_0, \tilde{T}_{i+1}, C \right) = L \cdot \left(\tilde{T}_{i+1} - \tilde{T}_i \right)_{fix} \cdot e^{-\int_t^{\tilde{T}_{i+1}} r(s)ds}$$
$$\cdot [y(t, T_0, T_N) \cdot \mathcal{N}(d_1) - C \cdot \mathcal{N}(d_2)],$$

with

$$d_1 := \frac{\ln \left(\frac{y(t, T_0, T_N)}{C} \right) + \frac{1}{2} \cdot \sigma^2 \cdot (T_1 - t)}{\sigma \cdot \sqrt{T_0 - t}}, \quad d_2 := d_1 - \sigma \cdot \sqrt{T_0 - t}.$$

Hence, the *Black price* of the *payer swaption* is given by[23]

$$PSwaption_F^{Black}(t, T_0, T_N, C) = \sum_{i=0}^{N_{fix}-1} PSwaplet_F^{Black}\left(t, T_0, \widetilde{T}_{i+1}, C\right)$$

$$= L \cdot PVBP(t, T_0, T_N) \qquad (5.32)$$
$$\cdot \left[y(t, T_0, T_N) \cdot \mathcal{N}(d_1) - C \cdot \mathcal{N}(d_2)\right],$$

since, as assumed in the Black model,

$$P\left(t, \widetilde{T}_{i+1}\right) = e^{-\int_t^{\widetilde{T}_{i+1}} r(s)ds}, \ i = 0, ..., N_{fix} - 1.$$

The *Black price* of the corresponding *receiver swaption* is given by[24]

$$RSwaption_F^{Black}(t, T_0, T_N, C) = L \cdot PVBP(t, T_0, T_N) \cdot \qquad (5.33)$$
$$\cdot \left[C \cdot \mathcal{N}(-d_2) - y(t, T_1, T_N) \cdot \mathcal{N}(-d_1)\right].$$

Traders usually quote swaption prices in terms of Black volatilities for different maturities. Using equations (5.32) and (5.33), these volatilities can be easily transformed to Black market prices. According to the quadratic minimization procedure described above, these market prices can then be used to estimate the parameters of a specific interest-rate model.

---

[23] Sometimes this equation can be found in a slightly different notation. If we define

$$PVNOM(t, T_0, T_N) := L \cdot PVBP(t, T_0, T_N) \cdot y(t, T_0, T_N)$$

and

$$PVINT(t, T_N, C) := L \cdot PVBP(t, T_0, T_N) \cdot C$$

for $t \in [t_0, T_0]$, equation (5.32) can be rewritten as

$$PSwaption_F^{Black}(t, T_0, T_N, C) = PVNOM(t, T_0, T_N) \cdot \mathcal{N}(d_1)$$
$$-PVINT(t, T_N, C) \cdot \mathcal{N}(d_2)$$

with

$$d_1 := \frac{\ln\left(\frac{PVNOM(t,T_0,T_N)}{PVINT(t,T_N,C)}\right) + \frac{1}{2} \cdot \sigma^2 \cdot (T_0 - t)}{\sigma \cdot \sqrt{T_0 - t}}, \ d_2 := d_1 - \sigma \cdot \sqrt{T_0 - t}.$$

[24] Note that, because of

$$C \cdot \mathcal{N}(-d_2) - y(t, T_0, T_N) \cdot \mathcal{N}(-d_1) = C \cdot [1 - \mathcal{N}(d_2)] - y(t, T_0, T_N) \cdot [1 - \mathcal{N}(d_1)]$$
$$= y(t, T_0, T_N) \cdot \mathcal{N}(d_1) - C \cdot \mathcal{N}(d_2)$$
$$+C - y(t, T_0, T_N)$$

we have

$$PSwaption_F^{Black}(t, T_0, T_N, C) = RSwaption_F^{Black}(t, T_0, T_N, C)$$

if we set

$$C := y(t, T_0, T_N).$$

Since this is done at time t_0 once we set up a swaption, the price of a payer swaption is always equal to the price of a receiver swaption at the trade date t_0.

5.8.4 Estimation of the Hull-White Model Parameters

In this section we present an illustrating case study to show how the parameters of the Hull-White model can be derived using Black forward volatilities. On December 12, 2000 the following zero rates[25], cap rates, and Black forward volatilities were available on the Reuters information system

Maturity	Black volatility	Zero rate	Cap rate
1Y	12.00%	4.94%	4.80%
3Y	15.30%	5.10%	4.92%
5Y	15.90%	5.27%	5.10%
7Y	15.50%	5.48%	5.28%

Using equation (5.28), the Black forward volatilities can be transformed to the following Black market prices Cap_i^{market}, $i = 1, ..., 4$ (years)

Maturity	Black market price	Zero rate	Cap rate
1Y	0.24	4.94%	4.80%
3Y	1.13	5.10%	4.92%
5Y	2.29	5.27%	5.10%
7Y	3.56	5.48%	5.28%

Solving optimization problem $\left(P_{a,\sigma}^{Cap}\right)$ we get the Hull-White parameters

$$a = 0.097 \quad \text{and} \quad \sigma = 1.01\%.$$

Inserted in the theoretical cap price formula for the Hull-White model of Section 5.6.2 we receive the following price table for the Hull-White and Black prices

Maturity	Black market price	Hull-White price	Difference
1Y	0.24	0.15	+0.09
3Y	1.13	1.13	±0.00
5Y	2.29	2.30	−0.01
7Y	3.56	3.55	+0.01

□

---

[25] Up to one year the zero rates are quoted *act/360*, for more than one year the quote is *30E/360*.

Part III

Measuring and Managing Interest-Rate Risk

6
Risk Measures

Since the introduction of the Treasury bill and Treasury bond futures in the U.S. in the mid 70's, many innovative interest-rate derivatives were invented all over the world. Usually these derivatives are sold by a financial institution or trader to one of its clients over-the-counter. If they are traded at an exchange the financial institution or trader can neutralize or hedge the position by buying the same instrument at the exchange. The resulting position is called *covered*. Unfortunately, most of these derivatives are tailored to the specific needs of a client with no equivalent exchange-traded product available for hedging purposes. This significantly complicates the hedging process. The trader has to buy or sell other financial instruments to synthetically create the same exposure as the derivative generates, which he or the financial institution sold to the client. If, and that is the predominating case, both portfolios do not fully coincide the trader is exposed to the risk that both positions evolve differently over time. This risk has to be quantified and updated by an adequate risk-management system. The main problem such a system has to solve is to supply the trader or risk manager with adequate risk numbers and, if advanced, to give assistance within the process of structuring and restructuring the portfolio according to the desired exposure or level of risk. Before we approach the question of how to define an adequate risk number we have to define the risk we are talking about a little more closely.

Following the directions of the Basle Comittee on Banking Supervision for the risk management in derivatives trading from 1991, *market risk* is defined to be the risk of a negative impact of changing market prices for the financial situation of an institution. For interest-rate instruments this risk is due to changes of the yield or zero-rate curve. Traditionally, it was measured by a parallel shift of the yield curve, but there is also the possibility of non-parallel movements of the yield curve, sometimes called *shape risk*. It is the focus of this work to measure and manage market risk. Nevertheless, a brief overview of other sources of risk seems to be appropriate.

Volatility risk is, roughly speaking, the risk resulting from changing volatilities of a financial instrument or the underlying variables, in our case the zero rates of different maturities. *Credit risk* is the risk that a debtor may not fullfil all of his duties. Rating agencies such as *Standard & Poors* or *Moodies* publish ratings as a measure of the credit-worthiness of a country or company. One way of controlling credit risk is to set an upper bound for the investment into the coupon bonds or shares of a specific country or company. Currently, a lot of research focuses on the modelling of credit risk. To investigate what properties a good credit risk model should have is important not only for the management of credit risk, but also for pricing credit derivatives such as default swaps or options on defaultable bonds. The interested reader may refer to Section 4.8 for an overview of this topic. The risk that a whole industry will be negatively influenced by changes in the economy is called *sector risk*, and may also be controlled by setting an upper bound on the total investment into this sector. If a portfolio is actively managed, the portfolio manager or trader may suffer under the risk of changing bid-ask spreads due to the changing liquidity and volume of the corresponding market. This risk is usually called *liquidity risk*, and may be controlled by trading financial instruments with a minimum trading volume only. Such instruments are, for example, the coupon-bond futures discussed in Section 5.4.2. If the portfolio consists of products quoted in different currencies there is the risk that the price of a foreign currency may change, the so-called *currency risk*. There are special instruments to control that risk. Often traders build portfolios of different currencies which are managed separately but are also controlled by an overall risk number such as the value at risk (see Section 6.2.2 for more details). Because the value at risk is a measure which considers the tail of a distribution, it may make sense to directly model extreme market movements. The reader interested in details on this so-called *extreme value theory* may refer to Embrechts, Klüppelberg, and Mikosch [EKM97].

In Section 6.1 we give an overview of the different risk measures based on small movements and small periods of time such as the first- and second-order sensitivities, especially Black or key-rate deltas and gammas or duration and convexity. In Section 6.2 we define the risk measures based on large movements of the risk factors or based on a longer time horizon such as the lower partial moments or the value at risk. An interesting question is what properties a somehow good risk measure should have. We give an answer to this question in Section 6.8. Because the simulation of a risk measure can be very time-consuming, it is important to carefully select the underlying interest-rate model. The simulation time can be reduced dramatically if the chosen model allows for closed-form solutions of the derivatives prices which have to be risk measured. It may also make sense to concentrate on just a few explanatory risk factors. This idea is discussed in Section 6.4.

6.1   Sensitivity Measures

The first and probably simplest thing a trader or risk manager can do is to measure the sensitivity of a financial instrument or derivative to small changes in the price of an underlying asset, i.e., roughly speaking, an asset which mainly determines the price of the derivative, in the next small interval of time. This sensitivity measure is known as the *delta* of a financial instrument. Correspondingly, immunizing a financial instrument or portfolio with respect to small price-changes of an underlying asset is known as *delta hedging*. Immunized this way for a short period of time, there appears the question of how often the hedged position has to be restructured. To put it another way, how sensitive is the delta to small changes in the price of an underlying asset? The risk measure which gives an answer to that question is called the *gamma* of a financial instrument or portfolio. But the portfolio or derivative is also exposed to small changes of other risk factors or parameters which have to be entered into the pricing formula. Such parameters may be the volatility, with the corresponding risk measure called *vega*, time, with the corresponding risk measure called *theta*, or an interest rate, with the corresponding risk measure called *rho*. This section is devoted to defining and discussing these sensitivity measures, also known as the *Greeks*, and the corresponding hedging idea. Throughout this and the following chapters we assume that all zero rates are quoted for an *act/act* daycount convention. For ease of exposition we therefore suppress the daycount notation for the upcoming time differences.

6.1.1   First- and Second-Order Sensitivity Measures

For a mathematical definition of the sensitivity measures, let $F = (F_1, ..., F_m)$, $m \in I\!N$, denote a vector of risk factors or parameters entering the price of a financial instrument or derivative $D(F)$.

Definition 6.1 (First- and Second-Order Sensitivity)
The first-order sensitivity $\Delta_{F_j}^D(F)$ of a financial instrument or derivative to small changes of the risk factor F_j is defined by

$$\Delta_{F_j}^D(F) := \frac{\partial}{\partial F_j} D(F) \approx \frac{\Delta_j D(F)}{\Delta F_j}$$

with $\Delta_j D\,(\mathsf{F})$ defined by

$$\Delta_j D\,(\mathsf{F}) := D\,(\mathsf{F} + e_j \cdot \Delta \mathsf{F}_j) - D\,(\mathsf{F})$$

and e_j denoting the $m-$dimensional unit vector with 1 in the $j-$th component and 0 else. Correspondingly, the second-order sensitivity $\Gamma^D_{\mathsf{F}_j \mathsf{F}_l}\,(\mathsf{F})$ of a financial instrument or derivative to small changes of the risk factors F_j and F_l is defined by

$$\Gamma^D_{\mathsf{F}_j \mathsf{F}_l}\,(\mathsf{F}) := \frac{\partial^2}{\partial \mathsf{F}_j \partial \mathsf{F}_l} D\,(\mathsf{F}) = \frac{\partial}{\partial \mathsf{F}_j} \Delta^D_{\mathsf{F}_l}\,(\mathsf{F}) \approx \frac{\Delta_j\,(\Delta_l D\,(\mathsf{F}))}{\Delta \mathsf{F}_j \cdot \Delta \mathsf{F}_l}.$$

*If a risk factor is considered to be a stochastic process underlying the pricing model, the first-order sensitivity of a derivatives price with respect to that risk factor is called the **delta** and the second-order sensitivity with respect to that risk factor is called the **gamma** of the derivative.*

Using a Taylor series expansion we know that the change of the derivatives price in relation to changing risk factors is approximately given by

$$\Delta D\,(\mathsf{F}) \approx \sum_{j=1}^m \Delta^D_{\mathsf{F}_j}\,(\mathsf{F}) \cdot \Delta \mathsf{F}_j + \frac{1}{2} \cdot \sum_{j=1}^m \sum_{l=1}^m \Gamma^D_{\mathsf{F}_j \mathsf{F}_l}\,(\mathsf{F}) \cdot \Delta \mathsf{F}_j \cdot \Delta \mathsf{F}_l. \qquad (6.1)$$

To derive the first- and second-order sensitivities of a portfolio let $V\,(\mathsf{F}, \varphi)$ be the price of the portfolio $\varphi = (\varphi_1, ..., \varphi_n)$ of financial instruments or derivatives with prices $D_1\,(\mathsf{F}), ..., D_n\,(\mathsf{F})$ depending on the risk factors F, i.e.

$$V\,(\varphi) = V\,(\mathsf{F}, \varphi) = \sum_{i=1}^n \varphi_i \cdot D_i\,(\mathsf{F}).$$

The first-order sensitivities of the portfolio φ are given by

$$\Delta^{V(\varphi)}_{\mathsf{F}_j}\,(\mathsf{F}) = \sum_{i=1}^n \varphi_i \cdot \Delta^{D_i}_{\mathsf{F}_j}\,(\mathsf{F}), \, j = 1, ..., m,$$

the second-order sensitivities of the portfolio φ are given by

$$\Gamma^{V(\varphi)}_{\mathsf{F}_j \mathsf{F}_l}\,(\mathsf{F}) = \sum_{i=1}^n \varphi_i \cdot \Gamma^{D_i}_{\mathsf{F}_j \mathsf{F}_l}\,(\mathsf{F}), \, j, l = 1, ..., m.$$

Using (6.1), the price-change $\Delta V\,(\mathsf{F}, \varphi)$ is approximately given by

$$\Delta V\,(\mathsf{F}, \varphi) \approx \sum_{j=1}^m \Delta^{V(\varphi)}_{\mathsf{F}_j}\,(\mathsf{F}) \cdot \Delta \mathsf{F}_j + \frac{1}{2} \cdot \sum_{j=1}^m \sum_{l=1}^m \Gamma^{V(\varphi)}_{\mathsf{F}_j \mathsf{F}_l}\,(\mathsf{F}) \cdot \Delta \mathsf{F}_j \cdot \Delta \mathsf{F}_l.$$

Hence, if the trader or risk manager manages to set up a portfolio with

$$\Delta^{V(\varphi)}_{\mathsf{F}_j}\,(\mathsf{F}) = 0 \text{ and } \Gamma^{V(\varphi)}_{\mathsf{F}_j \mathsf{F}_l}\,(\mathsf{F}) = 0 \text{ for all } j, l = 1, ..., m$$

his portfolio is nearly insensitive to small movements of the risk factors $F_1, ..., F_m$. This is, roughly speaking, the idea which is at the bottom of hedging.

Let us now look at a few *examples* of first and second-order sensitivities. As described in Section 5.3, the forward price at time t of a coupon bond with payments

$$C\left(T_i\right) \text{ at time } T_i, \ i = 1, ..., n, \ t := t_0 \leq T_1 < T_2 < \cdots < T_n = T_B \leq T^*,$$

and a present value of the coupon payments up to the maturity time T of the forward contract

$$I\left(t, T\right) = \sum_{i=1}^{k} C\left(T_i\right) \cdot P\left(t, T_i\right)$$

with $k := \max \left\{i \in \left\{1, ..., n\right\} : T_i \leq T\right\}$ is given by

$$Forward\left(t, T\right) = \left(Bond\left(t, T_B, C\right) - I\left(t, T\right)\right) \cdot P^{-1}\left(t, T\right).$$

Furthermore, the value at time t of a forward agreement on this coupon bond is given by

$$V_{Forward}\left(t, T, X\right) = Bond\left(t, T_B, C\right) - I\left(t, T\right) - X \cdot P\left(t, T\right).$$

If we consider the bond price to be the underlying and risk factor F_1, the delta of $V_{Forward}$ with respect to $F_1 = Bond\left(t, T_B, C\right)$ can be easily calculated by[1]

$$\Delta_{F_1}^{V_{Forward}}\left(F\right) = 1. \tag{6.2}$$

On the other hand, the delta of the forward price with respect to the coupon-bond price is given by

$$\Delta_{F_1}^{Forward}\left(F\right) = P^{-1}\left(t, T\right). \tag{6.3}$$

Let us consider the cheapest-to-deliver (CTD) bond $Bond\left(t, T_B^*, C^*\right)$ as a risk factor $F_1 = Bond\left(t, T_B^*, C^*\right) =: CTD$. Let us furthermore assume that the CTD bond doesn't change by a small change of the coupon-bond price and that there are no coupon payments in the time-period $[t, T]$. From Section 5.4.2 we then know that the futures price is given by

$$
\begin{aligned}
F\left(t, T\right) &= F\left(t, T, T_B^*, C^*\right) \\
&= \frac{Bond\left(t, T_B^*, C^*\right) \cdot \left[1 + R_L\left(t, T\right) \cdot \left(T - t\right)\right] - AI\left(t_0, T, C^*\right)}{Conv\left(T_B^*, C^*\right)} \\
&= \frac{CTD \cdot \left[1 + R_L\left(t, T\right) \cdot \left(T - t\right)\right] - AI\left(t_0, T, C^*\right)}{Conv\left(T_B^*, C^*\right)},
\end{aligned}
$$

---

[1] For ease of exposition we omit writing the parameters on the left hand side of the sensitivity equations here and in the sequel where they are clear from context.

giving a first-order sensitivity with respect to CTD of

$$\Delta^F_{CTD}(\mathsf{F}) = \frac{1 + R_L(t,T) \cdot (T-t)}{Conv(T_B^*, C^*)} \qquad (6.4)$$

and a second-order sensitivity with respect to CTD of 0. Note that the forward price, the futures price, and their deltas are not equal per se, because of the coupon-bond futures specifications such as the conversion to a fictitious underlying coupon bond. What we know is that they coincide under the assumptions and definitions of the Black model. So we will have a deeper look at this model and its resulting sensitivities in the next section.

6.1.2 Black Deltas and Gammas

While the prices and deltas of the examples in Section 6.1.1 were derived by arbitrage arguments, we need a special pricing model to evaluate the prices and therefore the first- and second-order sensitivities of options. So we will derive such sensitivities for the Black model in this section. As we have already seen, the assumptions of the Black model are

1. The probability distribution of the underlying stochastic process $S(T)$ at time $T \in [t, T^*]$, as seen from time $t \geq t_0$, is lognormal with a standard deviation of $\sigma \cdot \sqrt{T-t}$.

2. Interest rates are supposed to be deterministic for discounting purposes.

3. The forward price of S for time T, evaluated at time t is equal to the futures price, i.e.

$$F(t,T) = Forward(t,T) = S(t) \cdot P^{-1}(t,T). \qquad (6.5)$$

To derive the forward price of a coupon bond we have to set

$$S(t) = Bond(t, T_B, C) - I(t,T)$$

with T_B, C and $I(t,T)$ given as in Section 6.1.1, i.e.

$$F(t,T) = Forward(t,T) = (Bond(t, T_B, C) - I(t,T)) \cdot P^{-1}(t,T). \quad (6.6)$$

Hence, using equation (6.3), the Black-delta of the futures price, i.e. the delta of the futures price with respect to $\mathsf{F}_1 = Bond(t, T_B, C)$ in the Black model, is given by

$$\Delta^F_{\mathsf{F}_1}(\mathsf{F}) = \Delta^{Forward}_{\mathsf{F}_1}(\mathsf{F}) = P^{-1}(t,T). \qquad (6.7)$$

The corresponding second-order sensitivities with respect to F_1 are 0. From Section 5.8.3 the price at time t of a (European) futures call option on an interest-rate instrument described by the underlying stochastic process $(S(t))_{t \in [t_0, T^*]}$, with $T \in [t_0, T^*]$ denoting the maturity of the option, $F(t,T)$ the futures price of $S(t)$ for maturity time T, evaluated at time $t \in [t_0, T]$, X the strike price of the option, and σ the volatility of the futures price in Black's model, is given by

$$Call_F^{Black}(t,T,X) = e^{-\int_t^T r(s)ds} \cdot [F(t,T) \cdot \mathcal{N}(d_1) - X \cdot \mathcal{N}(d_2)],$$

with

$$d_1 := \frac{\ln\left(\frac{F(t,T)}{X}\right) + \frac{1}{2} \cdot \sigma^2 \cdot (T-t)}{\sigma \cdot \sqrt{T-t}}, \quad d_2 := d_1 - \sigma \cdot \sqrt{T-t}.$$

It can be easily shown that the first-order sensitivity of the (European) futures call option in Black's model with respect to the risk factor $F_2 = F(t,T)$, i.e. the future or forward price of the underlying security, is given by

$$\Delta_{F_2}^{Call_F^{Black}}(F) = e^{-\int_t^T r(s)ds} \cdot \mathcal{N}(d_1). \tag{6.8}$$

The price of a (European) futures put option is given by

$$Put_F^{Black}(t,T,X) = e^{-\int_t^T r(s)ds} \cdot [X \cdot \mathcal{N}(-d_2) - F(t,T) \cdot \mathcal{N}(-d_1)],$$

with a first-order sensitivity with respect to the risk factor $F_2 = F(t,T)$ of

$$\Delta_{F_2}^{Put_F^{Black}}(F) = e^{-\int_t^T r(s)ds} \cdot (\mathcal{N}(d_1) - 1). \tag{6.9}$$

The corresponding second-order sensitivities for the futures call and the futures put option are given by

$$\Gamma_{F_2}^{Call_F^{Black}}(F) = \Gamma_{F_2}^{Put_F^{Black}}(F) = e^{-\int_t^T r(s)ds} \cdot \frac{\mathcal{N}'(d_1)}{F(t,T) \cdot \sigma \cdot \sqrt{T-t}} \tag{6.10}$$

with $\mathcal{N}'(d_1)$ denoting the densitiy of the standard normal distribution evaluated at d_1. Using equations (6.5) and (6.6) and considering $F_1(t) := S(t)$ or $F_1(t) := Bond(t, T_B, C)$ as the underlying process and risk factor, we can derive the Black-deltas and Black-gammas with respect to F_1 as follows:

$$\begin{aligned} \Delta_{F_1}^{Call_F^{Black}}(F) &= \Delta_{F_2}^{Call_F^{Black}}(F) \cdot \Delta_{F_1}^{F_2}(F) \\ &= e^{-\int_t^T r(s)ds} \cdot \mathcal{N}(d_1) \cdot P^{-1}(t,T) \end{aligned}$$

and

$$\begin{aligned} \Delta_{F_1}^{Put_F^{Black}}(F) &= \Delta_{F_2}^{Put_F^{Black}}(F) \cdot \Delta_{F_1}^{F_2}(F) \\ &= e^{-\int_t^T r(s)ds} \cdot (\mathcal{N}(d_1) - 1) \cdot P^{-1}(t,T) \end{aligned}$$

for the Black-deltas, and

$$
\begin{aligned}
\Gamma^{Call_F^{Black}}_{F_1}(F) &= \Gamma^{Put_F^{Black}}_{F_1}(F) \\
&= \Gamma^{Call_F^{Black}}_{F_2}(F) \cdot \Delta^{F_2}_{F_1}(F) + \Delta^{Call_F^{Black}}_{F_2}(F) \cdot \Gamma^{F_2}_{F_1}(F) \\
&= \Gamma^{Call_F^{Black}}_{F_2}(F) \cdot \Delta^{F_2}_{F_1}(F) \\
&= e^{-\int_t^T r(s)ds} \cdot \frac{\mathcal{N}'(d_1)}{F(t,T) \cdot \sigma \cdot \sqrt{T-t}} \cdot P^{-1}(t,T)
\end{aligned}
$$

for the Black-gammas of the futures call and futures put options. Because interest rates are considered to be deterministic for discounting purposes in the Black model, we have

$$
e^{-\int_t^T r(s)ds} = P(t,T)
$$

and so the Black-deltas are given by

$$
\Delta^{Call_F^{Black}}_{F_1}(F) = \mathcal{N}(d_1), \quad \Delta^{Put_F^{Black}}_{F_1}(F_1) = \mathcal{N}(d_1) - 1
$$

and the Black-gammas by

$$
\Gamma^{Call_F^{Black}}_{F_1}(F) = \Gamma^{Put_F^{Black}}_{F_1}(F_1) = \frac{\mathcal{N}'(d_1)}{F(t,T) \cdot \sigma \cdot \sqrt{T-t}}.
$$

Example. Let us return to the coupon-bond option example of Sections 5.6.3 and 5.8.3. Our actual time is $t_0 = t = 10/20/2000$ and we consider the 6.5% German government bond maturing at $T_B = 10/14/2005$ with annual coupon payments. The forward price (dirty) on this bond for the future time $T = 06/20/2002$ is 108.141, the time to expiration of the forward is $T - t = 1.6657$, and the discount factor is $P(t,T) = 0.920342233$. We want to calculate the delta and the gamma of an option on this coupon bond with an exercise clean price of 102.00 and maturing at time T using Black's model for coupon-bond options. The dirty exercise price was calculated to be $X = 106.4342$ and the Black volatility is given by a quote of $\sigma = 4.6914\%$. Assuming that

$$
e^{-\int_t^T r(s)ds} = P(t,T) = 0.920342233
$$

and applying equations (6.8), (6.9), and (6.10), we get

$$
\begin{aligned}
\Delta^{Call_F^{Black}}_{F_2}(F) &= e^{-\int_t^T r(s)ds} \cdot \mathcal{N}(d_1) = 0.92034 \cdot 0.61523 \\
&= 0.56622
\end{aligned}
$$

and

$$
\Delta^{Put_F^{Black}}_{F_2}(F) = e^{-\int_t^T r(s)ds} \cdot (\mathcal{N}(d_1) - 1) = -0.35412.
$$

The corresponding gamma for the put and the call option is given by

$$\Gamma_{F_2}^{Call_F^{Black}}(F) = \Gamma_{F_2}^{Put_F^{Black}}(F) = e^{-\int_t^T r(s)ds} \cdot \frac{\mathcal{N}'(d_1)}{F(t,T)\cdot\sigma\cdot\sqrt{T-t}}$$

$$= 0.92034 \cdot \frac{0.38218}{108.1408\cdot 4.6914\%\cdot 1.6657}$$

$$= 0.05372.$$

Figures 6.1-6.3 show the variation of the Black deltas and gamma with respect to varying forward prices.

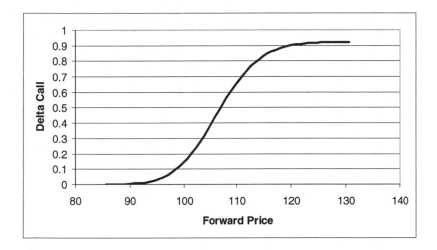

FIGURE 6.1. Variation of the Black delta of a call option with forward price

☐

Beside the problems with the assumptions given above, especially assumption 2 claiming that interest rates are supposed to be deterministic for discounting purposes, the problem appears that Black-deltas and -gammas of options on different coupon bonds cannot be compared directly, since they are sensitivities with respect to different risk factors. On the other hand, if we allow every coupon-bond price to be a risk factor, the number of risk factors will become too large for a reasonable risk management process. So we have to look for better or comparable risk numbers. We will do this by going one step back, remembering that interest-rate derivatives prices change because of changing interest rates. This raises the idea of breaking down the risk of these financial instruments to the zero rates of different maturities defined as risk factors.

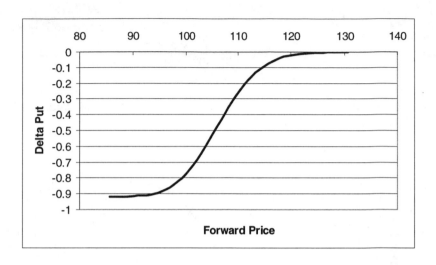

FIGURE 6.2. Variation of the Black delta of a put option with forward price

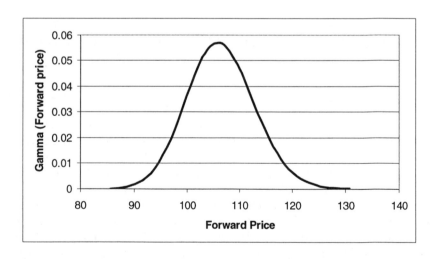

FIGURE 6.3. Variation of the Black gamma with forward price

6.1.3 Duration and Convexity

Remembering the interest-rate market model we defined in Section 4, the primary traded assets in this market are the zero-coupon bonds of different maturities T and prices $P(t,T)$ at time t with $t_0 \le t \le T \le T^*$. These zero-coupon bond prices are usually translated into the (continuous) zero rates or spot rates $R(t,T)$ at time t for the maturity time T or for the time to maturity $T - t$ by the relation

$$P(t,T) = e^{-R(t,T)\cdot(T-t)} \quad \text{or} \quad R(t,T) = -\frac{\ln P(t,T)}{T-t}.$$

It is convenient to consider every financial instrument traded in the interest-rate market to be given as a function of a finite number of these zero-coupon bonds or zero rates. Following this assumption, the natural risk factors for this market are the zero rates of different maturities $T \in [t_0, T^*]$. Interest-rate derivatives depending on different underlyings but indirectly, via the underlyings, depending on the same zero rates can be made comparable by calculating sensitivities with respect to these zero rates. Let us apply this idea to derive the sensitivities of a coupon bond with coupon payments $C = (C(T_1), ..., C(T_n))$, $t_0 \le T_1 < \cdots < T_n \le T^*$, and maturity $T_B = T_n$. The (dirty) price of this coupon bond at time $t \in [t_0, T_1]$ is given by

$$Bond\,(t, T_B, C) = \sum_{i=1}^{n} C(T_i) \cdot P(t, T_i) = \sum_{i=1}^{n} C(T_i) \cdot e^{-R(t,T_i)\cdot(T_i-t)}.$$

Choosing the risk factors $F_j(t) := R(t, T_j)$, $j = 1, ..., n$, the first- and second-order sensitivities of this coupon bond with respect to these risk factors F_j, $j = 1, ..., n$, are given by

$$\Delta_{F_j}^{Bond}(F) = -(T_j - t) \cdot C(T_j) \cdot e^{-R(t,T_j)\cdot(T_j-t)} \tag{6.11}$$

and

$$\Gamma_{F_j,F_l}^{Bond}(F) = \begin{cases} (T_j - t)^2 \cdot C(T_j) \cdot e^{-R(t,T_j)\cdot(T_j-t)} & \text{, if } j = l \\ 0 & \text{, if } j \neq l. \end{cases} \tag{6.12}$$

Using these sensitivities, the approximate price-change $\Delta Bond\,(t, T_B, C)$ of the coupon bond, depending on small changes of the risk factors F_j,

$j = 1, ..., n$, is given by

$$\Delta Bond\,(t, T_B, C) \approx \sum_{j=1}^{n} \Delta_{\mathsf{F}_j}^{Bond}\,(\mathsf{F}) \cdot \Delta \mathsf{F}_j$$

$$+ \frac{1}{2} \cdot \sum_{j=1}^{n} \sum_{l=1}^{n} \Gamma_{\mathsf{F}_j \mathsf{F}_l}^{Bond}\,(\mathsf{F}) \cdot \Delta \mathsf{F}_j \cdot \Delta \mathsf{F}_l$$

$$= \sum_{j=1}^{n} \Delta_{\mathsf{F}_j}^{Bond}\,(\mathsf{F}) \cdot \Delta R\,(t, T_j)$$

$$+ \frac{1}{2} \cdot \sum_{j=1}^{n} \Gamma_{\mathsf{F}_j \mathsf{F}_j}^{Bond}\,(\mathsf{F}) \cdot \Delta R^2\,(t, T_j).$$

It can be easily seen that the number of risk factors explodes with the number of different coupon payment and maturity dates as more coupon bonds enter the picture. On the other hand, we probably do not want to have different risk factors and risk sensitivities for payment dates or maturities which lie close together. To put it on a hedging ground, we want the risk sensitivities to show us that a coupon bond with maturity T_B may be hedged using another coupon bond with maturity T_B' if the maturities T_B and T_B' lie within a previously defined time segment or bucket. The idea is to suppose that zero rates within the same time or maturity segment move exactly the same way, i.e. with the same delta. The simplest way to realize this idea is to define only one maturity segment and assume that all zero rates move by

$$\Delta R\,(t, T) := \Delta \mathsf{F}\,(t) \quad \text{for all } T \in [t, T^*],$$

which is equivalent to assuming that the zero-rate curve moves by *parallel shifts only*. Doing this, we get

$$\Delta Bond\,(t, T_B, C) \approx \sum_{j=1}^{n} \Delta_{\mathsf{F}_j}^{Bond}\,(\mathsf{F}) \cdot \Delta \mathsf{F} + \frac{1}{2} \cdot \sum_{j=1}^{n} \Gamma_{\mathsf{F}_j \mathsf{F}_j}^{Bond}\,(\mathsf{F}) \cdot \Delta \mathsf{F}^2$$

$$= \Delta^{Bond}\,(\mathsf{F}) \cdot \Delta \mathsf{F} + \frac{1}{2} \cdot \Gamma^{Bond}\,(\mathsf{F}) \cdot \Delta \mathsf{F}^2 \qquad (6.13)$$

with

$$\Delta^{Bond}\,(\mathsf{F}) := \sum_{j=1}^{n} \Delta_{\mathsf{F}_j}^{Bond}\,(\mathsf{F}) \quad \text{and} \quad \Gamma^{Bond}\,(\mathsf{F}) := \sum_{j=1}^{n} \Gamma_{\mathsf{F}_j \mathsf{F}_j}^{Bond}\,(\mathsf{F}).$$

Multiplied by 1 basis point (bp) which is $0.01\% = 10^{-4}$ the expression

$$\Delta^{Bond}\,(\mathsf{F}) \cdot 1bp = \Delta^{Bond(t, T_B, C)}\,(\mathsf{F}) \cdot 1bp \qquad (6.14)$$

is known as the *price value of a basis point* and gives an information on how much a coupon-bond price would move if the zero-rate curve underlies a parallel shift of one basis point. The expression $\Gamma^{Bond}(F) = \Gamma^{Bond(t,T_B,C)}(F)$ is known as the *convexity*[2] of the coupon bond at time $t \in [t_0, T_B]$, also denoted by *convexity* (t, T_B, C), and is due to the non-linear reaction of the coupon-bond price on a parallel shift of the zero-rate curve. Portfolio managers are also interested in the relative change of a coupon-bond price, which is approximately given by

$$\frac{\Delta Bond(t,T_B,C)}{Bond(t,T_B,C)} \approx \frac{\Delta^{Bond(t,T_B,C)}(F)}{Bond(t,T_B,C)} \cdot \Delta F(t) + \frac{1}{2} \cdot \frac{\Gamma^{Bond(t,T_B,C)}(F)}{Bond(t,T_B,C)} \cdot \Delta F^2(t).$$

Definition 6.2 *The expression*

$$duration(t, T_B, C) := -\frac{\Delta^{Bond(t,T_B,C)}(F)}{Bond(t,T_B,C)} \tag{6.15}$$

is called the **duration** *of the bond* (T_B, C) *at time* $t \in [t_0, T_B]$.

The duration tells us how much the relative bond price would move if the zero-rate curve undergoes a small parallel shift. The negative sign shows that the coupon-bond price will fall as interest rates go up. It also gives us the probably most popular version of equation (6.13):

$$\Delta Bond(t, T_B, C) \approx -duration(t, T_B, C) \cdot Bond(t, T_B, C) \cdot \Delta F(t)$$
$$+\frac{1}{2} \cdot convexity(t, T_B, C) \cdot \Delta F^2(t).$$

Inserting for $\Delta^{Bond(t,T_B,C)}(F)$ into equation (6.15) we get another interpretation for the duration:

$$duration(t, T_B, C) = -\frac{\Delta^{Bond(t,T_B,C)}(F)}{Bond(t,T_B,C)} = -\sum_{j=1}^{n} \frac{\Delta_{F_j}^{Bond(t,T_B,C)}(F)}{Bond(t,T_B,C)}$$

$$= \sum_{j=1}^{n} (T_j - t) \cdot \frac{C(t_j) \cdot e^{-R(t,T_j)\cdot(T_j-t)}}{Bond(t,T_B,C)}$$

$$= \sum_{j=1}^{n} (T_j - t) \cdot \alpha_j$$

with

$$\alpha_j := \frac{C(t_j) \cdot e^{-R(t,T_j)\cdot(T_j-t)}}{Bond(t,T_B,C)}, \quad j = 1, ..., n.$$

---

[2] *Sometimes the factor $\frac{1}{2}$ of the Taylor expansion is already included in the convexity. Also the convexity may be already multiplied by $\Delta F^2(t) = (1bp)^2$ to make it comparable to the price value of a basis point.*

Since $\alpha_j \geq 0$ for all $j = 1,...,n$ and $\sum_{j=1}^n \alpha_j = 1$, the duration is the weighted average of the time to maturity of each payment where the weights are the proportion of the present value at time t of this payment relative to the present value of the coupon bond. It can be easily seen that the duration of a zero-coupon bond is equal to the time to maturity of the zero coupon bond. The duration $duration(t, \varphi)$ of a portfolio $\varphi = (\varphi_1,...,\varphi_n)$ of coupon bonds with dirty prices $Bond(t, T_B^i, C^i)$ and durations denoted by $duration(t, T_B^i, C^i)$, $i = 1,...,n$, at time $t \in [t_0, \min_{i=1,...,n} T_B^i]$ is given by

$$duration(t, \varphi) = -\frac{\Delta^{V(\varphi)}(F)}{V(\varphi, t)} \tag{6.16}$$

$$= -\frac{1}{V(\varphi, t)} \cdot \sum_{i=1}^n \varphi_i(t) \cdot \Delta^{Bond(t, T_B^i, C^i)}(F)$$

$$= \sum_{i=1}^n \frac{\varphi_i(t) \cdot Bond(t, T_B^i, C^i)}{V(\varphi, t)} \cdot duration(t, T_B^i, C^i),$$

where $V(\varphi, t)$ denotes the dirty price of the portfolio. Hence, the duration of a coupon-bond portfolio is the dirty price weighted average of the single coupon-bond durations. Because of this interpretation, the duration is still a very popular risk measure for bond portfolio managers who try to keep the duration, i.e. the relative price-change, of a portfolio within a given interval or close to that of a given market index or benchmark portfolio they want to duplicate (see, e.g. Elton and Gruber [EG91] for more details).

Definition 6.3 *Given the yield-to-maturity $y(t, T_B, C)$ of a coupon bond (T_B, C) at time $t \in [t_0, T_1]$, the so-called* **Macaulay duration** *of that coupon bond at time t is defined by*

$$duration_{Mac}(y, t, T_B, C) = \sum_{i=1}^n (T_i - t) \cdot \frac{C(T_i) \cdot (1+y)^{-(T_i-t)}}{Bond(t, T_B, C)}$$

and the **modified duration** *by*

$$duration_{mod}(y, t, T_B, C) = \frac{duration_{Mac}(y, t, T_B, C)}{1 + y(t, T_B, C)}.$$

The first-order sensitivity with respect to a small shift of the yield-to-maturity as a risk factor is given by

$$\Delta^{Bond(t, T_B, C)}(y) = -\sum_{i=1}^n (T_i - t) \cdot C(T_i) \cdot (1 + y(t, T_B, C))^{-(T_i-t)-1}$$

$$= -\frac{duration_{Mac}(y, t, T_B, C)}{1 + y(t, T_B, C)} \cdot Bond(t, T_B, C)$$

$$= -duration_{mod}(y, t, T_B, C) \cdot Bond(t, T_B, C).$$

Ignoring the second-order sensitivities, we get

$$\Delta Bond\left(t, T_B, C\right) \approx \Delta^{Bond(t,T_B,C)}\left(y\right) \cdot \Delta y\left(t, T_B, C\right)$$

$$\approx -duration_{mod}\left(y, t, T_B, C\right) \cdot Bond\left(t, T_B, C\right) \cdot \Delta y\left(t, T_B, C\right)$$

$$\approx -duration_{Mac}\left(y, t, T_B, C\right) \cdot Bond\left(t, T_B, C\right) \cdot \frac{\Delta y\left(t, T_B, C\right)}{1 + y\left(t, T_B, C\right)}$$

as an approximate change of the coupon-bond price with respect to a changing yield-to-maturity. Depending on the assumption that the absolute or relative changes of the yield-to-maturities for different coupon bonds or coupon-bond portfolios are equal, either the Macaulay or the modified duration is controlled.

Example. Let us revisit the coupon-bond example of Section 5.2. Our actual time is $t_0 = t = 10/20/2000$ and we are dealing with a 6.5% German government bond maturing at $T_B = 10/14/2005$ and annual coupon payments. The discount factors are given from market data and the dirty price of the coupon bond was calculated to be $Bond\left(t, T_B, C\right) = 105.71737$ for a notional of 100 Euro. Using equations (6.11) and (6.12) we get the following table:

T_j	$T_j - t$	$P(t, T_j)$	coupon	$\Delta_{F_j}^{Bond(t,T_B,C)}$	$\Gamma_{F_j,F_j}^{Bond(t,T_B,C)}$
10/14/01	0.9836	0.9524	6.5	-6.089057	5.988963
10/14/02	1.9836	0.9056	6.5	-11.676660	23.161376
10/14/03	2.9836	0.8604	6.5	-16.686871	49.785833
10/14/04	3.9836	0.8179	6.5	-21.178415	84.366474
10/14/05	4.9836	0.7768	106.5	-412.292845	2054.686806

From this we get

$$\Delta^{Bond(t,T_B,C)}\left(F\right) = \sum_{j=1}^{5} \Delta_{F_j}^{Bond(t,T_B,C)}\left(F\right) = -467.9237$$

with a duration of

$$duration\left(t, T_B, C\right) = -\frac{\Delta^{Bond(t,T_B,C)}\left(F\right)}{Bond\left(t, T_B, C\right)} = \frac{467.9237}{105.71737} = 4.4262$$

and a convexity of

$$convexity\left(t, T_B, C\right) = \Gamma^{Bond(t,T_B,C)}\left(F\right) = \sum_{j=1}^{5} \Gamma_{F_j,F_j}^{Bond(t,T_B,C)}\left(F\right)$$

$$= 2217.9895.$$

A parallel shift of the zero-rate curve of $1bp = 10^{-4}$ therefore results in an approximate price-change of

$$\Delta Bond\,(\mathsf{F}) = \Delta^{Bond(t,T_B,C)}\,(\mathsf{F}) \cdot 1bp = -0.0468.$$

The yield-to-maturity can be calculated (see example of Section 5.2) to be $y\,(t,T_B,C) = 5.193\%$, which gives us a Macaulay duration of

$$duration_{Mac}\,(y,t,T_B,C) = 4.4268,$$

a modified duration of

$$duration_{\mathrm{mod}}\,(y,t,T_B,C) = 4.2082,$$

and an approximate price-change resulting from a $1bp$ parallel shift of the yield curve of

$$
\begin{aligned}
\Delta Bond\,(y) &= -duration_{\mathrm{mod}}\,(y,t,T_B,C) \cdot Bond\,(t,T_B,C) \cdot 1bp \\
&= -0.0445.
\end{aligned}
$$

$\Box$

6.1.4 Key-Rate Deltas and Gammas

It is not surprising that one serious problem with the duration concept is the assumption that all zero rates move by exactly the same amount. In practice it is quite possible that short- and long-term zero rates move in opposite directions changing the whole slope of the zero-rate curve. For this reason, risk managers often feel more comfortable using more than one time to maturity segment. To do this let us divide the time to maturity interval $[0,T^* - t_0]$ into $m \in I\!N$ non-overlapping subintervals $KB_1, ..., KB_m$ with $[0,T^* - t] \subseteq \bigcup_{j=1,...,m} KB_j = [0,T^* - t_0]$. Dividing the time to maturity interval is preferred to dividing the maturity time interval $[t,T^*]$ because the latter changes as time t passes by while the first one stays the same still keeping the property of always starting relative to the actual time t. Since this simultaneously gives us m segments of the zero-rate curve we call $KB_1, ..., KB_m$ zero-rate or *key-rate buckets*. At time t, all zero rates having a time to maturity within the same bucket KB_j are supposed to move by exactly the same amount $\Delta\mathsf{F}_j\,(t)$, $j = 1, ..., m$. The risk factor $\mathsf{F}_j\,(t)$ therefore is the *key* for all zero-*rate* movements within the *bucket* KB_j, $j = 1, ..., m$, which explains the name of these buckets. Now let $D\,(R\,(t)\,,t)$ be the price of a financial instrument or derivative depending on (some elements of) the vector of zero rates[3]

$$R\,(t) := (R_1\,(t)\,,...,R_n\,(t))' := (R\,(t,T_1)\,,...,R\,(t,T_n))'$$

---

[3] Here, for convenience, we suppose that the vector $R = (R_1,...,R_n)$ with $R_i\,(t) := R\,(t,T_i)$, $i = 1, ..., n$, of zero rates includes all zero rates on which any of the

and time $t \in [t_0, T^*]$ with $t_0 \le t \le T_1 < \cdots < T_n \le T^*$.

Definition 6.4 (Key-Rate Delta and Gamma) *The key-rate delta $\Delta_{KB_j}^D (R)$ of a financial instrument or derivative with price $D = D (R)$ and with respect to a small change of size $\delta > 0$ of the risk factor F_j is defined by*

$$\Delta_{KB_j}^D (R) := \frac{\Delta_j D (R)}{2 \cdot \delta},$$

with $\Delta_j D (R)$ defined by[4]

$$\Delta_j D (R) := D \left(R + \sum_{i \in KB_j} e_i \cdot \delta \right) - D \left(R - \sum_{i \in KB_j} e_i \cdot \delta \right)$$

and e_i denoting the $n-$dimensional unit vector with 1 in the $i-$th component and 0 else. Correspondingly, the key-rate gamma $\Gamma_{KB_j, KB_l}^D (R)$ of a financial instrument or derivative with price $D = D (R)$ and with respect to small changes of the risk factors KB_j and KB_l is defined by

$$\Gamma_{KB_j KB_l}^D (R) := \frac{\Delta_j (\Delta_l D (R))}{4 \cdot \delta^2},$$

with $\Delta_j (\Delta_l D (R))$ defined by

$$\begin{aligned}
\Delta_j (\Delta_l D (R)) \quad := \quad & D \left(R + \sum_{i \in KB_j} e_i \cdot \delta + \sum_{i \in KB_l} e_i \cdot \delta \right) \\
- \quad & D \left(R - \sum_{i \in KB_j} e_i \cdot \delta + \sum_{i \in KB_l} e_i \cdot \delta \right) \\
- \quad & D \left(R + \sum_{i \in KB_j} e_i \cdot \delta - \sum_{i \in KB_l} e_i \cdot \delta \right) \\
+ \quad & D \left(R - \sum_{i \in KB_j} e_i \cdot \delta - \sum_{i \in KB_l} e_i \cdot \delta \right).
\end{aligned}$$

---

derivatives we deal with may depend. This is not really a problem. If we want to deal with a low number n of zero rates we can define any other interest rate $Y \notin \{R_1, ..., R_n\}$ with $R_i < Y < R_{i+1}$ to be the linear interpolation $Y = \lambda \cdot R_i + (1 - \lambda) \cdot R_{i+1}$ with a suitable $\lambda \in (0, 1)$.

[4]Note that here and in the sequel, we write $i \in KB_j$ instead of $i \in \{1, ..., n\}$: $t_i \in KB_j$, $j = 1, ..., m$, for ease of exposition.

Using a Taylor series expansion of $\Delta_j D(R)$ and $\Delta_j(\Delta_l D(R))$, a few steps of calculation show that

$$\Delta_{KB_j}^D(R) \approx \sum_{i \in KB_j} \Delta_{R_i}^D(R) \quad \text{and} \quad \Gamma_{KB_j KB_l}^D(R) \approx \frac{1}{2} \cdot \sum_{i \in KB_j} \sum_{k \in KB_l} \Gamma_{R_i R_k}^D(R).$$

$$(6.17)$$

As an *example* it is interesting to note that the key-rate gamma for a coupon bond with price $Bond(t, T_B, C)$ is approximately given by

$$\Gamma_{KB_j KB_l}^{Bond}(R) \approx \begin{cases} \frac{1}{2} \cdot \displaystyle\sum_{i \in KB_j} \Gamma_{R_i R_i}^{Bond}(R), & \text{if } j = l \\ 0, & \text{if } j \neq l, \end{cases}$$

which can be easily seen by combining equations (6.12) and (6.17). Hence, thinking of key-rate gammas as a matrix in $j, l = 1, ..., m$, coupon bonds have key-rate gamma exposure only on the diagonal elements of the matrix.

The approximate price-change $\Delta D(R)$ of a derivative, depending on small changes of the risk factors F, is given by

$$\begin{aligned} \Delta D(R) &\approx \sum_{i=1}^{n} \Delta_{R_i}^D(R) \cdot \Delta R_i + \frac{1}{2} \cdot \sum_{i=1}^{n} \sum_{k=1}^{n} \Gamma_{R_i R_k}^D(R) \cdot \Delta R_i \cdot \Delta R_k \\ &= \sum_{j=1}^{m} \sum_{i \in KB_j} \Delta_{R_i}^D(R) \cdot \Delta F_j \\ &\quad + \frac{1}{2} \cdot \sum_{j=1}^{m} \sum_{l=1}^{m} \sum_{i \in KB_j} \sum_{k \in KB_l} \Gamma_{R_i R_k}^D(R) \cdot \Delta F_j \cdot \Delta F_l \\ &\approx \sum_{j=1}^{m} \Delta_{KB_j}^D(R) \cdot \Delta F_j + \sum_{j=1}^{m} \sum_{l=1}^{m} \Gamma_{KB_j KB_l}^D(R) \cdot \Delta F_j \cdot \Delta F_l. \end{aligned}$$

Multiplied by 1 basis point (bp), the expression $\Delta_{KB_j}^D(R) \cdot 1bp$ gives information[5] on how much the price of the derivative would move if the zero-rate curve increased by a parallel shift of one basis point within key-rate bucket $KB_j, j = 1, ..., m$, as a linear approximation, while the expression

$$\Gamma_{KB_j KB_l}^D(R) \cdot (1bp)^2 = \Gamma_{KB_j KB_l}^D(R) \cdot 10^{-8}$$

gives the add-on to the linear approximation $\Delta_{KB_j}^D(R) \cdot 1bp$ with respect to a non-linear (quadratic) approximation of the derivatives price-change

---

[5] This is the information which is plotted for practical applications, as is done in the case studies of Section 7.1.3.

if the zero-rate curve underlies a parallel shift of one basis point within key-rate buckets KB_j and KB_l, $j, l = 1, ..., m$. If we choose $m = 1$ as a special case, we suppose that the zero-rate curve is changing by parallel shifts only, and we get

$$\Delta D\left(R\right) \approx \Delta_{KB_1}^{D}\left(R\right) \cdot \Delta \mathsf{F}_1 + \Gamma_{KB_1 KB_1}^{D}\left(R\right) \cdot \left(\Delta \mathsf{F}_1\right)^2.$$

If we choose the derivative to be a coupon bond with price $Bond\left(t, T_B, C\right)$, we learn from (6.17) that the resulting approximation can be considered as a generalization of equation (6.13). To derive the key-rate delta and gamma of a portfolio, let $V\left(\varphi, R\right)$ be the price of a portfolio $\varphi = \left(\varphi_1, ..., \varphi_n\right)$ of financial instruments or derivatives, with prices $D_1\left(R\right), ..., D_n\left(R\right)$ depending on the vector of zero rates R, i.e.

$$V\left(\varphi\right) = V\left(\varphi, R\right) = \sum_{i=1}^{n} \varphi_i \cdot D_i\left(R\right).$$

Then the key-rate deltas of the portfolio φ are given by

$$\Delta_{KB_j}^{V\left(\varphi\right)}\left(R\right) = \sum_{i=1}^{n} \varphi_i \cdot \Delta_{KB_j}^{D_i}\left(R\right), \, j = 1, ..., m,$$

and the key-rate gammas of the portfolio φ are given by

$$\Gamma_{KB_j KB_l}^{V\left(\varphi\right)}\left(R\right) = \sum_{i=1}^{n} \varphi_i \cdot \Gamma_{KB_j KB_l}^{D_i}\left(R\right), \, j, l = 1, ..., m.$$

Furthermore, the price-change $\Delta V\left(\varphi, R\right)$ of the portfolio $\varphi = \left(\varphi_1, ..., \varphi_n\right)$ is approximately given by

$$\Delta V\left(\varphi, R\right) \approx \sum_{j=1}^{m} \Delta_{KB_j}^{V\left(\varphi\right)}\left(R\right) \cdot \Delta \mathsf{F}_j + \sum_{j=1}^{m} \sum_{l=1}^{m} \Gamma_{KB_j KB_l}^{V\left(\varphi\right)}\left(R\right) \cdot \Delta \mathsf{F}_j \cdot \Delta \mathsf{F}_l.$$

Hence, if the trader or risk manager manages to set up a portfolio with

$$\Delta_{KB_j}^{V\left(\varphi\right)}\left(R\right) = 0 \text{ and } \Gamma_{KB_j KB_l}^{V\left(\varphi\right)}\left(R\right) = 0 \text{ for all } j, l = 1, ..., m,$$

his portfolio is nearly insensitive to parallel movements within the key-rate buckets $KB_1, ..., KB_m$. By using key-rate buckets a financial institution or portfolio manager is able to examine the effect of small changes in all the zero rates with maturities lying in a special key-rate bucket while leaving all other zero rates unchanged. If this exposure would be uncomfortable selected instruments would be used to reduce it carefully without changing the exposure with respect to the other key-rate buckets. In the context of managing the assets and liabilities of a portfolio, this approach is sometimes referred to as *GAP management* (see, e.g. Hull [Hul00], p. 114).

6.1.5 Other Sensitivity Measures

Interest-rate derivatives prices usually depend not only on the prices of underlying coupon bonds, futures or zero rates. Prices also change just because of time to maturity effects or because of changes in the volatility structure. So the time t, the volatility parameter σ and, for the Hull-White model, the mean reversion rate a as well as the short rate r are included in the vector of risk factors F. The *theta* of a portfolio $\varphi = (\varphi_1, ..., \varphi_n)$ of financial instruments or derivatives with prices $D_1(\mathsf{F}), ..., D_n(\mathsf{F})$ depending on the risk factors F is defined to be the first-order sensitivity of the portfolio value $V(\mathsf{F}, \varphi)$ with respect to a small change in time, with all other risk factors remaining the same, i.e.

$$\Theta^{V(\varphi)}(\mathsf{F}) = \Delta_t^{V(\varphi)}(\mathsf{F}) := \frac{\partial}{\partial t} V(\mathsf{F}, \varphi).$$

It is also referred to as the *time decay* of the derivative. As an *example* the theta of a (European) futures call option with maturity time T and exercise price X in Black's model, holding the futures price $\mathsf{F}_2 = F(t, T) =: F$ fixed, is given by

$$\Theta^{Call_F^{Black}}(\mathsf{F}) = r(t) \cdot Call_F^{Black}(t, T, X) - e^{-\int_t^T r(s)ds} \cdot \frac{F \cdot \mathcal{N}'(d_1) \cdot \sigma}{2 \cdot \sqrt{T-t}}.$$

The corresponding theta of a (European) futures put option is given by

$$\Theta^{Put_F^{Black}}(\mathsf{F}) = r(t) \cdot Put_F^{Black}(t, T, X) - e^{-\int_t^T r(s)ds} \cdot \frac{F \cdot \mathcal{N}'(d_1) \cdot \sigma}{2 \cdot \sqrt{T-t}}.$$

The first-order sensitivity with respect to small changes in the volatility or the mean reversion, if relevant, are referred to as *vega$_1$* and *vega$_2$*, i.e.

$$\mathcal{V}_1^{V(\varphi)}(\mathsf{F}) = \Delta_\sigma^{V(\varphi)}(\mathsf{F}) := \frac{\partial}{\partial \sigma} V(\mathsf{F}, \varphi)$$

and

$$\mathcal{V}_2^{V(\varphi)}(\mathsf{F}) = \Delta_a^{V(\varphi)}(\mathsf{F}) := \frac{\partial}{\partial a} V(\mathsf{F}, \varphi).$$

As an *example* the vega$_1$ of a (European) futures call or put option with maturity time T and exercise price X in Black's model is given by

$$\mathcal{V}_1^{Call_F^{Black}}(\mathsf{F}) = \mathcal{V}_1^{Put_F^{Black}}(\mathsf{F}) = e^{-\int_t^T r(s)ds} \cdot F(t, T) \cdot \mathcal{N}'(d_1) \cdot \sqrt{T-t}.$$

Even if we have assumed in some of our pricing models that the volatility or volatility structure of an underlying asset, forward or short rate is constant, volatility does change over time in practice. So it does make sense to control the potential changes of a portfolio as a consequence of changing volatility. Another first-order sensitivity, called the *rho* of a portfolio, is defined as the change of the portfolio value with respect to small changes in the short rate, i.e.

$$\mathcal{R}^{V(\varphi)}\left(\mathsf{F}\right) = \Delta_r^{V(\varphi)}\left(\mathsf{F}\right) := \frac{\partial}{\partial r} V\left(\mathsf{F}, \varphi\right).$$

As an *example* the rho of a (European) futures call option with maturity time T and exercise price X in Black's model with constant interest rate $r(s) \equiv r$, $s \in [t, T]$, and holding the futures price $\mathsf{F}_2 = F(t, T) =: F$ fixed, is given by

$$\mathcal{R}^{Call_F^{Black}}\left(\mathsf{F}\right) = -\left(T - t\right) \cdot Call_F^{Black}\left(t, T, X\right).$$

The corresponding rho of a (European) futures put option is given by

$$\mathcal{R}^{Put_F^{Black}}\left(\mathsf{F}\right) = -\left(T - t\right) \cdot Put_F^{Black}\left(t, T, X\right).$$

Rho is not only of interest for the Black model with constant interest rate r but also for short-rate models such as the Hull-White model, where r can be considered as the underlying of the derivatives prices or portfolio value. In this case, rho can be considered as the delta of the portfolio value with respect to the underlying short rate. It could also make sense to control the second-order sensitivity with respect to r as the gamma within a short-rate model. Nevertheless, we already captured the risk with respect to the short rate by measuring the key-rate delta of the first, probably small, key-rate bucket. Examples for the sensitivity measures theta, vega and rho are given, e.g., in Hull [Hul00], p. 312-329.

6.2   Downside Risk Measures

In practice, traders, risk and portfolio managers are not continuously rebalancing their portfolios to ensure first- and second-order measures of zero, also called first- or second-order neutrality, over time. Since this would be too expensive, they rather decide whether a calculated first or second-order sensitivity is acceptable or not, indicating a short-term risk due to small movements of the risk factors inherent in their portfolio. Only in the latter case do they adjust their portfolio. Beside this short-term sensitivity risk, portfolio managers change their portfolio if the risk of the portfolio return falling below a given benchmark return is too high. This considers large movements of the risk factors as well as longer-term horizons, and is also known as *downside risk*. It is usually carried out by a scenario analysis and involves calculating the portfolio's profit or loss over a specified period under a variety of different scenarios. The time horizon may depend on the liquidity of the instruments or the planning period of the portfolio manager. The scenarios can either be chosen by the management or generated by a scenario simulation. One possibility is thus to use a *Monte Carlo simulation* based on a specific interest-rate market model which enables the trader, risk or portfolio manager to derive a complete probability distribution of the future profits and losses. Using this distribution, different measures of downside risk such as the *lower partial moments* or the *value at risk* can be calculated. These are the measures we will discuss in the following sections.

6.2.1 Lower Partial Moments

Throughout the traditional portfolio theory introduced by Markowitz [Mar52] and Sharpe [Sha64], the risk of a portfolio of financial products is measured by the variance or standard deviation of the portfolio return over a given planning horizon $T \in [t_0, T^*]$, with t_0 and T^* given by our interest-rate market model. Given this model, the prices at time $t \in [t_0, T]$ of the interest-rate derivatives we consider may be calculated as $D_1(t), ..., D_n(t)$. At each time t, the future value $V(\varphi, t, T)$ of a portfolio $\varphi = (\varphi_1, ..., \varphi_n)$ of these derivatives at the end of the planning horizon T is a random variable, and depends on the probability background of the financial market model, especially on the filtration $\mathcal{F}$, the probability measure Q, and the pricing model. Hence, the portfolio return $R(\varphi, t, T)$, usually quoted for discrete compounding, is random and defined by[6]

$$R(\varphi, t, T) := \frac{V(\varphi, t, T) - V(\varphi, t)}{V(\varphi, t)}, \tag{6.18}$$

with $V(\varphi, t)$ denoting the portfolio value at time t. It is important to note that for the definitions and calculations in this section we consider the portfolio φ to be fixed from the time of evaluation t to the planning horizon T. Given these definitions, the variance of the portfolio return at time t is defined by

$$Var_Q[R(\varphi, t, T)|\mathcal{F}_t] := E_Q[R^2(\varphi, t, T)|\mathcal{F}_t] - (E_Q[R(\varphi, t, T)|\mathcal{F}_t])^2.$$

Bookstaber and Clarke [BC81] and [BC83], Pelsser and Vorst [PV95], and Scheuenstuhl and Zagst [SZ96] and [SZ97] show that adding options to a portfolio may lead to rather asymmetric return distributions. But the use of variance as a measure of risk for asymmetric return distributions has already been questioned by financial theorists such as, e.g., Markowitz

---

[6]Note that the return is for the period from t_0 to T and is therefore not annualized. The only reason we do not work with annualized returns here is to simplify the transformation between returns and absolute portfolio values, which we will consider in Sections 6.2.2 and 7.2.1.

[Mar70] and [Mar91], p. 188-201. He states that risk measured by semi-variance tends to produce better results for portfolio analyses than risk measured by variance. The so-called lower partial moments can be considered as a generalization of the semi-variance (see, e.g. Harlow and Rao [HR89] or Harlow [Har91]). Following Harlow [Har91], we give the following definition.

Definition 6.5 (Lower Partial Moment) *Let $\varphi = (\varphi_1, ..., \varphi_n)$ be a portfolio of interest-rate derivatives, $[t, T]$ be the considered time-period, and $R(\varphi, t, T)$ be the portfolio return over that period as defined in equation (6.18). Then the lower partial moment $LPM_l(\varphi, R, b, t, T)$ of order $l \in \mathbb{N}$ for the portfolio return, with respect to a given investor-specific benchmark or threshold return $b(t, T) \in \mathbb{R}$, is defined by*

$$LPM_l(\varphi, R, b, t, T) = E_Q\left[1_{(-\infty, b(t,T))}\left(R(\varphi, t, T)\right) \cdot (b(t, T) - R(\varphi, t, T))^l \,|\mathcal{F}_t\right].$$

The corresponding lower partial moment $LPM_l(\varphi, V, B, t, T)$ of order $l \in \mathbb{N}$ for the future portfolio value, with respect to a given investor-specific benchmark or threshold $B(t, T) \in \mathbb{R}$, is defined by

$$LPM_l(\varphi, V, B, t, T) = E_Q\left[1_{(-\infty, B(t,T))}\left(V(\varphi, t, T)\right) \cdot (B(t, T) - V(\varphi, t, T))^l \,|\mathcal{F}_t\right]$$

with

$$B(t, T) := V(t, T) \cdot (1 + b(t, T))$$

defining the gateway between the lower partial moments for the portfolio return and for the future portfolio value.

Since

$$
\begin{aligned}
B(t, T) - V(\varphi, t, T) &= V(t, T) \cdot (1 + b(t, T)) - V(t, T) \cdot (1 + R(\varphi, t, T)) \\
&= V(t, T) \cdot (b(t, T) - R(\varphi, t, T)),
\end{aligned}
$$

it can be easily seen that

$$LPM_l(\varphi, V, B, t, T) = [V(t, T)]^l \cdot LPM_l(\varphi, R, b, t, T)$$

if $V(t, T) > 0$. The lower partial moment only considers realizations of the portfolio return or future value below the investor-specific benchmark, measured to a power of l. For $l = 0$ this is just the probability that the random portfolio return or future value falls below the given benchmark, which is referred to as *shortfall probability*. Setting the benchmark equal to 0, this is the probability of loss. For $l = 1$, the lower partial moment is the expected deviation of the portfolio returns or future values below the benchmark, sometimes called (expected) *regret*. For $l = 2$, the lower partial moment is weighting the squared deviations below the benchmark and thus is the *semi-variance* if the benchmark is set equal to the expected value of the portfolio return or future value.

For ease of exposition, we concentrate on the case $t = t_0$, knowing that, $Q - a.s.$,

$$LPM_l \left(\varphi, R, b, t_0, T \right) = E_Q \left[1_{(-\infty, b(t_0, T))} \left(R \left(\varphi, t_0, T \right) \right) \cdot \left(b \left(t_0, T \right) - R \left(\varphi, t_0, T \right) \right)^l \right].$$
(6.19)

If we also know the distribution function of the portfolio return $R \left(\varphi, t_0, T \right)$ and denote it by $F_{R(\varphi, t_0, T)}$, equation (6.19) is equivalent to

$$LPM_l \left(\varphi, R, b, t_0, T \right) = \int_{-\infty}^{b(t_0, T)} \left(b \left(t_0, T \right) - x \right)^l \, dF_{R(\varphi, t_0, T)} \left(x \right). \quad (6.20)$$

The problem with equations (6.19) and (6.20) is that the distribution function of the portfolio return is not always known. Especially, when derivatives are included in the portfolio, the return distribution may become asymmetric, and cannot be described by a normal distribution. In this case it may be profitable to use a simulation technique such as a Monte Carlo simulation. This is done by assuming that the portfolio return is dependent on the risk factors listed in the risk vector $\mathsf{F} = (\mathsf{F}_1, ..., \mathsf{F}_m)$ and that the joint distribution of these parameters is known, as well as the (approximate) functional relation between the portfolio return and the risk vector, i.e. $R \left(\varphi, t_0, T \right) = R \left(\mathsf{F}, \varphi, t_0, T \right)$, or equivalently the future value of the portfolio

$$V \left(\varphi, t_0, T \right) = V \left(\mathsf{F}, \varphi, t_0, T \right) = V \left(\varphi, t_0 \right) \cdot \left(1 + R \left(\mathsf{F}, \varphi, t_0, T \right) \right). \quad (6.21)$$

Under this assumption, we simulate[7] the risk vector F, getting the simulations

$$\mathsf{F}^k = \left(\mathsf{F}_1^k, ..., \mathsf{F}_m^k \right), \ k = 1, ..., K.$$

Inserting these simulations into the return or portfolio value function, we get the simulations for the portfolio returns and future portfolio values:

$$R^k \left(\varphi, t_0, T \right) := R \left(\mathsf{F}^k, \varphi, t_0, T \right) \text{ and } V^k \left(\varphi, t_0, T \right) := V \left(\mathsf{F}^k, \varphi, t_0, T \right),$$

$k = 1, ..., K$. In this case, a probability of $p_k = 1/K$ would be assessed to each of the simulated returns. Other methods such as the one proposed by Jamshidian and Zhu [JZ97] derive scenarios $R^k \left(\varphi, t_0, T \right)$ and probabilities p_k which need not be equal for all scenarios. However, the discrete version of equation (6.20) is given by

$$LPM_l \left(\varphi, R, b, t_0, T \right) = \sum_{\substack{k=1,...,K \\ R^k(\varphi, t_0, T) < b(t_0, T)}} p_k \cdot \left(b \left(t_0, T \right) - R^k \left(\varphi, t_0, T \right) \right)^l.$$

---

[7] For Monte Carlo simulations based on normally distributed risk factors see, e.g. Boyle [Boy77], Hammersley and Handscomb [HH64] or Hull [Hul00]. For other distributions the so-called acceptance-rejection method can be used, which is explained in, e.g., Moskowitz and Caflish [MC95b]. Other methods such as the variance reduction or the Quasi-Monte Carlo method help to reduce the computer cost, especially for large portfolios, and are described in Morokoff and Caflish [MC95a].

The corresponding formulations for the lower partial moment $LPM_l\left(\varphi, V, B, t_0, T\right)$ of the future portfolio value are straightforward. As described above, the risk or portfolio manager controls the downside risk of his portfolio by setting an upper limit on specific lower partial moments probably evaluated for different benchmark returns.

6.2.2 Value at Risk

One of the most popular risk numbers nowadays is the value at risk (VaR), indicating the maximum possible loss of a portfolio or trading book with respect to a given time horizon, a given significance level, measured by a change of the portfolios market price, and relative to its expected value over the specified time horizon. One of the targets in calculating the VaR of a portfolio is the derivation of risk numbers for different trading books that can be aggregated to a global risk number. This number may then be used to derive the amount of capital that has to be allocated by a banking institution to cover its trading risk, or even to set up a limit system for the traders. The various methods for calculating the VaR could be distinguished either by using an exact (full valuation) or approximate (approximate valuation) mathematical pricing formula for the future portfolio value $V\left(\varphi, t_0, T\right)$ of equation (6.21) or the change $\Delta V\left(\varphi, t_0, T\right) := V\left(\varphi, t_0, T\right) - V\left(\varphi, t_0\right)$ of the portfolio value from time t_0 to time T. Methods with full valuation include the Empirical method, the Monte Carlo method, or the Approximate Full Valuation method introduced by Zagst [Zag97a]. Methods with approximate valuation could be classified by the Portfolio Normal method and the Delta methods, where the last include the widely used RiskMetrics^{TM} method. A detailed overview and definition of the different methods is given, e.g., in Wilson [Wil96] or Zagst [Zag97a]; a comparison of the methods can be found, e.g., in Smithson and Minton [SM96], Leong [Leo96] or Zagst [Zag97b] and [Zag97c]. While the methods with approximate valuation can be considered to be numerically efficient because of their assumptions, such as normally distributed risk factors or a linear dependence of the portfolio price from the vector of risk factors F, the methods with full valuation usually come with a high cost for data mining or simulation. Nevertheless, for portfolios with rather asymmetric return distributions the methods with approximate valuation may not be adequate for calculating the VaR of those portfolios. Since we want to include complex products such as options we have to deal with asymmetric return distributions (see, e.g., Bookstaber and Clarke [BC81] and [BC83]). For this reason we concentrate on using a method with full valuation dependent on a set of scenarios derived by Monte Carlo simulation methods as described in Section 6.2.1.

Under the assumptions and definitions of Section 6.2.1, let $V\left(\varphi, t_0, T\right)$ be the future value of a portfolio $\varphi = \left(\varphi_1, ..., \varphi_n\right)$ of interest-rate derivatives at the end of the planning horizon T measured at time t_0, $R\left(\varphi, t_0, T\right)$

the portfolio return, and $F_{R(\varphi,t_0,T)}$ the distribution function of $R(\varphi,t_0,T)$. Then the change $\Delta V(\varphi,t_0,T)$ of the portfolio value from time t_0 to time T is given by

$$\Delta V(\varphi,t_0,T) = V(\varphi,t_0) \cdot R(\varphi,t_0,T). \tag{6.22}$$

Let $F_{\Delta V(\varphi,t_0,T)}$ be the distribution function of $\Delta V(\varphi,t_0,T)$, which may be derived from $F_{R(\varphi,t_0,T)}$ using equation (6.22). Furthermore, let the $\alpha-quantile$ $c_{\Delta V(\varphi,t_0,T)}(\alpha)$ of $F_{\Delta V(\varphi,t_0,T)}$ be defined by

$$c_{\Delta V(\varphi,t_0,T)}(\alpha) := \sup\{x \in \mathbb{R} : F_{\Delta V(\varphi,t_0,T)}(x) \le \alpha\}, \tag{6.23}$$

and let

$$F_{\Delta V(\varphi,t_0,T)}(x-) := \lim_{n\to\infty} F_{\Delta V(\varphi,t_0,T)}\left(x - \frac{1}{n}\right) \text{ for } x \in \mathbb{R}.$$

Then we can state the mathematical definition of the VaR.

Definition 6.6 (Value at Risk) *Let* $\varphi = (\varphi_1, ..., \varphi_n)$ *be a portfolio or trading book of interest-rate derivatives,* $[t_0, T]$ *be the considered time period, and* $1 - \alpha \in (0,1)$ *be a given confidence level. Then the value at risk* $VaR(\alpha,\varphi,t_0,T)$ *is defined by*

$$VaR(\alpha,\varphi,t_0,T) := E_Q[\Delta V(\varphi,t_0,T)] - \sup\{x \in \mathbb{R} : F_{\Delta V(\varphi,t_0,T)}(x-) \le \alpha\}.$$

Popular confidence levels are 95% or 99%. From a practical point of view the following lemma is quite interesting.

Lemma 6.7 *Let* $c \in \mathbb{R}$*, possibly dependent on* φ, t_0*, and* T*. Then the following statements are equivalent:*

a) $Q(\Delta V(\varphi,t_0,T) < c) \le \alpha$,

b) $c \le c_{\Delta V(\varphi,t_0,T)}(\alpha)$.

Proof.

(i) Let $Q(\Delta V(\varphi,t_0,T) < c) \le \alpha$, and suppose that $c > c_{\Delta V(\varphi,t_0,T)}(\alpha)$. Then

$$c_{\Delta V(\varphi,t_0,T)}(\alpha) < \frac{1}{2} \cdot \left(c_{\Delta V(\varphi,t_0,T)}(\alpha) + c\right) < c,$$

and because of Definition 6.23 we get

$$Q(\Delta V(\varphi,t_0,T) < c) \ge Q\left(\Delta V(\varphi,t_0,T) \le \frac{1}{2} \cdot \left(c_{\Delta V(\varphi,t_0,T)}(\alpha) + c\right)\right)$$
$$> \alpha,$$

which contradicts the assumption. Hence, $c \le c_{\Delta V(\varphi,t_0,T)}(\alpha)$.

(ii) Let $c \leq c_{\Delta V(\varphi, t_0, T)}(\alpha)$. Then

$$
\begin{aligned}
Q\left(\Delta V\left(\varphi, t_0, T\right) < c\right) &\leq Q\left(\Delta V\left(\varphi, t_0, T\right) < c_{\Delta V(\varphi, t_0, T)}(\alpha)\right) \\
&= F_{\Delta V(\varphi, t_0, T)}\left(c_{\Delta V(\varphi, t_0, T)}(\alpha) -\right) \\
&\leq \alpha.
\end{aligned}
$$

$\square$

As an immediate consequence of this lemma we can easily see that the value at risk is equal to the difference between the expected portfolio value and the α−quantile of the portfolio's distribution function, i.e.

$$
VaR\left(\alpha, \varphi, t_0, T\right) = E_Q\left[\Delta V\left(\varphi, t_0, T\right)\right] - c_{\Delta V(\varphi, t_0, T)}(\alpha). \tag{6.24}
$$

If the distribution function of $\Delta V\left(\varphi, t_0, T\right)$ is unknown, we have to simulate it. Under the assumption of Section 6.2.1, we get the simulations $\mathsf{F}^k = \left(\mathsf{F}_1^k, ..., \mathsf{F}_m^k\right)$, $k = 1, ..., K$, for the risk vector F and, inserting these simulations into the return or portfolio value function, we get the simulations for the portfolio returns and future portfolio values as

$$
R^k\left(\varphi, t_0, T\right) := R\left(\mathsf{F}^k, \varphi, t_0, T\right)
$$

and

$$
\Delta V^k\left(\varphi, t_0, T\right) := V\left(\varphi, t_0\right) \cdot R\left(\mathsf{F}^k, \varphi, t_0, T\right),
$$

$k = 1, ..., K$. Let p_k be the probabilities of the scenarios or corresponding values assigned by index k. Then the distribution function of $\Delta V^k\left(\varphi, t_0, T\right)$ can be simulated by

$$
F_{\Delta V(\varphi, t_0, T)}(x) = \sum_{\substack{k=1,...,K \\ \Delta V^k(\varphi, t_0, T) \leq x}} p_k.
$$

The value at risk is straightforward to simulate using the previous definitions and the simulated distribution function instead of the real one. To control his maximum possible loss, the risk or portfolio manager may be interested in setting an upper bound on the value at risk of his portfolio.

6.3 Coherent Risk Measures

Value at risk is often critizied for not satisfying specific desirable properties one would expect from a risk measure. For example, the diversification of a portfolio is not always adequately represented by the VaR (see example below). Also, adding a fixed (constant) amount of money will not reduce the value at risk of the portfolio or trading book. Hence, the question had to be answered as is how a "good" risk measure should be characterized.

It was not before the year 2000 when Artzner, Delbaen, Eber, and Heath [ADEH99] presented a set of desirable properties which a risk measure should satisfy and called all risk measures holding their conditions *coherent*. They consider a finite probability space $(\Omega, \mathcal{F}, Q)$, i.e. the set Ω is finite, which may be interpreted as the set of all simulated vectors of risk factors $\mathsf{F}^k = \left(\mathsf{F}_1^k, ..., \mathsf{F}_m^k\right)$ and their probabilities p_k, $k = 1, ..., K$, used in the previous sections. Furthermore, we let $\mathcal{X}$ denote the set of all random variables $X : \Omega \to I\!\!R$ with

$$\|X\|_\infty := \sup_{\omega \in \Omega} X\left(\omega\right) \le M \in I\!\!R.$$

Given a specific portfolio or trading book φ, the set $\mathcal{X}$ can be interpreted as the set of all simulated changes of the portfolio value from time t_0 to the end of a specific planning horizon or trading period T, i.e.

$$\mathcal{X} = \left\{\Delta V^k\left(\varphi, t_0, T\right) := V\left(\mathsf{F}^k, \varphi, t_0, T\right) - V\left(\varphi, t_0\right), \; k = 1, ..., K\right\},$$

with $V\left(\varphi, t_0\right)$ and $V\left(\mathsf{F}^k, \varphi, t_0, T\right)$ denoting the initial portfolio value at time t_0 and the simulated portfolio value for time T as seen from time t_0. For ease of notation, we simply call X a *risk position* (with value X). A *risk measure* is then defined as a measurable mapping $\rho : \mathcal{X} \to I\!\!R$. For ease of notation we simply call $\rho\left(X\right)$ the *risk* of a risk position X. If we interpret $\rho\left(X\right)$ to be the (additional) amount of capital which has to be allocated by a banking institution to cover the risk position X, a negative value of $\rho\left(X\right)$ indicates that the capital being allocated is higher than necessary and may be reduced. For ease of exposition we also assume that the interest on allocated capital is zero. However, it should be noted that this assumption can be easily relaxed.

6.3.1 Characterization of Coherent Risk Measures

In this section we state the properties which a risk measure should hold, following the axiomatic approach of Artzner, Delbaen, Eber, and Heath. The first condition is the monotonicity axiom, which claims that a risk position with a lower value should be assigned a higher risk.

Axiom 1 [Monotonicity]. For all $X, Y \in \mathcal{X}$ with $X \le Y$, we have

$$\rho\left(X\right) \ge \rho\left(Y\right).$$

The second property is called translation-invariance, and assures that an additional position of riskless capital should reduce the risk by exactly this amount of money.

Axiom 2 [Translation-Invariance]. For all $X \in \mathcal{X}$ and for all $c \in \mathbb{R}$, we have

$$\rho(X + c) = \rho(X) - c.$$

The axiom of positive homogenity claims that multiplication of an actual risk position should simply result in a corresponding multiplication of risk.

Axiom 3 [Positive Homogenity]. For all $X \in \mathcal{X}$ and for all $\lambda \geq 0$, we have

$$\rho(\lambda \cdot X) = \lambda \cdot \rho(X).$$

The fourth property can be viewed as an aspect of diversification and tells us that adding two risk positions should lead to correlation effects and so reduce risk.

Axiom 4 [Subadditivity]. For all $X, Y \in \mathcal{X}$, we have

$$\rho(X + Y) \leq \rho(X) + \rho(Y).$$

Using these axioms, we can now state the following definition.

Definition 6.8 (Coherent Risk Measure) *A risk measure satisfying the axioms 1-4 is called coherent.*

Examples. It can be easily checked that $\rho : \mathcal{X} \to \mathbb{R}$, defined by

$$\rho(X) := - \inf_{\omega \subset \Omega} X(\omega),$$

is a coherent risk measure. Let $\mathbb{P} = \mathbb{P}(\Omega)$ be the set of all probability measures on the measure space $(\Omega, \mathcal{F})$ and $\mathbb{P}_\rho \subseteq \mathbb{P}$. Then it is straightforward to show that $\rho : \mathcal{X} \to \mathbb{R}$, defined by

$$\rho(X) := - \inf_{Q \in \mathbb{P}_\rho} E_Q[X],$$

is a coherent risk measure. $\qquad\qquad\qquad\qquad\qquad\qquad\qquad\qquad\qquad\square$

The last example gives rise to the following theorem.

Theorem 6.9 (Characterization of Coherent Risk Measures)
A risk measure ρ is coherent if and only if there is a set $\mathbb{P}_\rho$ of probability measures on the measure space $(\Omega, \mathcal{F})$ such that for all $X \in \mathcal{X}$ we have

$$\rho(X) := - \inf_{Q \in \mathbb{P}_\rho} E_Q[X].$$

Proof. As already mentioned, it can be easily verified that

$$\rho(X) := - \inf_{Q \in \mathbb{P}_\rho} E_Q[X] = \sup_{Q \in \mathbb{P}_\rho} E_Q[-X]$$

is a coherent risk measure. For the opposite direction let $E^*[X] := \rho(-X)$ for all $X \in \mathcal{X}$. Then by axioms 1-4, we have

(i) for all $X, Y \in \mathcal{X}$ with $X \leq Y$:

$$E^*[X] = \rho(-X) \leq \rho(-Y) = E^*[Y],$$

(ii) for all $X \in \mathcal{X}$ and for all $\lambda \geq 0$, $c \in \mathbb{R}$:

$$\begin{aligned} E^*[\lambda \cdot X + c] &= \rho(\lambda \cdot (-X) + (-c)) = \rho(\lambda \cdot (-X)) - (-c) \\ &= \lambda \cdot \rho(-X) + c = \lambda \cdot E^*[X] + c, \end{aligned}$$

(iii) for all $X, Y \in \mathcal{X}$:

$$\begin{aligned} E^*[X + Y] &= \rho((-X) + (-Y)) \leq \rho(-X) + \rho(-Y) \\ &= E^*[X] + E^*[Y]. \end{aligned}$$

Hence, assumptions 2.7-2.9 in Huber [Hub81], p. 255 are satisfied and it follows from Proposition 2.1 in Huber [Hub81], p. 256, that for all $X \in \mathcal{X}$ we have

$$E^*[X] = \sup_{Q \in \mathbb{P}_\rho} E_Q[X]$$

with

$$\mathbb{P}_\rho = \{Q \in \mathbb{P} : E_Q[X] \leq E^*[X] \text{ for all } X \in \mathcal{X}\}.$$

Hence, for all $X \in \mathcal{X}$,

$$\rho(X) = E^*[-X] = \sup_{Q \in \mathbb{P}_\rho} E_Q[-X] = - \inf_{Q \in \mathbb{P}_\rho} E_Q[X]$$

with

$$\mathbb{P}_\rho = \{Q \in \mathbb{P} : E_Q[X] \leq \rho(-X) \text{ for all } X \in \mathcal{X}\}.$$

$\square$

The central result we used in the previous proof was Proposition 2.1 in Huber [Hub81]. It deals with the so-called upper expectation $E^*[X]$ of a random variable X induced by a subset $\mathbb{P}_\rho$ of the set of all probability measures $\mathbb{P}$ on a measurable space $(\Omega, \mathcal{F})$. Huber showed that for any measurable mapping $\rho : \mathcal{X} \to \mathbb{R}$ which satisfies a set of conditions ((i),...,(iii) in the previous proof) there is a subset $\mathbb{P}_\rho$ such that ρ can be represented by an upper expectation induced by $\mathbb{P}_\rho$.

6.3.2 What About Value at Risk?

Although it is one of the most popular risk numbers, we now show that the value at risk is not a coherent risk measure. However, we will also show that if we slightly modify the VaR to what we call VaR_0, we still get a risk measure which, while in general not coherent, is nevertheless closely related to the set of coherent risk measures. Indeed, we will show that $VaR_0(\alpha, X)$, $\alpha \in (0,1)$, is the infimum risk of a given risk position X we get by applying all coherent risk measures which are uniformly (i.e. for each possible risk position) greater or equal than $VaR_0(\alpha, \cdot)$. But let us start by restating the definition of $VaR(\alpha, X)$, i.e. the value at risk of the risk position X at a confidence level of $\alpha \in (0,1)$, according to Definition 6.6 and equation (6.24):

$$
\begin{aligned}
VaR(\alpha, X) \quad &:= \quad E_Q[X] - \sup\{x \in \mathbb{R} : Q(\{\omega : X(\omega) < x\}) \le \alpha\} \\
&= \quad E_Q[X] - c_X(\alpha).
\end{aligned}
$$

Correspondingly, we set

$$
\begin{aligned}
VaR_0(\alpha, X) \quad &:= \quad VaR(\alpha, X) - E_Q[X] \\
&= \quad -\sup\{x \in \mathbb{R} : Q(\{\omega : X(\omega) < x\}) \le \alpha\} \\
&= \quad -c_X(\alpha).
\end{aligned}
$$

With respect to axiom 4 we give the following example.

Example. Let X and Y be defined by

$$
X(\omega) = Y(\omega) = \begin{cases} -1, & \text{with probability } 5\% \\ 0, & \text{with probability } 90\% \\ 1, & \text{with probability } 5\%. \end{cases}
$$

Then $E_Q[X] = E_Q[Y] = 0$, and $VaR_0(7.5\%, X) = VaR_0(7.5\%, Y) = 0$. Furthermore,

$$
X(\omega) + Y(\omega) = \begin{cases} -2 & \text{with probability } 0.25\% \\ -1 & \text{with probability } 9\% \\ 0 & \text{with probability } 81.5\% \\ 1 & \text{with probability } 9\% \\ 2 & \text{with probability } 0.25\%, \end{cases}
$$

with $E_Q[X + Y] = 0$ and

$$
\begin{aligned}
VaR(7.5\%, X+Y) \quad &= \quad VaR_0(7.5\%, X+Y) \\
&= \quad -\sup\{x \in \mathbb{R} : Q(\{\omega : X(\omega) + Y(\omega) < x\}) \le 7.5\%\} \\
&= \quad -(-1) = 1 \\
&> \quad VaR_0(7.5\%, X) + VaR_0(7.5\%, Y) \\
&= \quad VaR(7.5\%, X) + VaR(7.5\%, Y).
\end{aligned}
$$

We thus learn that the value at risk does not in general satisfy axiom 4. $\square$

Despite the fact that the VaR is not in general a coherent risk measure, the following lemma shows that the modification to VaR_0 leads to a risk measure which is, under specific assumptions, coherent. This is the reason why VaR_0 is often used for calculating the overnight risk of a portfolio or trading book.

Lemma 6.10 Let $\alpha \in (0,1)$. Then VaR_0 satisfies axioms 1-3. If all $X, Y \in \mathcal{X}$ are jointly normal distributed and $\alpha < \frac{1}{2}$, then VaR_0 satisfies axiom 4. In this case, VaR_0 is a coherent risk measure.

Proof. To prove axiom 1, let $X, Y \in \mathcal{X}$ with $X \leq Y$. Then

$$Q\left(\{\omega : X(\omega) < x\}\right) \geq Q\left(\{\omega : Y(\omega) < x\}\right),$$

and hence,

$$
\begin{aligned}
VaR_0(\alpha, X) &= -\sup\{x \in \mathbb{R} : Q(\{\omega : X(\omega) < x\}) \leq \alpha\} \\
&\geq -\sup\{x \in \mathbb{R} : Q(\{\omega : Y(\omega) < x\}) \leq \alpha\} \\
&= VaR_0(\alpha, Y).
\end{aligned}
$$

For axiom 2, let $X \in \mathcal{X}$ and $c \in \mathbb{R}$. Then

$$
\begin{aligned}
VaR_0(\alpha, X + c) &= -\sup\{x \in \mathbb{R} : Q(\{\omega : X(\omega) + c < x\}) \leq \alpha\} \\
&= -\sup\{x \in \mathbb{R} : Q(\{\omega : X(\omega) < x - c\}) \leq \alpha\} \\
&= -\sup\{y + c \in \mathbb{R} : Q(\{\omega : X(\omega) < y\}) \leq \alpha\} \\
&= -\sup\{y \in \mathbb{R} : Q(\{\omega : X(\omega) < y\}) \leq \alpha\} - c \\
&= VaR_0(\alpha, X) - c.
\end{aligned}
$$

For axiom 3, let $X \in \mathcal{X}$ and for all $\lambda \geq 0$. Then

$$
\begin{aligned}
VaR_0(\alpha, \lambda \cdot X) &= -\sup\{x \in \mathbb{R} : Q(\{\omega : \lambda \cdot X(\omega) < x\}) \leq \alpha\} \\
&= -\sup\left\{x \in \mathbb{R} : Q\left(\left\{\omega : X(\omega) < \frac{x}{\lambda}\right\}\right) \leq \alpha\right\} \\
&= -\sup\{\lambda \cdot y \in \mathbb{R} : Q(\{\omega : X(\omega) < y\}) \leq \alpha\} \\
&= \lambda \cdot (-\sup\{y \in \mathbb{R} : Q(\{\omega : X(\omega) < y\}) \leq \alpha\}) \\
&= \lambda \cdot VaR_0(\alpha, X).
\end{aligned}
$$

Now let $X, Y \in \mathcal{X}$ be jointly normal distributed and $\alpha \leq \frac{1}{2}$. Furthermore, let $N_{0,1}^{-1}$ denote the inverse of the standard normal distribution function and $\sigma(X)$ and $\varrho(X, Y)$ be the standard deviation and correlation of the random variable(s) in the brackets. Then

$$
\begin{aligned}
\sigma^2(X + Y) &= \sigma^2(X) + \sigma^2(Y) + 2 \cdot \varrho(X, Y) \cdot \sigma(X) \cdot \sigma(Y) \\
&\leq \sigma^2(X) + \sigma^2(Y) + 2 \cdot \sigma(X) \cdot \sigma(Y) \\
&= (\sigma(X) + \sigma(Y))^2,
\end{aligned}
$$

and, since $N_{0,1}^{-1}(\alpha) \leq 0$,

$$
\begin{aligned}
VaR_0(\alpha, X+Y) &= -\left(E_Q[X+Y] + N_{0,1}^{-1}(\alpha) \cdot \sigma(X+Y)\right) \\
&\leq -\left(E_Q[X] + E_Q[Y] + N_{0,1}^{-1}(\alpha) \cdot (\sigma(X) + \sigma(Y))\right) \\
&= VaR_0(\alpha, X) + VaR_0(\alpha, Y).
\end{aligned}
$$

$\square$

Remark. *From the proof of Lemma 6.10 we learn that for $\alpha \in (0,1)$, $X \in X$, and $c \in \mathbb{R}$ we have*

$$
\begin{aligned}
VaR(\alpha, X+c) &= E_Q[X+c] + VaR_0(\alpha, X+c) \\
&= E_Q[X] + c + VaR_0(\alpha, X) - c \\
&= VaR(\alpha, X),
\end{aligned}
$$

i.e. VaR doesn't satisfy axiom 2. Furthermore, it can be easily checked that, in general, VaR also fails to satisfy axiom 1. $\square$

Even though VaR_0 is not in general a coherent risk measure, we can show that it is closely related to the set of coherent risk measures. To do this, we make use of the following lemma.

Lemma 6.11 *Let $\alpha \in (0,1)$ and ρ be a coherent risk measure defined by the set of probability measures $\mathbb{P}_\rho$ via*

$$
\rho(X) = - \inf_{Q \in \mathbb{P}_\rho} E_Q[X].
$$

Then the following statements are equivalent.

a) $\rho(\cdot) \geq VaR_0(\alpha, \cdot)$, *i.e.* $\rho(X) \geq VaR_0(\alpha, X)$ *for all $X \in \mathcal{X}$.*

b) *For each $A \in \mathcal{F}$ with $Q(A) > \alpha$ and each $\varepsilon > 0$ there is a $Q_\varepsilon \in \mathbb{P}_\rho$ such that $Q_\varepsilon(A) > 1 - \varepsilon$.*

Proof. Let ρ be a coherent risk measure as supposed with $\rho(\cdot) \geq VaR_0(\alpha, \cdot)$. Furthermore, let $A \in \mathcal{F}$ with $Q(A) > \alpha \in (0,1)$ and $X_A := -1_A$. Then

$$
\begin{aligned}
Q(\{\omega \in \Omega : X_A(\omega) \leq x\}) &= Q(\{\omega \in \Omega : -1_A(\omega) \leq x\}) \\
&= \begin{cases} 0, & \text{if } x < -1 \\ Q(A) & \text{if } -1 \leq x < 0 \\ 1 & \text{if } x \geq 0 \end{cases}
\end{aligned}
$$

and so because $Q(A) > \alpha$,

$$
\begin{aligned}
VaR_0(\alpha, X_A) &= -\sup\{x \in \mathbb{R} : Q(\{\omega \in \Omega : X_A(\omega) < x\}) \leq \alpha\} \\
&= 1.
\end{aligned}
$$

Hence, using a),

$$
\begin{aligned}
1 &= VaR_0\left(\alpha, X_A\right) \le \rho\left(X_A\right) = -\inf_{Q \in \mathbb{P}_\rho} E_Q\left[X_A\right] \\
&= \sup_{Q \in \mathbb{P}_\rho} E_Q\left[-X_A\right] = \sup_{Q \in \mathbb{P}_\rho} Q\left(A\right),
\end{aligned}
$$

i.e. for each $\varepsilon > 0$ there is a $Q_\varepsilon \in \mathbb{P}_\rho$ such that $Q_\varepsilon\left(A\right) > 1 - \varepsilon$.
For the opposite direction, let b) be satisfied and $X \in \mathcal{X}$. By definition of the $\alpha-$quantile $c_X\left(\alpha\right)$, $\alpha \in \left(0,1\right)$, we know that for each $\varepsilon > 0$,

$$
Q\left(\left\{\omega \in \Omega : X\left(\omega\right) \le c_X\left(\alpha\right) + \varepsilon\right\}\right) > \alpha.
$$

With $c_\varepsilon := c_X\left(\alpha\right) + \varepsilon$ and $A_\varepsilon := \left\{\omega \in \Omega : X\left(\omega\right) \le c_\varepsilon\right\}$ we know that $Q\left(A_\varepsilon\right) > \alpha$ and thus, using b), there is a $Q_\varepsilon \in \mathbb{P}_\rho$ such that $Q_\varepsilon\left(A_\varepsilon\right) > 1-\varepsilon$, i.e. $Q_\varepsilon\left(A_\varepsilon^c\right) < \varepsilon$. Hence, for each $\varepsilon > 0$ there is a $A_\varepsilon \in \mathcal{F}$ and a $Q_\varepsilon \in \mathbb{P}_\rho$ such that

$$
0 \le Q_\varepsilon\left(A_\varepsilon^c\right) < \varepsilon, \text{ i.e. } 1 - \varepsilon < Q_\varepsilon\left(A_\varepsilon\right) \le 1.
$$

Now, because $\|X\|_\infty := \sup_{\omega \in \Omega} X\left(\omega\right) \le M$,

$$
\begin{aligned}
E_{Q_\varepsilon}\left[-X\right] &= \int_{A_\varepsilon} -X\, dQ_\varepsilon + \int_{A_\varepsilon^c} -X\, dQ_\varepsilon \\
&\ge -c_\varepsilon \cdot Q_\varepsilon\left(A_\varepsilon\right) - Q_\varepsilon\left(A_\varepsilon^c\right) \cdot \sup_{\omega \in A_\varepsilon^c} X\left(\omega\right) \\
&\ge -c_\varepsilon \cdot Q_\varepsilon\left(A_\varepsilon\right) - Q_\varepsilon\left(A_\varepsilon^c\right) \cdot M.
\end{aligned}
$$

By definition of ρ, we get for each $\varepsilon > 0$,

$$
\begin{aligned}
\rho\left(X\right) &= \sup_{Q \in \mathbb{P}_\rho} E_Q\left[-X\right] \ge E_{Q_\varepsilon}\left[-X\right] \\
&\ge -c_\varepsilon \cdot Q_\varepsilon\left(A_\varepsilon\right) - Q_\varepsilon\left(A_\varepsilon^c\right) \cdot M
\end{aligned}
$$

and thus, since $\lim_{\varepsilon \to 0} c_\varepsilon = c_X\left(\alpha\right)$,

$$
\begin{aligned}
\rho\left(X\right) &\ge \lim_{\varepsilon \to 0}\left\{-c_\varepsilon \cdot Q_\varepsilon\left(A_\varepsilon\right) - Q_\varepsilon\left(A_\varepsilon^c\right) \cdot M\right\} \\
&= -c_X\left(\alpha\right) \cdot \lim_{\varepsilon \to 0} Q_\varepsilon\left(A_\varepsilon\right) - M \cdot \lim_{\varepsilon \to 0} Q_\varepsilon\left(A_\varepsilon^c\right) \\
&= -c_X\left(\alpha\right) \cdot 1 - M \cdot 0 = -c_X\left(\alpha\right),
\end{aligned}
$$

i.e. $\rho\left(X\right) \ge VaR_0\left(\alpha, X\right)$. $\square$

We will now apply Lemma 6.11 to prove the following theorem.

Lemma 6.12 Let $\alpha \in \left(0,1\right)$. Then for all $X \in \mathcal{X}$,

$$
VaR_0\left(\alpha, X\right) = \inf\left\{\rho\left(X\right) : \rho\left(\cdot\right) \ge VaR_0\left(\alpha, \cdot\right) \text{ and } \rho \text{ is coherent}\right\}.
$$

Proof. Let $\alpha \in (0,1)$ and $X_0 \in \mathcal{X}$ be arbitrary but fixed. Furthermore, let $A \in \mathcal{F}$ with $Q(A) > \alpha$. By definition of the $\alpha-$quantile $c_{X_0}(\alpha)$, we know that $Q(\{\omega : X_0(\omega) \geq c_{X_0}(\alpha)\}) \geq 1 - \alpha$, and thus

$$
\begin{aligned}
1 \;\geq\; & Q(A \cup \{\omega : X_0(\omega) \geq c_{X_0}(\alpha)\}) \\
=\; & Q(A) + Q(\{\omega : X_0(\omega) \geq c_{X_0}(\alpha)\}) - Q(A \cap \{\omega : X_0(\omega) \geq c_{X_0}(\alpha)\}) \\
>\; & \alpha + 1 - \alpha - Q(A \cap \{\omega : X_0(\omega) \geq c_{X_0}(\alpha)\}) \\
=\; & 1 - Q(A \cap \{\omega : X_0(\omega) \geq c_{X_0}(\alpha)\}),
\end{aligned}
$$

i.e.

$$
Q(A \cap \{\omega : X_0(\omega) \geq c_{X_0}(\alpha)\}) > 0.
$$

Now let $Q_A = Q_{A,X_0} \in \mathbb{P}$ be defined by

$$
\begin{aligned}
Q_A(B) \;:=\; & \int_B \frac{\mathbf{1}_{A \cap \{\omega : X_0(\omega) \geq c_{X_0}(\alpha)\}}(\omega)}{Q(A \cap \{\omega : X_0(\omega) \geq c_{X_0}(\alpha)\})} dQ \\
=\; & \frac{Q(B \cap A \cap \{\omega : X_0(\omega) \geq c_{X_0}(\alpha)\})}{Q(A \cap \{\omega : X_0(\omega) \geq c_{X_0}(\alpha)\})} \\
=\; & Q(B | A \cap \{\omega : X_0(\omega) \geq c_{X_0}(\alpha)\})
\end{aligned}
$$

for all $B \in \mathcal{F}$. Then

$$
Q_A(A) = 1 > 1 - \varepsilon \text{ for each } \varepsilon > 0.
$$

Now let

$$
\mathbb{P}_{\rho_0} = \mathbb{P}_{\rho_0, X_0} := \{Q_A : A \in \mathcal{F}, Q(A) > \alpha\},
$$

and $\rho_0 = \rho_{X_0}$ be defined by

$$
\rho_0(X) := \sup_{Q_A \in \mathbb{P}_{\rho_0}} E_{Q_A}[-X].
$$

Then we have shown that for each $A \in \mathcal{F}$ with $Q(A) > \alpha$ and each $\varepsilon > 0$ there is a $Q_\varepsilon \in \mathbb{P}_{\rho_0}$ such that $Q_\varepsilon(A) > 1 - \varepsilon$. So by Theorem 6.9 and Lemma 6.11, we know that ρ_0 is a coherent risk measure with

$$
\rho_0(\cdot) \geq VaR_0(\alpha, \cdot).
$$

Furthermore, for all $Q_A \in \mathbb{P}_{\rho_0}$,

$$
\begin{aligned}
E_{Q_A}[-X_0] \;=\; & -\int \frac{X_0 \cdot \mathbf{1}_{A \cap \{\omega : X_0(\omega) \geq c_{X_0}(\alpha)\}}(\omega)}{Q(A \cap \{\omega : X_0(\omega) \geq c_{X_0}(\alpha)\})} dQ \\
\leq\; & -\int \frac{c_{X_0}(\alpha) \cdot \mathbf{1}_{A \cap \{\omega : X_0(\omega) \geq c_{X_0}(\alpha)\}}(\omega)}{Q(A \cap \{\omega : X_0(\omega) \geq c_{X_0}(\alpha)\})} dQ \\
=\; & -c_{X_0}(\alpha) \cdot 1 = VaR_0(\alpha, X_0),
\end{aligned}
$$

and thus, for our arbitrary but fixed $X_0 \in \mathcal{X}$,

$$\rho_0\left(X_0\right) = \sup_{Q_A \in \mathbb{P}_{\rho_0}} E_{Q_A}\left[-X_0\right] \leq VaR_0\left(\alpha, X_0\right),$$

i.e.

$$\rho_0\left(X_0\right) = VaR_0\left(\alpha, X_0\right).$$

We conclude that under all coherent risk measures ρ with $\rho(\cdot) \geq VaR_0\left(\alpha, \cdot\right)$ and for arbitrary but fixed $X_0 \in \mathcal{X}$ there is at least one risk measure ρ_0 with $\rho_0\left(X_0\right) = VaR_0\left(\alpha, X_0\right)$. Hence, we have for all $X \in \mathcal{X}$,

$$VaR_0\left(\alpha, X\right) = \inf\left\{\rho\left(X\right) : \rho\left(\cdot\right) \geq VaR_0\left(\alpha, \cdot\right) \text{ and } \rho \text{ is coherent}\right\}.$$

□

6.3.3 Worst and Tail Conditional Expectations

In Section 6.2 we defined the lower partial moments of order $l \in \mathbb{N}$. As we have seen, the lower partial moment of order $l = 0$ was closely related to the value at risk which, in general, is not a coherent risk measure. The lower partial moment of order $l = 1$ was defined as the expected deviation below a given benchmark. Especially with respect to the tails of a distribution, the conditional expectation of a risk position X, given that X falls below some critical (extreme) boundary such as the value at risk, is a measure of great interest, not only for insurance companies (see, e.g. Hogg and Klugman [HK84] or Embrechts, Klüppelberg, and Mikosch [EKM97] for more details). In this section we concentrate on such conditional expectations and their relation to coherent risk measures. We therefore give the following definition.

Definition 6.13 (Worst Conditional Expectation) *For $A \in \mathcal{F}$ with $Q\left(A\right) > 0$ let $Q_A \in \mathbb{P}$ be defined by*

$$Q_A\left(B\right) := Q\left(B|A\right) := \frac{Q\left(A \cap B\right)}{Q\left(A\right)} \text{ for all } B \in \mathcal{F}.$$

Furthermore, let the set $\mathbb{P}_\alpha$ be defined for arbitrary but fixed $\alpha \in (0,1)$ by

$$\mathbb{P}_\alpha := \left\{Q_A : Q\left(A\right) > \alpha, A \in \mathcal{F}\right\}.$$

Then the so-called worst conditional expectation $WCE_\alpha = WCE\left(\alpha, \cdot\right)$ is defined by

$$WCE\left(\alpha, X\right) := - \inf_{Q_A \in \mathbb{P}_\alpha} E_{Q_A}\left[X\right]$$

for all $X \in \mathcal{X}$.

Using Theorem 6.9 we can directly see that WCE_α is a coherent risk measure. Note that $E_{Q_A}[X]$ is the ordinary conditional expectation of X under $A \in \mathcal{F}$ on the probability space $(\Omega, \mathcal{F}, Q)$, usually denoted by $E_Q[X|A]$. This motivates the definition of the tail conditional expectation.

Definition 6.14 (Tail Conditional Expectation) *Let $\alpha \in (0, 1)$. Then the tail conditional expectation or tail value at risk $TCE_\alpha = TCE(\alpha, \cdot)$ is defined by*

$$TCE(\alpha, X) := -E_Q[X|X \leq -VaR_0(\alpha, X)]$$

for all $X \in \mathcal{X}$.

By the previous definition we can easily see that $TCE(\alpha, \cdot) > VaR_0(\alpha, \cdot)$. If we compare the tail conditional expectation with the worst conditional expectation, we get the following theorem.

Lemma 6.15 *Let $\alpha \in (0, 1)$. Then we have*

$$VaR_0(\alpha, \cdot) \leq TCE(\alpha, \cdot) \leq WCE(\alpha, \cdot).$$

Proof. Let $\alpha \in (0, 1)$, $X \in \mathcal{X}$, and $c_X(\alpha)$ denote the α–quantile of X. Then

$$
\begin{aligned}
-TCE(\alpha, X) &= E_Q[X|X \leq -VaR_0(\alpha, X)] \\
&= \int \frac{1_{(-\infty, -VaR_0(\alpha, X)]}(X) \cdot X}{Q(\{\omega \in \Omega : X(\omega) \leq -VaR_0(\alpha, X)\})} dQ \\
&\leq -VaR_0(\alpha, X) \cdot \frac{\int 1_{(-\infty, -VaR_0(\alpha, X)]}(X) \, dQ}{Q(\{\omega \in \Omega : X(\omega) \leq -VaR_0(\alpha, X)\})} \\
&= -VaR_0(\alpha, X),
\end{aligned}
$$

i.e. $VaR_0(\alpha, X) \leq TCE(\alpha, X)$. For the right inequality, we know because of equation (6.24) that

$$VaR_0(\alpha, X) = -c_X(\alpha),$$

and so

$$TCE(\alpha, X) = -E_Q[X|X \leq -VaR_0(\alpha, X)] = -E_Q[X|X \leq c_X(\alpha)].$$

Now, for $n \in \mathbb{N}$ let

$$A_n := \left\{\omega \in \Omega : X(\omega) \leq c_X(\alpha) + \frac{1}{n}\right\}$$

and

$$A = A_\infty := \{\omega \in \Omega : X(\omega) \leq c_X(\alpha)\}.$$

(i) If $Q(A) > \alpha$, then

$$
\begin{aligned}
-TCE(\alpha, X) &= E_Q[X|X \le c_X(\alpha)] = E_Q[X|A] \\
&\ge \inf\{E_Q[X|A] : Q(A) > \alpha\} \\
&= -WCE(\alpha, X),
\end{aligned}
$$

i.e.

$$
TCE(\alpha, X) \le WCE(\alpha, X).
$$

(ii) If $Q(A) = \alpha$, then $Q(A_n) > \alpha$ for all $n \in I\!N$ and hence,

$$
\begin{aligned}
-WCE(\alpha, X) &= \inf\{E_Q[X|A] : Q(A) > \alpha\} \\
&\le E_Q[X|A_n] = \frac{E_Q[X \cdot 1_{A_n}]}{Q(A_n)}
\end{aligned}
$$

for all $n \in I\!N$. Therefore,

$$
\begin{aligned}
-WCE(\alpha, X) &\le \lim_{n\to\infty} \frac{E_Q[X \cdot 1_{A_n}]}{Q(A_n)} = \frac{\lim_{n\to\infty} E_Q[X \cdot 1_{A_n}]}{\lim_{n\to\infty} Q(A_n)} \\
&= \frac{E_Q[X \cdot 1_A]}{Q(A)} = E_Q[X|A] \\
&= E_Q[X|X \le c_X(\alpha)] \\
&= -TCE(\alpha, X),
\end{aligned}
$$

i.e.

$$
TCE(\alpha, X) \le WCE(\alpha, X).
$$

$\square$

Especially for Monte Carlo and Quasi Monte Carlo simulations we often deal with uniform distributions. Under the additional assumption that all simulated values are different, we can show that the worst conditional expectation WCE_α and the tail conditional expectation TCE_α, $\alpha \in (0,1)$, coincide. In this case the tail conditional expectation is a coherent risk measure. The result is stated in the following lemma.

Lemma 6.16 *Let* $\alpha \in (0,1)$ *and let* Q *be the uniform distribution on* $(\Omega, \mathcal{F})$. *Furthermore, suppose that for all* $X \in \mathcal{X}$, $X(\omega) \ne X(\omega')$ *for all* $\omega, \omega' \in \Omega$ *with* $\omega \ne \omega'$. *Then*

$$
TCE(\alpha, \cdot) = WCE_\alpha(\cdot).
$$

Especially, $TCE(\alpha, \cdot)$ *is a coherent risk measure.*

Proof. Let $\alpha \in (0,1)$, $X \in \mathcal{X}$, and $c_X(\alpha)$ denote the $\alpha-$quantile of X. Then for each $\varepsilon > 0$,

$$
Q(\{\omega \in \Omega : X(\omega) \le c_X(\alpha)\}) \ge \alpha, \ Q(\{\omega \in \Omega : X(\omega) < c_X(\alpha) + \varepsilon\}) > \alpha
$$

and
$$c_X(\alpha) = -VaR_0(\alpha, X).$$

Now let $\mathcal{X} = \{x_1, ..., x_K\}$, $K \in I\!N$, with $x_1 < \cdots < x_K$. By assumption we then know that $\omega_k \in \Omega$ with $X(\omega_k) = x_k$ is uniquely defined for all $k \in \{1, ..., K\}$ and

$$Q(\omega_k) := Q(\{\omega_k\}) = Q(\{\omega \in \Omega : X(\omega) = x_k\}) = \frac{1}{K} \text{ for all } k \in \{1, ..., K\}.$$

Suppose that $\alpha \in \left[\frac{k}{K}, \frac{k+1}{K}\right)$ for a $k \in \{1, ..., K-1\}$. We then know that $c_X(\alpha) = (k+1)/K$, and thus

$$TCE(\alpha, X) = -E_Q[X|X \le c_X(\alpha)] = -\frac{1}{k+1} \cdot \sum_{i=1}^{k+1} x_k,$$

as well as

$$Q(A) = \sum_{\substack{k=1 \\ \omega_k \in A}}^{K} \frac{1}{K} \text{ for all } A \in \mathcal{F}.$$

Hence, for all $A \in \mathcal{F}$ we know that

$$Q(A) > \alpha \quad \Leftrightarrow \quad Q(A) = \sum_{\substack{k=1 \\ \omega_k \in A}}^{K} \frac{1}{K} \ge \frac{k+1}{K}$$
$$\Leftrightarrow \quad K \cdot Q(A) \ge k+1,$$

and so

$$
\begin{aligned}
WCE(\alpha, X) &= -\inf\{E_Q[X|A] : Q(A) > \alpha\} \\[2mm]
&= -\inf\left\{ \sum_{\substack{k=1 \\ \omega_k \in A}}^{K} \frac{x_k}{K \cdot Q(A)} : Q(A) > \alpha \right\} \\[2mm]
&= -\inf\left\{ \sum_{\substack{k=1 \\ \omega_k \in A}}^{K} \frac{x_k}{K \cdot Q(A)} : K \cdot Q(A) \ge k+1 \right\} \\[2mm]
&= -\frac{1}{k+1} \cdot \sum_{i=1}^{k+1} x_k = TCE(\alpha, X).
\end{aligned}
$$

$\square$

Remark. *Let us slightly extend the definition of the lower partial moment $LPM_l(X, B)$ of order $l \in I\!N$ for a risk position $X \in X$ and a given bechmark B at time t_0 (see Definition 6.5) to allow for equality by*

$$LPM_l^*(X, B) = E_Q\left[1_{(-\infty, B]}(X) \cdot (B - X)^l\right].$$

Then

$$-E_Q\left[X|X \le B\right] = \frac{-E_Q\left[1_{(-\infty,B]}(X) \cdot X\right]}{Q\left(\{\omega \in \Omega : X(\omega) \le B\}\right)}$$

$$= \frac{E_Q\left[1_{(-\infty,B]}(X) \cdot (B-X)\right] - E_Q\left[1_{(-\infty,B]}(X) \cdot B\right]}{LPM_0^*(X,B)}$$

$$= \frac{LPM_1^*(X,B) - B \cdot LPM_0^*(X,B)}{LPM_0^*(X,B)}$$

$$= \frac{LPM_1^*(X,B)}{LPM_0^*(X,B)} - B.$$

Setting $B := -VaR_0(\alpha, X)$ we get

$$
\begin{aligned}
TCE(\alpha, X) &= -E_Q\left[X|X \le -VaR_0(\alpha, X)\right] \\
&= \frac{LPM_1^*(X, -VaR_0(\alpha, X))}{LPM_0^*(X, -VaR_0(\alpha, X))} + VaR_0(\alpha, X),
\end{aligned}
$$

and thus for every $S \in [VaR_0(\alpha, X), \infty)$,

$$TCE(\alpha, X) \le S \Leftrightarrow \frac{LPM_1^*(X, -VaR_0(\alpha, X))}{LPM_0^*(X, -VaR_0(\alpha, X))} + VaR_0(\alpha, X) \le S,$$

which is equivalent to

$$LPM_1^*(X, -VaR_0(\alpha, X)) \le (S - VaR_0(\alpha, X)) \cdot LPM_0^*(X, -VaR_0(\alpha, X)). \tag{6.25}$$

As we have already seen,

$$LPM_0^*(X, -VaR_0(\alpha, X)) = Q\left(\{\omega \in \Omega : X(\omega) \le c_X(\alpha)\}\right) \ge \alpha.$$

So inequality (6.25) is satisfied if

$$LPM_1^*(X, -VaR_0(\alpha, X)) \le \alpha \cdot (S - VaR_0(\alpha, X)). \tag{6.26}$$

If we would like to set a limit on the tail conditional expectation for risk or portfolio management purposes, we could use inequality (6.26) as a sufficient condition which is a limit on the lower partial moment of order $l = 1$ combined with the value at risk. This topic is discussed in Section 7.2. $\quad\square$

6.4   Reducing Dimensions

As we have seen in the previous sections it is quite crucial for sensitivity analysis as well as for simulations to assume that the prices of derivatives or the rate of return over a given planning horizon are dependent on a set of risk factors. Within our interest-rate market model the primary traded assets, the zero-coupon bonds of different maturities, can be considered to be dependent on the zero-rate curve. Hence, neglecting side effects such as the aging of a portfolio because of a decreasing time to maturity, changes of the zero-coupon bond prices are thus mainly due to changes of the yield curve. Interest-rate models such as those included in the Heath-Jarrow-Morton framework were developed to describe the behaviour of the zero-rate curve via forward rates. The one-factor models of Vasicek [Vas77], Cox, Ingersoll, and Ross [CIR85], Ho and Lee [HL86], Black, Derman, and Toy [BDT90] or Hull and White [HW90],[HW93], and [HW94b] explain the development of the zero-rate curve using a stochastic process for the short rate. Brennan and Schwartz [BS82] use a two-factor model of the short rate and a long-term rate, also called consol rate. Longstaff and Schwartz [LS92] developed a two-factor model for the short rate and the volatility which, in their model, is considered to be stochastic. Hull and White [HW94a] invented a two-factor model for the short rate and a stochastic drift factor. An overview of these and other one- or two-factor models was given in Sections 4.5 and 4.6. The so-called (linear) multi-factor models try to explain the behaviour of the zero-rate curve by a set of factors $\mathsf{F} = (\mathsf{F}_1, ..., \mathsf{F}_m)$, $m \in I\!N$. Usually they describe the changes $\Delta R(t, T_i)$ of a selected set of zero rates $R(t, T_i)$, $i = 1, ..., n$, over a given time-period from t to $t + \Delta t$, $t \in [t_0, T^* - \Delta t]$, by the linear equation

$$\Delta R(t, T_i) = \sum_{j=1}^{m} a_{ij} \cdot \Delta \mathsf{F}_j(t) + \varepsilon_i(t), i = 1, ..., n,$$

where it is assumed that the random variables or error terms $\varepsilon_1(t), ..., \varepsilon_n(t)$ follow a joint normal distribution. Dependent on the individual model these error terms may be correlated or not. Most models assume that their expectation is zero and their distribution independent of time (see, e.g., Greene [Gre93] for more details). Writing $R_i(t) := R(t, T_i)$, $i = 1, ..., n$, the zero-rate curve at time t is described by the vector $R(t) := (R_1(t), ..., R_n(t))$ with all other zero rates $Y(t) \notin \{R_1(t), ..., R_n(t)\}$ with $R_i(t) < Y(t) < R_{i+1}(t)$ being defined by the linear interpolation $Y(t) = \lambda \cdot R_i(t) + (1 - \lambda) \cdot R_{i+1}(t)$, $\lambda \in (0, 1)$. It is supposed that the vector $R(t)$ is chosen to give a sufficiently good approximation of the true zero-rate curve. We also assume that the factors F are chosen to justify an approximate representation of the zero-rate curve changes[8] by

$$\Delta R_i(t) \approx \sum_{j=1}^{m} a_{ij} \cdot \Delta F_j(t), i = 1, ..., n. \tag{6.27}$$

The numbers $a_{ij}, j = 1, ..., m$, are referred to as the *factor loadings* for the zero rate $R_i(t), i = 1, ..., n$. Let us consider now the price D of an interest-rate derivative which is assumed to depend on the zero-rate curve $R(t)$ at any time $t \in [t_0, T^*]$, i.e. $D(t) = D(R(t))$ or briefly $D = D(R)$. Using (6.1), the price-change $\Delta D(R)$ is approximately given by

$$\Delta D(R) \approx \sum_{i=1}^{n} \Delta_{R_i}^{D}(R) \cdot \Delta R_i + \frac{1}{2} \cdot \sum_{i=1}^{n} \sum_{k=1}^{n} \Gamma_{R_i R_k}^{D}(R) \cdot \Delta R_i \cdot \Delta R_k.$$

Inserting approximation (6.27), we get

$$\Delta D(R) \approx \sum_{j=1}^{m} \Delta_{F_j}^{D}(F) \cdot \Delta F_j + \frac{1}{2} \cdot \sum_{j=1}^{m} \sum_{l=1}^{m} \Gamma_{F_j F_l}^{D}(F) \cdot \Delta F_j \cdot \Delta F_l,$$

where we set

$$\Delta_{F_j}^{D}(F) = \sum_{i=1}^{n} \Delta_{R_i}^{D}(R) \cdot a_{ij},$$

and

$$\Gamma_{F_j F_l}^{D}(F) = \sum_{i=1}^{n} \sum_{k=1}^{n} \Gamma_{R_i R_k}^{D}(R) \cdot a_{ij} \cdot a_{kl}$$

for $j,l = 1, ..., m$. $\Delta_{F_j}^{D}(F)$ is called the *factor delta* and $\Gamma_{F_j F_l}^{D}(F)$ the *factor gamma* of derivative D with respect to factor F_j or factors F_j and F_l, respectively. It can be easily seen that, setting $m := 1$ and $a_{i1} := 1$ for all $i = 1, ..., n$, we get

$$\Delta R_i(t) \approx \Delta F_1(t), i = 1, ..., n,$$

which is a parallel shift of the zero-rate curve as we used it for the duration and convexity in Section 6.1.3. Unfortunately, it can be shown (see Dahl [Dah93]) that a simple parallel shift of the zero-rate curve induces arbitrage possibilities and therefore rather rarely happens in reality. Another possibility is to define the key-rate buckets $KB_1, KB_2, ..., KB_m$ and

$$a_{ij} := \begin{cases} 1, & \text{if } i \in KB_j \\ 0, & \text{else} \end{cases}$$

---

[8]Within RiskMetrics^{TM}, for example, the zero rates for the maturities $1M(onth), 3M, 6M$ as well as $1Y(ear), 2Y, 3Y, 4Y, 5Y, 7Y, 9Y$, and $10Y$ are chosen to sufficiently explain the zero curve changes from 1 to 10 years.

for $i = 1, ..., n$ and $j = 1, ..., m$. Using (6.27), this implies that

$$\Delta R_i(t) \approx \sum_{j=1}^{m} a_{ij} \cdot \Delta F_j(t) = \begin{cases} \Delta F_j(t), & \text{if } i \in KB_j, j = 1, ..., m \\ 0, & \text{else,} \end{cases}$$

i.e. all zero rates within one key-rate bucket move by a parallel shift. Also, it can be directly seen from approximation (6.17) that

$$\sum_{i=1}^{n} a_{ij} \cdot \Delta_{R_i}^{D}(R) = \sum_{i \in KB_j} \Delta_{R_i}^{D}(R) \approx \Delta_{KB_j}^{D}(R)$$

and

$$\frac{1}{2} \cdot \sum_{i=1}^{n} \sum_{k=1}^{n} a_{ij} \cdot a_{kl} \cdot \Gamma_{R_i R_k}^{D}(R) = \frac{1}{2} \cdot \sum_{i \in KB_j} \sum_{k \in KB_l} \Gamma_{R_i R_k}^{D}(R) \approx \Gamma_{KB_j KB_l}^{D}(R).$$

The key question when using key-rate buckets for risk management is how many of them should be used and how they should be defined along the time to maturity axes to manage a portfolio successfully. The answer largely depends on the portfolio itself. If the trader or risk manager deals with a portfolio consisting of assets with short-term maturities, he may be interested in key-rate buckets up to the maximum time to maturity of the portfolio assets only, and will probably divide this part of the time to maturity axes into a greater number of segments than a portfolio manager managing assets with cash flows and maturities widely spread between one and ten years using yearly key-rate buckets. The choice of the key-rate buckets also depends on the cash flows and time to maturity of the hedge instruments.

However, if we look at the RiskMetrics^{TM} correlation (Correl.) data of the price volatilities (in %), i.e. the relative daily changes of the zero-coupon bond prices or discount factors, for maturities from 1 month (money market (M.M.)) to ten years (bond market (B.M.)), we see that there are almost perfect correlations. Let us take a closer look at the correlation matrix extracted from the data set available on the internet ($http://www.riskmetrics.com$) on October 23, 2000 and rounded to two decimals:

Correl.	1M	3M	6M	1Y	2Y	3Y	4Y	5Y	7Y	9Y	10Y
M. M. 1M	1.00	0.78	0.85	0.75	0.18	0.16	0.10	0.07	0.06	0.01	0.01
3M		1.00	0.80	0.73	0.17	0.10	0.06	-0.001	-0.05	-0.09	-0.11
6M			1.00	0.93	0.26	0.23	0.16	0.12	0.07	-0.01	-0.01
1Y				1.00	0.26	0.24	0.18	0.13	0.10	0.03	0.01
B. M. 2Y					1.00	0.96	0.93	0.88	0.79	0.67	0.63
3Y						1.00	0.97	0.94	0.85	0.74	0.72
4Y							1.00	0.96	0.89	0.80	0.78
5Y								1.00	0.92	0.85	0.83
7Y									1.00	0.95	0.93
9Y										1.00	0.99
10Y											1.00
1.65*Vol.	0.02	0.05	0.09	0.22	0.68	1.01	1.26	1.52	1.87	2.29	2.47
Volat. (%)	0.01	0.03	0.05	0.13	0.42	0.61	0.77	0.92	1.13	1.39	1.50

As we can see, the correlation between the maturities of $3, 4$, and 5 years is pairwise greater than 0.94. The correlation between the maturities of 9 and 10 years is even greater than 0.98. So it seems appropriate to define the key-rate buckets $KB_1 := [0M, 3Y]$, $KB_2 := (3Y, 6Y]$, $KB_3 := (6Y, 8Y]$, and $KB_4 := (8Y, 10Y]$. We may even divide the first key-rate buckets into smaller ones if we include hedge instruments with specific exposure in this bucket and if we especially want to deal with this time to maturity segments in more detail. Since this will not be the case in most of our applications we either deal with KB_1, KB_2, KB_3, and KB_4 or yearly buckets for a detailed risk exposure.

It should be mentioned that the factor loadings can also be evaluated empirically. This is usually done by a so-called *principal component analysis* (see, e.g. Greene [Gre93] for more details). The advantage of this method is that the explanatory factors are uncorrelated and that even a small number of three or four risk factors already explain a high percentage of the empirical term structure variation. Dahl [Dah89] showed that, in Denmark, three factors are sufficient to explain 99.6% of all term structure movements within the chosen sample set. Comparable results have been found by Garbade [Gar86] and [Gar89] and Litterman and Scheinkman [LS88] for the U.S. Treasury bond market, and by Newton and Chau [NC91], who give an overview of a great variety of markets. Figure 6.4 shows the factor loadings of the three factors which explain most of the term structure changes in the German bond market evaluated by using daily yield changes of the REX[9] sub indices with maturities ranging from one to ten years and a sample set of data from October 27, 1998 to December 11, 2000.

---

[9]REX denotes the German bond (in German: *Renten*) *index*.

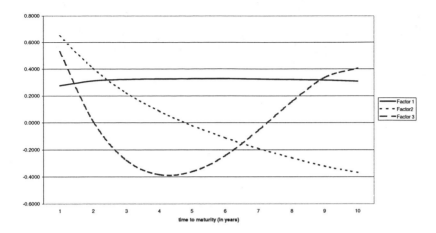

FIGURE 6.4. Factor loadings derived by a principal component analysis of the German bond market

The first factor corresponds to a parallel shift of the term structure, the second changes the *steepness*, and the third changes the *curvature* of the yield curve. These factors explain 99.3% of all yield curve movements within the chosen sample set. The problem arising with the use of these factors is that the factor loadings vary over time and have to be evaluated regularly on the basis of an updated sample set. Furthermore, these factors and their importance are not identical over different markets. For example, the factor explaining most movements in the Danish market was the steepness factor while the corresponding one in Germany was a parallel shift or level factor. Most of the time, however, the set of the most important three factors is identical over different markets. But even here the third factor for the Danish sample set is neither a level, nor a steepness, nor a curvature factor. This is somehow unsatisfactory and basically the reason why we have chosen the key-rate buckets for risk management purposes here.

7
Risk Management

Basically, risk management deals with the problem of protecting a portfolio or trading book against unexpected changes of market prices or other parameters. It therefore expresses the desire of a portfolio manager or trader to guarantee a minimum holding period return or to create a portfolio which helps him to fullfil specific liabilities over time. Risk management may help to avoid extreme events, to reduce the tracking error or even the trading costs. However, there are different possibilities for setting up a risk management or hedging process. The method which should be applied may well depend on the time horizon of the risk manager. If he is interested in controlling short-term risk, or if he would like to hedge against small movements in market prices, he may decide for a sensitivity-based risk management. This method is described in Section 7.1. If he has a longer time horizon and wants to be safe against large market movements he may decide for a downside risk management, which is discussed in Section 7.2.

7.1 Sensitivity-Based Risk Management

Sensitivity-based risk management deals with the problem of controlling a portfolio's sensitivity with respect to a given set of risk factors. It concentrates on hedging against small movements of the risk factors in a small period of time. This is of special interest to traders in charge of controlling the intraday or overnight market risk of their trading book. Section 7.1.1

gives a general definition of first- and second-order hedging which is applied to special risk measures such as the duration measure in Section 7.1.2 or the key-rate deltas and gammas in Section 7.1.3. For ease of exposition we omit the index for the specific daycount convention in writing the length of a time interval. The reader interested in more details on this topic refer to Section 5.1.

7.1.1   First- and Second-Order Hedging

Having defined the sensitivity measures of first- and second-order, we now discuss how these measures can be used for risk management or hedging purposes. To do so, let $V(\mathsf{F}, \varphi)$ be the price of a portfolio $\varphi = (\varphi_1, ..., \varphi_n)$ of financial instruments or derivatives with prices $D_1(\mathsf{F}), ..., D_n(\mathsf{F})$ depending on the vector of risk factors $\mathsf{F} = (\mathsf{F}_1, ..., \mathsf{F}_m)$, i.e.

$$V(\varphi) = V(\mathsf{F}, \varphi) = \sum_{i=1}^{n} \varphi_i \cdot D_i(\mathsf{F}),$$

and let $H_1(\mathsf{F}), ..., H_K(\mathsf{F})$, $K \in \mathbb{N}$, be the prices of the financial instruments which the trader or risk manager would allow for hedging purposes and which we will call hedge instruments. Furthermore, let $h = (h_1, ..., h_K)$ be a portfolio consisting of these hedge instruments, called a hedge portfolio, with a portfolio price given by

$$V(h) = V(\mathsf{F}, h) = \sum_{k=1}^{K} h_k \cdot H_k(\mathsf{F}).$$

The first-order sensitivities of the portfolios φ and h are given by

$$\Delta_{\mathsf{F}_j}^{V(\varphi)} = \sum_{i=1}^{n} \varphi_i \cdot \Delta_{\mathsf{F}_j}^{D_i}(\mathsf{F})$$

and

$$\Delta_{\mathsf{F}_j}^{V(h)} = \sum_{k=1}^{K} h_k \cdot \Delta_{\mathsf{F}_j}^{H_k}(\mathsf{F}), \, j = 1, ..., m.$$

For a fixed vector $\alpha = (\alpha_1, ..., \alpha_m) \in [0, 1]^m$, the idea behind the so-called *first-order hedging* is to find a portfolio $h^* = (h_1^*, ..., h_K^*)$, sometimes called the vector of the *first-order hedge ratios*, which solves the optimization problem

$$(P_1) \begin{cases} \sum_{j=1}^{m} \alpha_j \cdot \left(\Delta_{\mathsf{F}_j}^{V(\varphi)}(\mathsf{F}) - \Delta_{\mathsf{F}_j}^{V(h)}(\mathsf{F}) \right)^2 \to \min \\ h \in Z_1 \subseteq \mathbb{R}^K, \end{cases}$$

where Z_1 denotes the set of all possible hedge portfolios, which we assume to be set up by linear restrictions. This set may be equal to $\mathbb{R}^K$ if no trading restrictions are set by the trader, but there may be restrictions on the trading volume such as $s_{low} \leq h \leq s_{high}$ or even special corridors into which the risk manager or trader would like to drive the first-order sensitivities, i.e.

$$\Delta_{low} \leq \Delta_{\mathsf{F}}^{V(\varphi)}(\mathsf{F}) - \Delta_{\mathsf{F}}^{V(h)}(\mathsf{F}) \leq \Delta_{high}.$$

Such restrictions may be interesting especially for those first-order sensitivities for which α_j was set equal to 0 and which are therefore not included in the minimization process. Note that (P_1) is of the general form

$$(P) \begin{cases} h'Qh + c'h + d \to \min \\ h \in Z \end{cases}$$

for suitable vectors c and d, a symmetric matrix $Q \in \mathbb{R}^{K \times K}$, and a set of possible portfolios Z set up by linear restrictions. Especially, Q in the case of problem (P_1) is given by

$$Q = (q_{kl})_{k,l=1,\dots,K} \quad \text{with} \quad q_{kl} := \sum_{j=1}^{m} \alpha_j \cdot \Delta_{\mathsf{F}_j}^{H_k}(\mathsf{F}) \cdot \Delta_{\mathsf{F}_j}^{H_l}(\mathsf{F}), \ k,l = 1, \dots, K.$$

The goal function is convex if and only if Q is positive semi-definite, it is strictly convex if and only if Q is positive definite. In either case, the corresponding optimization problem is called quadratic. It is well-known from the theory of non-linear optimization that the quadratic optimization problem (P) has a solution if Q is positive semi-definite, $Z \neq \emptyset$, and the goal function is bounded below. (P) has a unique solution if Q is positive definite and $Z \neq \emptyset$. Very often traders and risk managers are only interested in first-order hedging with respect to a single risk factor, i.e. $\alpha = e_j$ for some $j \in \{1, \dots, m\}$ and $Z_1 = \mathbb{R}^K$. We will refer to this special case as *single-factor first-order hedging*. In this case, and if enough instruments with an exposure in the corresponding risk factor are made available for hedging, the value of the goal function in the previous optimization problem is zero, i.e.

$$\Delta_{\mathsf{F}_j}^{V(\varphi)}(\mathsf{F}) = \Delta_{\mathsf{F}_j}^{V(h^*)}(\mathsf{F}) = \sum_{k=1}^{K} h_k^* \cdot \Delta_{\mathsf{F}_j}^{H_k}(\mathsf{F}). \tag{7.1}$$

The resulting hedged portfolio $(\varphi, -h^*)$ derived by adding a short position in the hedge portfolio h^* to the portfolio φ is called *first-order neutral* (with respect to risk factor F_j). The second-order sensitivities of the portfolios φ and h are given by

$$\Gamma_{\mathsf{F}_j \mathsf{F}_l}^{V(\varphi)}(\mathsf{F}) = \sum_{i=1}^{n} \varphi_i \cdot \Gamma_{\mathsf{F}_j \mathsf{F}_l}^{D_i}(\mathsf{F})$$

and

$$\Gamma_{F_j F_l}^{V(h)} (F) = \sum_{k=1}^{K} h_k \cdot \Gamma_{F_j F_l}^{H_k} (F), \; j, l = 1, ..., m.$$

For a fixed matrix $\beta = (\beta_{jl})_{j,l=1,...,m} \in [0,1]^{m \times m}$, the idea behind the so-called *second-order hedging* is to find a portfolio $h^* = (h_1^*, ..., h_K^*)$, sometimes called the vector of the *second-order hedge ratios*, which solves the optimization problem

$$(P_2) \begin{cases} \displaystyle\sum_{j=1}^{m} \sum_{l=1}^{m} \beta_{jl} \cdot \left(\Gamma_{F_j F_l}^{V(\varphi)} (F) - \Gamma_{F_j F_l}^{V(h)} (F) \right)^2 \to \min \\ h \in Z_2 \subseteq I\!R^K. \end{cases}$$

Again, we suppose that the set of all possible hedge portfolios Z_2 is set up by linear restrictions. One possible restriction may be equation (7.1) to ensure that the residual portfolio will have a first-order sensitivity of zero with respect to factor F_j and second-order sensitivities as low as possible. Traders and risk managers are very often interested in second-order hedging only with respect to a single risk factor, i.e. $\beta_{jj} = 1$ for some $j \in \{1, ..., m\}$ and $\beta_{jl} = 0$ for all other possible pairs (j, l) as well as $Z_2 = I\!R^K$. We will refer to this special case as *single-factor second-order hedging*. In this case, and if enough instruments with an exposure in the corresponding risk factor are made available for hedging, the value of the goal function in the previous optimization problem is zero, i.e.

$$\Gamma_{F_j F_j}^{V(\varphi)} (F) = \Gamma_{F_j F_j}^{V(h^*)} (F) = \sum_{k=1}^{K} h_k^* \cdot \Gamma_{F_j F_l}^{H_k} (F). \tag{7.2}$$

The resulting hedged portfolio $(\varphi, -h^*)$ is called *second-order neutral* (with respect to risk factor F_j).

Advanced risk management often includes first- and second-order hedging. In this case traders are not only interested in reducing the sensitivity of their portfolio to given risk factors but also would like to reduce the frequency of restructuring the portfolio after small market changes. There are a few possibilities for how this can be done. The first one is to minimize a combination of the first- and second-order sensitivities. For a fixed number $\lambda \in [0,1]$, the idea is to find a portfolio $h^* = (h_1^*, ..., h_K^*)$ which solves the optimization problem

$$(P_\lambda) \begin{cases} \lambda \cdot \displaystyle\sum_{j=1}^{m} \alpha_j \cdot \left(\Delta_{F_i}^{V(\varphi)} (F) - \Delta_{F_i}^{V(h)} (F) \right)^2 \\ + (1 - \lambda) \cdot \displaystyle\sum_{j=1}^{m} \sum_{l=1}^{m} \beta_{jl} \cdot \left(\Gamma_{F_j F_l}^{V(\varphi)} (F) - \Gamma_{F_j F_l}^{V(h)} (F) \right)^2 \to \min \\ h \in Z_{2-\lambda}, \end{cases}$$

where the set $Z_{2-\lambda}$ of possible hedge portfolios is supposed to be set up by linear restrictions. Choosing $\lambda = 1$ the risk manager or trader is interested in first-order hedging only. Choosing $\lambda = 0$ as another extreme, he is interested in pure second-order hedging.

Let us look at a first *example*. In Section 6.1.1 we learned that the first-order sensitivity of the (static) coupon-bond futures price $F(t,T)$ with respect to the cheapest-to-deliver bond (CTD) with price $Bond(t, T_B^*, C^*)$, under the assumptions that the CTD bond doesn't change by a small change of the coupon-bond price, that there are no coupon payments in the time-period $[t, T]$, and that all hedge ratios are due to the corresponding notional amounts of the future and the CTD, is given by (see equation (6.4))

$$\Delta_{CTD}^F(\mathsf{F}) = \frac{1 + R_L(t,T) \cdot (T-t)}{Conv(T_B^*, C^*)}.$$

Since $\Delta_{CTD}^{CTD}(\mathsf{F}) = 1$, the first-order hedge ratio for hedging the future with the CTD to receive first-order neutrality, according to equation (7.1), is given by

$$\Delta_{CTD}^F(\mathsf{F}) = h_{CTD}^* \cdot \Delta_{CTD}^{CTD}(\mathsf{F}) = h_{CTD}^*.$$

The corresponding first-order hedge ratio for hedging the CTD with the future to receive first-order neutrality is given by

$$h_F^* = \frac{Conv(T_B^*, C^*)}{1 + R_L(t,T) \cdot (T-t)},$$

i.e. we have to buy h_F^* futures to hedge the CTD. Under the additional assumption that the repo rate $R_L(t,T)$ doesn't change if the price of the CTD changes[1], this is also the hedge ratio with respect to a changing zero-rate curve since, in this case, the future price-changes with changes of the zero-rate curve only by the price-changes of the CTD. Note that the hedge ratios have to be multiplied by the corresponding ratio of the notional amounts if these are different for the future and the CTD (see Section 7.1.2 for an example).

7.1.2 Duration-Based Hedging

Let us now drop the assumption that the repo rate $R_L(t,T)$ doesn't change with price-changes of the CTD, and assume that all price-changes are due to a parallel shift of the zero-rate curve. This is the assumption we made in Section 6.1.3 claiming that the zero rates of all maturities move by

$$\Delta R(t,T) := \Delta \mathsf{F}(t) \quad \text{for all } T \in [t, T^*]. \tag{7.3}$$

---

[1] Note, that this assumption is consistent with assumption 2 of the Black model, i.e. interest rates are supposed to be deterministic for discounting purposes.

Consequently, we assume that the linearly quoted repo rate moves according to a parallel shift of the continously quoted zero-rate curve. The continuously compounded equivalent $R_C(t,T)$ to the repo rate is implicitly defined by the equation

$$e^{R_C(t,T)\cdot(T-t)} = 1 + R_L(t,T) \cdot (T-t),$$

or, the other way round,

$$R_L(t,T) = \frac{1}{T-t} \cdot \left[e^{R_C(t,T)\cdot(T-t)} - 1 \right].$$

Hence, a small change of $R_C(t,T)$ will change the repo rate by

$$
\begin{aligned}
\Delta^{R_L}(\mathsf{F}) &:= \frac{\partial}{\partial R_C(t,T)} R_L(t,T) = e^{R_C(t,T)\cdot(T-t)} \\
&= 1 + R_L(t,T) \cdot (T-t).
\end{aligned}
$$

Furthermore, using equation (6.4), we know that a small change of the CTD will change the (static) futures price by

$$\Delta^F_{CTD}(\mathsf{F}) = \frac{1 + R_L(t,T) \cdot (T-t)}{Conv(T^*_B, C^*)},$$

with (T^*_B, C^*) characterizing the cheapest-to-deliver bond. Denoting the first-order sensitivities of the (static) future price $F = F(t,T)$ and the cheapest-to-deliver bond price $CTD = CDT(t, T^*_B, C^*)$ with respect to a parallel shift of the zero-rate curve by $\Delta^F(\mathsf{F})$ and $\Delta^{CTD}(\mathsf{F})$, we know by equation (6.15) that

$$\Delta^{CTD}(\mathsf{F}) = -duration(t, T^*_B, C^*) \cdot CTD.$$

To derive $\Delta^F(\mathsf{F})$ let us now look at the (static) futures price, again assuming that the CTD bond doesn't change by a small change of the coupon-bond price, that there are no coupon payments in the time-period $[t,T]$, and that all hedge ratios are due to the corresponding notional amounts of the future and the CTD. Following equation (5.4) it is given by

$$F(t,T) = CTD \cdot \frac{1 + R_L(t,T) \cdot (T-t)}{Conv(T^*_B, C^*)} - \frac{Accrued(t_0, T, C^*)}{Conv(T^*_B, C^*)}.$$

Hence,

$$
\begin{aligned}
\Delta^F(\mathsf{F}) &= \Delta^{CTD}(\mathsf{F}) \cdot \frac{1 + R_L(t,T) \cdot (T-t)}{Conv(T_B^*, C^*)} \\
&\quad + CTD \cdot \frac{1 + R_L(t,T) \cdot (T-t)}{Conv(T_B^*, C^*)} \cdot (T-t) \\
&= CTD \cdot \frac{1 + R_L(t,T) \cdot (T-t)}{Conv(T_B^*, C^*)} \cdot \\
&\quad \cdot \left[-duration(t, T_B^*, C^*) + T - t\right] \\
&= \left(F(t,T) + \frac{Accrued(t_0, T, C^*)}{Conv(T_B^*, C^*)}\right) \cdot \\
&\quad \cdot \left[-duration(t, T_B^*, C^*) + T - t\right] \\
&= F_{dirty}(t,T) \cdot \left[-duration(t, T_B^*, C^*) + T - t\right]
\end{aligned}
$$

with

$$
F_{dirty}(t,T) := F(t,T) + \frac{Accrued(t_0, T, C^*)}{Conv(T_B^*, C^*)}
$$

denoting the so-called *dirty price of the future*. If we may suppose that $T - t \approx 0$, we know that $F_{dirty}(t,T) \approx \frac{CTD(t, T_B^*, C^*)}{Conv(T_B^*, C^*)}$ which leads us to

$$
\begin{aligned}
\Delta^F(\mathsf{F}) &\approx -duration(t, T_B^*, C^*) \cdot F_{dirty}(t,T) \\
&\approx -duration(t, T_B^*, C^*) \cdot \frac{CTD(t, T_B^*, C^*)}{Conv(T_B^*, C^*)}.
\end{aligned}
$$

Having made this preparatory work, we can now turn our interest to the problem of hedging a portfolio of coupon bonds with a coupon-bond future to receive first-order neutrality with respect to a parallel shift of the zero-rate curve. The result is summarized in the following lemma.

Lemma 7.1 (Duration-Based Hedge Ratios) *Let* $V(\mathsf{F}, \varphi, t)$ *be the dirty price of a portfolio* $\varphi = (\varphi_1, ..., \varphi_n)$ *of coupon bonds depending on the risk factor* F *at time* $t \in [t_0, T]$. *Furthermore, let* $duration(\varphi, t)$ *denote the duration and* $\Delta^{V(\varphi)}(\mathsf{F}) = \Delta^{V(\mathsf{F}, \varphi, t)}(\mathsf{F})$ *denote the first-order sensitivity of the coupon-bond portfolio with respect to* F *at time* $t \in [t_0, T]$. *Under the assumption that the CTD bond doesn't change by a small change of the coupon-bond price and that there are no coupon payments in the time-period* $[t, T]$, *the first-order hedge ratio* $h_F^*(t)$ *at time* $t \in [t_0, T]$ *for hedging the coupon-bond portfiolio with the future to receive first-order neutrality is given by*

$$
\begin{aligned}
h_F^*(t) &= \frac{\Delta^{V(\varphi)}(\mathsf{F})}{\Delta^F(\mathsf{F})} \\
&= \frac{-duration(\varphi, t) \cdot V(\mathsf{F}, \varphi, t)}{F_{dirty}(t,T) \cdot \left[-duration(t, T_B^*, C^*) + T - t\right]}
\end{aligned}
\tag{7.4}
$$

for $t \in [t_0, T]$ and with $CTD = CTD(t, T_B^*, C^*)$. The hedge ratio $h_F^*(t)$ is sometimes called duration-based hedge ratio at time $t \in [t_0, T]$.

In practice, risk and portfolio managers often use reasonable approximations of equation (7.4). These are summarized in the following corollary.

Corollary 7.2 *Let the assumptions of Lemma 7.1 be satisfied. Furthermore, let $V_{clean}(\mathsf{F}, t, \varphi)$ denote the clean price of the portfolio φ at time $t \in [t_0, T]$.*

a) If $T - t \approx 0$, then

$$h_F^*(t) \approx \frac{duration(t, \varphi) \cdot V(\mathsf{F}, \varphi, t)}{duration(t, T_B^*, C^*) \cdot F_{dirty}(t, T)}$$
$$\approx \frac{duration(\varphi, t) \cdot V(\mathsf{F}, \varphi, t) \cdot Conv(T_B^*, C^*)}{duration(t, T_B^*, C^*) \cdot CTD}. \tag{7.5}$$

b) If, in addition to the assumptions of part a),

$$\frac{V(\mathsf{F}, \varphi, t)}{F_{dirty}(t, T)} \approx \frac{V_{clean}(\mathsf{F}, \varphi, t)}{F(t, T)},$$

then

$$h_F^*(t) \approx \frac{duration(\varphi, t) \cdot V_{clean}(\mathsf{F}, \varphi, t)}{duration(t, T_B^*, C^*) \cdot F(t, T)}. \tag{7.6}$$

Especially equation (7.6) is very popular among risk and portfolio managers since durations, clean coupon bond and (clean) futures prices are directly available in the market. Also, up to today, Macaulay and modified duration is available via commercial software and information systems rather than the zero-rate based duration of equation (7.6). So risk and portfolio managers tend to use one of these instead of the zero-rate based duration. The following example shows this practical application of equation (7.6) using the Macaulay duration. The problems arising with this application are analyzed in the case study of Section 7.1.3. For the practical application, note that h_F^* was calculated under the assumption that the futures price is evaluated relative to the same notional amount as the CTD. If the notional N_F of the future and the notional N_{CTD} of the CTD do not coincide, the duration-based hedge ratio of equations (7.4), (7.5) or (7.6) has to be adjusted to

$$h_F^{trading} = \frac{N_{CTD}}{N_F} \cdot h_F^*. \tag{7.7}$$

Case Study (Hedging Bond Portfolios with Futures)

Let us consider the following coupon-bond portfolio $\varphi = (\varphi_1, ..., \varphi_7)$ with a notional amount of 90 Mio. Euro, a portfolio clean price of $V_{clean}(\varphi, t) = 88,740,750$ (prices are already multiplied with $\frac{N_{CTD}}{100}$) at October 20, 2000 (t), and a Macaulay duration of $duration_{Mac}(\varphi, t) = 5.41$ years consisting of the following coupon bonds:

Notional amount (in Mio.)	Coupon (in %)	Maturity
10	4.050	05/17/02
10	4.125	08/27/04
15	7.500	11/11/04
5	6.250	04/26/06
20	3.750	01/04/09
10	4.500	07/04/09
20	5.250	07/04/10

Usually, the portfolio is divided for a better duration-based hedging result which we do by splitting into the portfolios $\varphi^1 = (\varphi_1, ..., \varphi_4)$ with coupon bonds having a time to maturity of up to 6 years and $\varphi^2 = (\varphi_5, ..., \varphi_7)$ with coupon bonds having a time to maturity of more than 6 years. The corresponding portfolio prices at time t were $V_{clean}(\varphi^1, t) = 41,092,750$ and $V_{clean}(\varphi^2, t) = 47,648,000$ (prices are already multiplied with $\frac{N_{CTD}}{100}$), the Macaulay durations were $duration_{Mac}(\varphi^1, t) = 3.19$ years and $duration_{Mac}(\varphi^2, t) = 7.33$ years. The idea is to hedge portfolio φ^1 with the Bobl future and portfolio φ^2 with the Bund future. The price for the futures at time t were $F_{Bobl}(t) := F(t, T_{Bobl}) = 103.38$ for the Bobl and $F_{Bund}(t) := F(t, T_{Bund}) = 105.58$ for the Bund. The CTD for the Bobl future at that time was a 6.5% government bond with maturity time 10/14/05 and a duration of 4.43 years, the CTD for the Bund future was a 5.375% government bond with maturity time 01/04/10 and a duration of 7.21 years. At October 20, 2000 both futures were for a notional of $N_F = 100,000$ Euro. Using equation (7.6) and the Macaulay duration as explained above, we get the following hedge ratios at time t for the Bobl and Bund futures:

$$h_{Bobl}^{trading}(t) \approx \frac{duration_{Mac}(\varphi^1, t) \cdot V_{clean}(\varphi^1, t)}{\frac{N_{Bobl}}{100} \cdot duration\left(t, T_{CTD(Bobl)}^*, C_{CTD(Bobl)}^*\right) \cdot F_{Bobl}(t)}$$

$$= \frac{3.19 \cdot 41,092,750}{\frac{100,000}{100} \cdot 4.43 \cdot 103.38} = 286.23 \approx 286$$

and

$$h_{Bund}^{trading}(t) \approx \frac{duration_{Mac}(\varphi^2, t) \cdot V_{clean}(\varphi^2, t)}{\frac{N_{Bund}}{100} \cdot duration\left(t, T^*_{CTD(Bund)}, C^*_{CTD(Bund)}\right) \cdot F_{Bund}(t)}$$

$$= \frac{7.33 \cdot 47,648,000}{1,000 \cdot 7.21 \cdot 105.58} = 458.81 \approx 459.$$

So we are hedging the coupon-bond portfolio by selling 286 Bobl futures and 459 Bund futures.   □

Sometimes it is also interesting to duration-hedge a coupon-bond portfolio with other coupon bonds. For this reason, let $V(\mathsf{F}, \varphi, t)$ and $duration(\varphi, t)$ be the price and duration of a portfolio $\varphi = (\varphi_1, ..., \varphi_n)$ of coupon bonds at time $t \in [t_0, T]$. Furthermore, let $h = (h_1, ..., h_K)$ be a portfolio consisting of the coupon bonds which are available for hedging and which we will briefly denote by hedging coupon bonds. Let $H_k(\mathsf{F}, t) = Bond\left(t, T^k_B, C^k\right)$ and $duration\left(t, T^k_B, C^k\right)$, $k = 1, ..., K \in I\!N$, be the prices and durations of these hedge instruments and

$$V(h, t) = V(\mathsf{F}, h, t) = \sum_{k=1}^{K} h_k \cdot H_k(\mathsf{F}, t) \text{ and } duration(h, t)$$

be the price and duration of the hedge portfolio as given in equation (6.16). Then the condition for the first-order hedge ratios $h^* = h^*(t)$ with $h^* = (h_1^*, ..., h_K^*)$ for hedging the coupon-bond portfolio with the hedging coupon bonds to receive first-order neutrality at time t is given by

$$\Delta^{V(\varphi)}(\mathsf{F}) = \Delta^{V(h^*)}(\mathsf{F}) = \sum_{k=1}^{K} h_k^* \cdot \Delta^{H_k}(\mathsf{F}), \tag{7.8}$$

or equivalently

$$-duration(\varphi, t) \cdot V(\mathsf{F}, \varphi, t) = -duration(h^*, t) \cdot V(\mathsf{F}, h^*, t).$$

If $K = 1$ with $T_B := T^1_B$ and $C := C^1$, this is

$$-duration(\varphi, t) \cdot V(\mathsf{F}, \varphi, t) = -duration(t, T_B, C) \cdot h^* \cdot Bond(t, T_B, C),$$

or

$$h^* = \frac{duration(\varphi, t) \cdot V(\mathsf{F}, \varphi, t)}{duration(t, T_B, C) \cdot Bond(t, T_B, C)}.$$

If $K > 1$, there is more than one possibility, and further equations have to be added, such as

$$duration(\varphi, t) = duration(h^*, t).$$

In this case we need a minimum of two coupon bonds for hedging. For a number of two hedging coupon bonds we always get a combination of a coupon bond with a shorter and a coupon bond with a longer duration than that of the portfolio φ. We will return to hedging coupon-bond portfolios with coupon bonds in the next section.

7.1.3 Key-Rate Delta and Gamma Hedging

In Section 6.1.4 we showed that the key-rate deltas are a natural generalization of the duration concept. To realize this concept, the time to maturity interval $[0, T^* - t]$ was divided into $m \in I\!N$ non-overlapping subintervals $KB_1, ..., KB_m$ called the key-rate buckets. At time t, all zero rates having a time to maturity within the same bucket KB_j are supposed to move by exactly the same amount $\Delta F_j(t)$, $j = 1, ..., m$. If $D(R(t), t)$ denotes the price of a financial instrument or derivative depending on (some elements of) the vector of zero rates $R(t) := (R(t, T_1), ..., R(t, T_n))'$ and time $t \in [t_0, T^*]$ with $t_0 \leq t \leq T_1 < \cdots < T_n \leq T^*$, the key-rate delta $\Delta_{KB_j}^D(R)$ is defined to be the first-order sensitivity of the derivatives price with respect to a small parallel shift of the zero-rate curve within key-rate bucket KB_j, $j = 1, ..., m$, and all other zero rates unchanged. The corresponding second-order sensitivity, the key-rate gamma with respect to the key-rate buckets KB_j and KB_l, $j, l = 1, ..., m$, is denoted by $\Gamma_{KB_j KB_l}^D(R)$. Using these definitions, the approximate price-change $\Delta D(R)$ of a derivative, depending on small changes of the risk factors F, is given by

$$\Delta D(R) \approx \sum_{j=1}^{m} \Delta_{KB_j}^D(R) \cdot \Delta F_j + \sum_{j=1}^{m} \sum_{l=1}^{m} \Gamma_{KB_j KB_l}^D(R) \cdot \Delta F_j \cdot \Delta F_l.$$

The price-change $\Delta V(\varphi, R)$ of the portfolio $\varphi = (\varphi_1, ..., \varphi_n)$ of financial instruments or derivatives with prices $D_1(R), ..., D_n(R)$ depending on the vector of zero rates R which is supposed to include all zero rates on which the derivatives may depend is approximately given by

$$\Delta V(\varphi, R) \approx \sum_{j=1}^{m} \Delta_{KB_j}^{V(\varphi)}(R) \cdot \Delta F_j + \sum_{j=1}^{m} \sum_{l=1}^{m} \Gamma_{KB_j KB_l}^{V(\varphi)}(R) \cdot \Delta F_j \cdot \Delta F_l,$$

where $V(\varphi, R)$ denotes the price of the portfolio, the key-rate deltas of the portfolio are given by

$$\Delta_{KB_j}^{V(\varphi)}(R) = \sum_{i=1}^{n} \varphi_i \cdot \Delta_{KB_j}^{D_i}(R), \, j = 1, ..., m,$$

and the key-rate gammas of the portfolio are given by

$$\Gamma_{KB_j KB_l}^{V(\varphi)}(R) = \sum_{i=1}^{n} \varphi_i \cdot \Gamma_{KB_j KB_l}^{D_i}(R), \, j, l = 1, ..., m.$$

Let $H_1(R), ..., H_K(R)$, $K \in \mathbb{N}$, be the prices of the hedge instruments with key-rate deltas and gammas denoted by $\Delta_{KB_j}^{H_k}(R)$ and $\Gamma_{KB_jKB_l}^{H_k}(R)$, $j, l = 1, ..., m$, $k = 1, ..., K$. Furthermore, let $h = (h_1, ..., h_K)$ be the hedge portfolio with price $V(h) = V(h, R)$ and key-rate delta and gamma given by $\Delta_{KB_j}^{V(h)}(R)$ and $\Gamma_{KB_jKB_l}^{V(h)}(R)$, $j, l = 1, ..., m$. For a fixed vector $\alpha = (\alpha_1, ..., \alpha_m) \in [0, 1]^m$, the so-called *key-rate delta hedging* is the search for a portfolio $h^* = (h_1^*, ..., h_K^*)$, sometimes called the vector of the *key-rate delta hedge ratios*, which solves the optimization problem

$$(P_1) \begin{cases} \sum_{j=1}^{m} \alpha_j \cdot \left(\Delta_{KB_j}^{V(\varphi)}(R) - \Delta_{KB_j}^{V(h)}(R) \right)^2 \to \min \\ h \in Z_1 \subseteq \mathbb{R}^K, \end{cases}$$

where Z_1 denotes the set of all possible hedge portfolios which we assume to be set up by linear restrictions. For a fixed matrix $\beta = (\beta_{jl})_{j,l=1,...,m} \in [0, 1]^{m \times m}$, the corresponding *key-rate gamma hedging* is the search for a portfolio $h^* = (h_1^*, ..., h_K^*)$, sometimes called the vector of the *key-rate gamma hedge ratios*, which solves the quadratic optimization problem

$$(P_2) \begin{cases} \sum_{j=1}^{m} \sum_{l=1}^{m} \beta_{jl} \cdot \left(\Gamma_{KB_jKB_l}^{V(\varphi)}(R) - \Gamma_{KB_jKB_l}^{V(h)}(R) \right)^2 \to \min \\ h \in Z_2 \subseteq \mathbb{R}^K. \end{cases}$$

Again, we suppose that the set of all possible hedge portfolios Z_2 is set up by linear restrictions. Combinations are possible as already mentioned in Section 7.1.1. Nevertheless, because of the complexity of the resulting optimization problems, key-rate delta hedging plays the dominant role in practice.

Case Study (Key-Rate Delta Hedging)[2]

In the *Duration-Based Hedging with Futures* case study of Section 7.1.2 we hedged, at time $t = 10/20/00$, a coupon-bond portfolio with a notional amount of 90 Mio. Euro using Bobl and Bund futures. We did this, using the simplest approximation of Corollary 7.2b), by selling $h_1^0 = 286$ Bobl and $h_2^0 = 459$ Bund futures which is a notional of 74.5 Mio. Euro. According-ing to the classification of the coupon bonds into two maturity segments in this case study we now define the key-rate buckets $KB_1^A := [0M, 6Y]$ and $KB_2^A := (6Y, 10Y]$. The zero-rate curve at time t and the key-rate deltas of the duration hedged portfolio $(\varphi_1, ..., \varphi_7, -h_1^0, -h_2^0)$ is shown[3] in

---

[2] All calculations and optimizations were done using the software tool Risk *Advisor* from risklab *germany*.

[3] Remember, that we plot the key rate deltas with respect to an increase of the corresponding zero rates by $1bp$.

figures 7.1-7.2. The two deltas do not completely net out because we split the portfolio, and so do not correctly consider coupon payments of the longer maturity coupon bonds which fall in the first key-rate bucket. Furthermore, the duration-based hedging formula of Corollary 7.2b) is just an approximation.

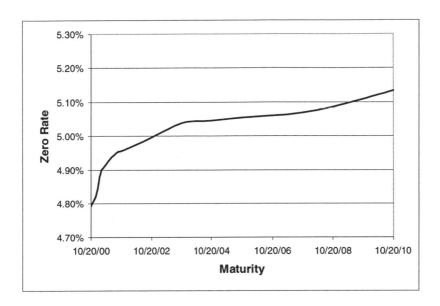

FIGURE 7.1. Continuous zero rate curve derived from German government bonds on October 20, 2000

On the other hand we apply the optimization problem (P_1) doing a key-rate delta hedging with respect to the same key-rate buckets and instruments. The optimal solution of this problem is given by selling $h_1^A = 360$ Bobl and $h_2^A = 442$ Bund futures which is a notional of 80.18 Mio. Euro. The corresponding key-rate deltas of the hedged portfolio $\left(\varphi_1, ..., \varphi_7, -h_1^A, -h_2^A\right)$ in figure 7.3 are both close to zero and by far smaller than those of portfolio $\left(\varphi_1, ..., \varphi_7, -h_1^0, -h_2^0\right)$.

However, this is not the full picture. Let us consider the key-rate buckets $KB_1^B := [0M, 3Y]$, $KB_2^B := (3Y, 6Y]$, $KB_3^B := (6Y, 8Y]$, and $KB_4^B := (8Y, 10Y]$ and let us examine the corresponding key-rate deltas of portfolio $\left(\varphi_1, ..., \varphi_7, -h_1^A, -h_2^A\right)$. The result is plotted in figure 7.4 and shows the risk inherent in what we thought a well hedged portfolio. It is indeed well hedged under the assumption that the zero-rate curve moves by parallel shifts only in the key-rate buckets KB_1^A and KB_2^A when the key-rate deltas of buckets KB_1^B and KB_2^B as well as those of buckets KB_3^B and KB_4^B net out. The hedge may be rather bad if the yield curve twists.

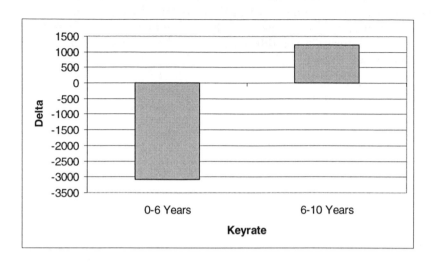

FIGURE 7.2. Key rate deltas of the duration hedged portfolio $(\varphi_1, ..., \varphi_7, -h_1^0, -h_2^0)$ with respect to the key rate buckets KB_1^A and KB_2^A

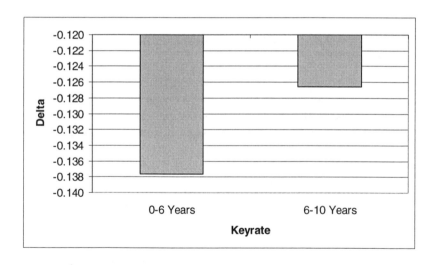

FIGURE 7.3. Key rate deltas of the key rate hedged portfolio $(\varphi_1, ..., \varphi_7, -h_1^A, -h_2^A)$ with respect to the key rate buckets KB_1^A and KB_2^A

One step further, let us use a 8% coupon bond with maturity time 01/21/02 and a 6% coupon bond with maturity time 07/04/07 as hedge instruments H_3 and H_4 and apply optimization problem (P_1) with respect to the key-rate buckets KB_1^B, KB_2^B, KB_3^B, and KB_4^B. The optimal solution is a hedge portfolio of $h_1^B = 259$ Bobl futures, $h_2^B = 445$ Bund futures which is a notional of 70.4 Mio., a notional of $h_3^B = -0.42$ Mio. of H_3, and a notional of $h_4^B = 33.62$ Mio. of H_4. This is a total notional amount of 103.6 Mio., i.e. the notional amount of our hedge portfolios increases the finer we hedge. Nevertheless, the risk numbers of the key-rate hedged portfolio $\left(\varphi_1, ..., \varphi_7, -h_1^B, -h_2^B, -h_3^B, -h_4^B\right)$, shown in figure 7.5, look much better than those of the hedged portfolio $\left(\varphi_1, ..., \varphi_7, -h_1^A, -h_2^A\right)$ as shown in figure 7.4.

Let us finally compare the risk of the starting coupon-bond portfolio with that of the key-rate hedged portfolio $\left(\varphi_1, ..., \varphi_7, -h_1^B, -h_2^B, -h_3^B, -h_4^B\right)$ for the yearly key-rate buckets $KB_1^C := [0M, 1Y]$, $KB_2^C := (1Y, 2Y]$, ..., $KB_{10}^C := (9Y, 10Y]$ as shown in figures 7.6-7.7. Notice the typically negative key-rate deltas of the long coupon-bond portfolio in figure 7.6, due to the fact that coupon-bond prices fall if we increase the zero rates by $1bp$. We see that the main risk is concentrated in the key-rate buckets $KB_4^C, ..., KB_6^C$, corresponding to time to maturities from 3 to 6 years, and in the key-rate buckets KB_9^C and KB_{10}^C, corresponding to time to maturities from 8 to 10 years. As the analysis of the RiskMetrics^{TM} correlations in Section 6.4 showed, it is rather plausible to assume that the zero-rate curve in each of these two time to maturity segments will move by parallel shifts. So we can assume that the key-rate deltas will net out making the hedged portfolio $\left(\varphi_1, ..., \varphi_7, -h_1^B, -h_2^B, -h_3^B, -h_4^B\right)$ a rather good one. We could now add additional conditions such as a limit for the notional amount of the hedged portfolio to meet more specific needs of the risk controller or portfolio manager.   □

One of the most important applications in portfolio management is the derivation of a portfolio that mirrors a given index portfolio. Such a portfolio is also called a *tracking portfolio* and the process of managing a portfolio to duplicate the index over time is called *index tracking*. Usually this is done by adjusting the duration or the key-rate delta of the tracking portfolio with respect to a single key-rate bucket to match that of the index. No wonder that the results can be very disappointing if the zero-rate movements are non-parallel. Therefore, we dedicated the following case study to compare the tracking portfolios for the J.P. Morgan government bond index Germany (JPMGBG), briefly denoted by J.P. Morgan index derived by a key-rate delta hedge with one (consistent with duration-based hedging) and with ten key-rate buckets.

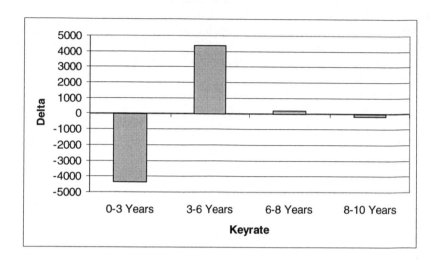

FIGURE 7.4. Key rate deltas of the key rate hedged portfolio $\left(\varphi_1, ..., \varphi_7, -h_1^A, -h_2^A\right)$ with respect to the key rate buckets KB_1^B, KB_2^B, KB_3^B, and KB_4^B

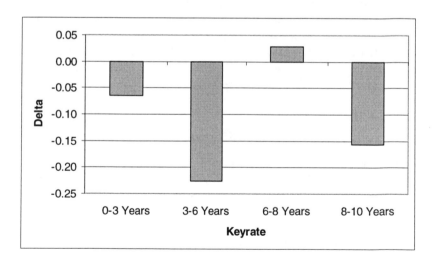

FIGURE 7.5. Key rate deltas of the key rate hedged portfolio $\left(\varphi_1, ..., \varphi_7, -h_1^B, -h_2^B, -h_3^B, -h_4^B\right)$ with respect to the key rate buckets KB_1^B, KB_2^B, KB_3^B, and KB_4^B

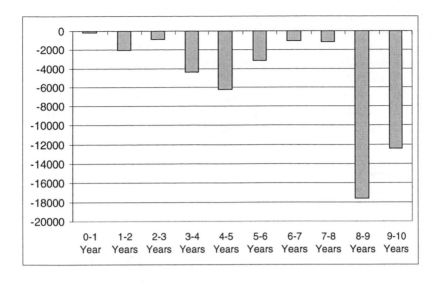

FIGURE 7.6. Key rate deltas of the initial bond portfolio $(\varphi_1, ..., \varphi_7)$ with respect to the key rate buckets $KB_1^C, KB_2^C, ..., KB_{10}^C$

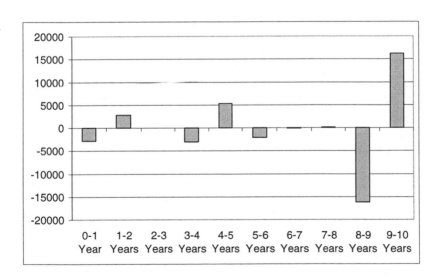

FIGURE 7.7. Key rate deltas of the key rate hedged portfolio $\left(\varphi_1, ..., \varphi_7, -h_1^B, -h_2^B, -h_3^B, -h_4^B\right)$ with respect to the key rate buckets $KB_1^C, KB_2^C, ..., KB_{10}^C$

Case Study (Index Tracking)[4]

Starting in April 1998, the J.P. Morgan index was hedged once a month. First this was done with respect to one key-rate bucket from 0 to 10 years due to a parallel shift of the yield curve, second with respect to the ten key-rate buckets $KB_1^C, KB_2^C, ..., KB_{10}^C$. In addition we claimed that the fair or clean prices of the J.P. Morgan index and those of the hedge portfolios should be equal. The corresponding hedge concepts are referred to as (fair price) duration tracking for the case of one key-rate bucket and (fair price) key-rate tracking for the case of ten key-rate buckets. The portfolios are compared after one month and then readjusted. Figure 7.8 shows the price behaviour of the duration tracking portfolio compared to the J.P. Morgan index. This happened while the correponding zero-rate curves changed as plotted in figure 7.9.

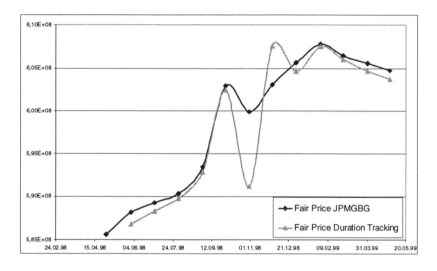

FIGURE 7.8. Price behaviour of the duration tracking portfolio compared to the J.P. Morgan index.

At the very beginning the duration tracking did quite well until the "big surprise" in October 1998 opened the eyes by a significant underperformance compared to the J.P. Morgan index which is shown numerically in the following table.

Time	J.P. Morgan	Duration tracking	change
8/31-09/30/98	$9,486,111$	$9,021,417$	parallel
9/30-10/31/98	$-3,031,613$	$-11,749,764$	twist

---

[4] All calculations and optimizations were done using the software tool Risk *Advisor* from risklab *germany*.

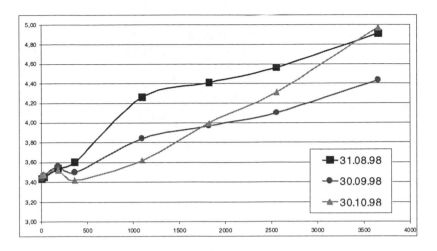

FIGURE 7.9. Changes of the zero rate curves.

Where did this underperformance in October 1998 come from? To answer this question let us dip into September 30, 1998. If we compare the price value of a basis point (see Section 6.1.3) of $-255,325.69$ for the J.P. Morgan index and $-255,326.21$ for the duration tracking portfolio at a clean price of $585,583,106.07$ for the first and $585,583,179.57$ for the latter we would have not expected such a development. This is underlined by the key-rate delta picture in figures 7.10-7.11 plotted by their days to maturity.

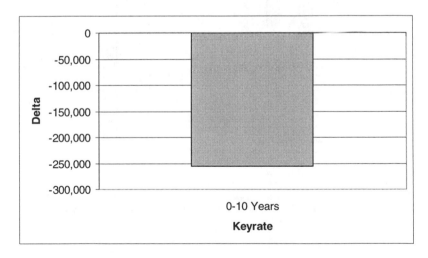

FIGURE 7.10. Key rate delta of the J.P. Morgan index with respect to one key rate bucket.

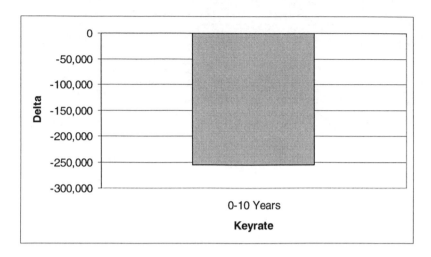

FIGURE 7.11. Key rate delta of the duration tracking portfolio with respect to one key rate bucket

However, with a little more insight, given by the key-rate delta pictures of figures 7.12-7.13 for a number of 10 key-rate buckets, we see that both portfolios have a completely different risk design.

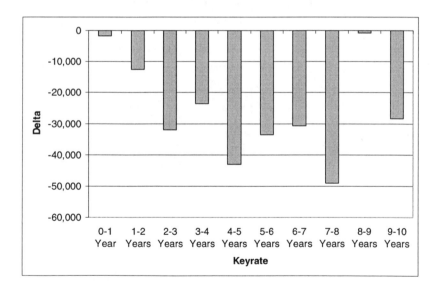

FIGURE 7.12. Key rate deltas of the J.P. Morgan index with respect to ten key rate buckets

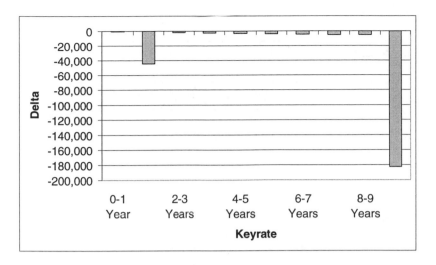

FIGURE 7.13. Key rate deltas of the duration tracking portfolio with respect to ten key rate buckets

More dramatically this is shown in figure 7.14, looking at the key-rate deltas of the portfolio derived by going long the duration tracking portfolio and short the J.P. Morgan index. We can see that the duration tracking portfolio extremely overweights the last maturity segment, resulting in a negative key-rate delta. A zero-rate curve moving up in this last maturity bucket may therefore have led to the underperformance we observed. And indeed, as already shown in figure 7.9, the zero-rate curve did increase in this bucket combined with a twist of the whole zero-rate curve.

If we use all ten key-rate buckets to solve hedging problem (P_1), the key-rate deltas of the resulting key-rate tracking portfolio pretty much equal those of the J.P. Morgan index (see figure 7.15). If we examine the portfolio resulting from going long the key-rate tracking portfolio and short the J.P. Morgan index we see that the key-rate deltas are much smaller than those of figure 7.14, especially in the last segment (see figure 7.16). Consequently, the corresponding price behaviour of the key-rate tracking portfolio is much closer to that of the J.P. Morgan index than that of the duration tracking portfolio which is shown in figure 7.17. As we can see, the key-rate tracking portfolio rather smoothly tracks the index, almost perfect compared to the duration tracking portfolio. This is documented by the tracking error, which is defined to be the square root of the sum of the squared deviations of the index price and the price of the corresponding tracking portfolio. It is calculated as $2,926,855$ or 0.50% for the duration tracking portfolio and $331,102$ or 0.05% for the key-rate tracking portfolio which is only 10% of the first value.   □

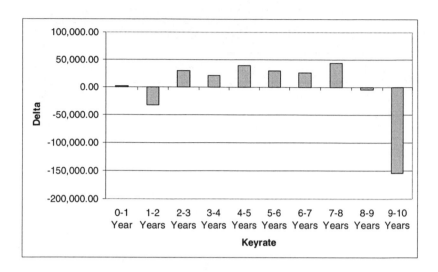

FIGURE 7.14. Key rate deltas of the portfolio resulting from going long the duration tracking portfolio and short the J.P. Morgan index with respect to ten key rate buckets

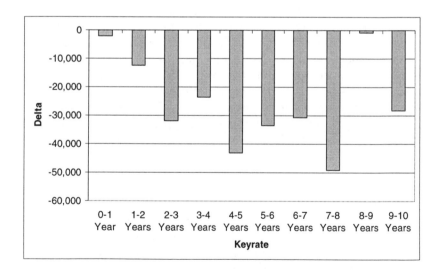

FIGURE 7.15. Key rate deltas of the key rate tracking portfolio with respect to ten key rate buckets

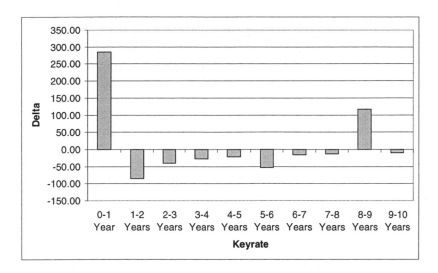

FIGURE 7.16. Key rate deltas of the portfolio resulting from going long the key rate tracking portfolio and short the J.P. Morgan index with respect to ten key rate buckets

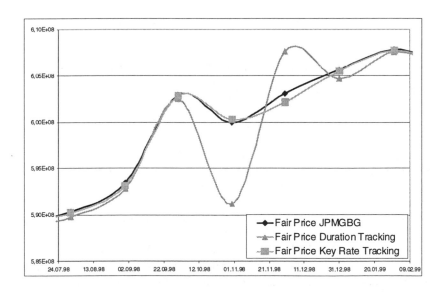

FIGURE 7.17. Price behaviour of the key rate tracking portfolio compared to the duration tracking portfolio and the J.P. Morgan index

Note that we optimized the tracking portfolios of the previous example without any linear restrictions but the price equality. Nevertheless, it could be interesting to add other restrictions such as a limit on the hedging volume or limits on the investment in instruments with a maturity falling in specific key-rate buckets to control the time structure of the portfolio. It could also be interesting to limit the sensitivity with respect to other risk factors such as the theta or vega, which could be easily implemented as shown in Section 7.1.1.

Another area of increasing interest is protecting a given portfolio φ of loans, considered to play the role of the index portfolio, with a portfolio h of coupon bonds, considered as playing the role of the tracking portfolio, against the risk of changing interest rates. This is already part of the so-called *asset liability management*. If we claim that both assets and liabilities have identical key-rate deltas, the price-change of the residual portfolio is approximately given by

$$\sum_{j=1}^{m}\sum_{l=1}^{m}\left(\Gamma_{KB_jKB_l}^{V(\varphi)}(R) - \Gamma_{KB_jKB_l}^{V(h)}(R)\right) \cdot \Delta F_j \cdot \Delta F_l.$$

Let us suppose that we model loans such as coupon bonds, and remember that coupon bonds have gamma exposure only in the diagonal elements of the gamma matrix. Then this price-change is equal to

$$\sum_{j=1}^{m}\left(\Gamma_{KB_jKB_j}^{V(\varphi)}(R) - \Gamma_{KB_jKB_j}^{V(h)}(R)\right) \cdot (\Delta F_j)^2.$$

If we claim that the key-rate gammas $\Gamma_{KB_jKB_j}^{V(h)}(R)$ of the assets are always larger than the key-rate gammas $\Gamma_{KB_jKB_j}^{V(\varphi)}(R)$ of the liability side, i.e.

$$\Gamma_{KB_jKB_j}^{V(h)}(R) \geq \Gamma_{KB_jKB_j}^{V(\varphi)}(R), \text{ for all } j = 1, ..., m,$$

the approximate price-change of the residual portfolio will always be negative. In other words, an increase in price because of changing zero rates in any of the key-rate buckets will always be greater, a decrease in price will always be less on the asset side than that on the liability side. This generalizes the restriction that, under a parallel movement of the zero-rate curve, the convexity of the asset side should always be larger than the convexity of the liability side which can be found, e.g., in Dahl [Dah93].

Having discussed different possibilities for setting restrictions, there is also the possibility of changing the goal function. Under limited first- or second-order sensitivities it could be interesting to minimize the transaction costs or to maximize the expected return of a portfolio. The latter is of special interest for portfolio managers, especially when the planning horizon is long-term rather than short-term. Since this fits the background of downside risk measures, we will discuss downside risk management with the intention of maximizing the expected return of a portfolio in the next section.

7.2   Downside Risk Management

In this section we deal with the problem of managing the risk of large movements of the risk factors or interest rates in a possibly longer period of time. This problem especially appears in portfolio management when the portfolio manager wants to avoid the return of his portfolio falling below some given benchmark return. This problem is adressed in Section 7.2.1. On the other hand, the portfolio manager or trader may be controlled by a limit set on the value at risk of his portfolio therefore trying to be safe against extreme events. We will present a solution to this problem in Section 7.2.2.

7.2.1   Risk Management Based on Lower Partial Moments

The process of performing an optimal asset allocation basically deals with the problem of finding a portfolio that maximizes the expected utility of the investor or portfolio manager. In other words, the portfolio manager aims to choose a portfolio with a distribution function that maximizes the *expected utility*. As long as it is supposed that the returns of the portfolio assets follow a normal distribution, the return distribution of any portfolio considered will also be normal. In this case, as is done throughout the traditional portfolio theory introduced by Markowitz [Mar52] and Sharpe [Sha64], the problem of finding an expected utility-maximizing portfolio or distribution function for a risk-averse trader or portfolio manager, represented by a concave utility function, can be restricted to finding an optimal combination of the two parameters mean and variance. This dramatically simplifies the whole asset allocation process and is known as *mean-variance analysis*. It is the aim of the portfolio manager to find a portfolio that maximizes his expected return under a given risk level or a portfolio that minimizes his risk under a given return level. Risk in this case is measured by the variance of the portfolio return.

Unfortunately, selection rules based on the two parameters mean and variance are of limited generality. Roughly speaking they are optimal if the utility function is quadratic or if it is concave and the return distribution is elliptical[5]. Quadratic utility functions imply increasing absolute risk aversion and are therefore rather implausible, as was pointed out by Arrow [Arr71] and Hicks [Hic62]. In an empirical study Lintner [Lin72] has shown that the world of asset returns is more likely to be lognormal than normal. Bookstaber and Clarke [BC81] and [BC83] show that adding options to a portfolio may lead to rather *asymmetric return distributions*. But the use of variance as a measure of risk for asymmetric return distributions has already been questioned by financial theorists including, e.g., Markowitz [Mar70] and [Mar91], p. 188-201. He states that portfolio analyses based on semi-variance tend to produce better portfolios than those based on variance. Unfortunately, the computing cost involved with the use of semi-variance is much higher than that with the analyses based on variance as we need the entire joint distribution for the first technique while we only need the covariance matrix for the latter.

Extensive research has been done to derive concepts for ordering uncertain prospects, resulting in principles such as the *stochastic dominance* of order 1, 2 or 3 (see, e.g., Bawa [Baw75] or Martin, Cox and MacMinn [MCM88]). V. Bawa argues that it is quite reasonable to assume that the average investor takes decisions with respect to a finite, increasing and concave utility function with decreasing absolute risk aversion and that there is a strong rationale for using mean-lower partial variance rules as an approximation for the portfolio selection of this kind of investor. Under this rule one portfolio is better than the other if its mean is not lower and its lower partial variance is not higher than that of the other portfolio at every possible benchmark. As the lower partial variance is a generalization of the semi-variance, this follows the suggestion of Markowitz, still leaving the computational problems because we have to calculate the lower partial variances for each distribution at every possible benchmark. For the special case of important two-parameter distribution classes V. Bawa shows that the optimal decision rule reduces to simple manageable forms involving only two parameters. One of those parameters is always the mean while the other one is the appropriate measure of dispersion usually represented by the scale parameter. Except for the case of symmetric distributions the scale parameter is not equal to the variance. For distributions not belonging to these two-parameter distribution classes it may be a good approximation to compare the lower partial variance at several benchmarks. The so-called lower partial moments of Section 6.2.1 can be considered as a generalization of the lower partial variance (see, e.g. Harlow and Rao [HR89] or Harlow [Har91]). Since we want to optimize interest-rate portfolios which include derivatives and which therefore may have rather asymmetric return distributions, we will use lower partial moments to control the downside risk of

---

[5] The normal distribution is an important member of this class of symmetric distributions. For more details and conditions see, e.g., Ingersoll [Ing87], p. 104-113.

the portfolio exposure. As proposed above this is done by calculating the lower partial moments at several benchmarks. Because we do not have any information on the distribution of the possible or optimal portfolios, we are not able to restrict the set of distributions to the set of two-parameter families defined by Bawa. So we have to approximate the distribution function and, as a consequence, the optimization problem based on a simulation model. For this simulation we assume that we are given a planning horizon $T \in [t_0, T^*]$, with t_0 and T^* defined by our interest-rate market model, and that the prices at any time $t \in [t_0, T]$ of the interest-rate derivatives $i = 1, ..., n$ we consider are given by $D_1(t), ..., D_n(t)$. Furthermore, let $R_i(t_0, T)$, $i = 1, ..., n$, denote the random return of an investment in derivative i, which is usually quoted for discrete compounding and defined by

$$R_i(t_0, T) := \frac{V_i(t_0, T) - V_i(t_0)}{V_i(t_0)}, \ i = 1, ..., n,$$

with $V_i(t_0)$ and $V_i(t_0, T)$ denoting the value of derivative i at time t_0 and time T as seen from time t_0. It is for the portfolio manager to decide whether he likes to set the derivatives values equal to their quoted prices or their dirty prices or he wants to add the future value of potential cash flows between t_0 and T to derive $V_i(t_0, T)$. However, as in Section 6.2.1, for our simulation purposes we assume that the returns are dependent on the vector $\mathsf{F} = (\mathsf{F}_1, ..., \mathsf{F}_m)$ of risk factors and that the joint distribution of these parameters is known, as well as the functional relation between the returns and the risk vector, i.e. $R_i(t_0, T) = R_i(\mathsf{F}, t_0, T)$, or equivalently the future value of the derivatives

$$V_i(t_0, T) = V_i(\mathsf{F}, t_0, T) = V_i(t_0) \cdot (1 + R_i(\mathsf{F}, t_0, T)), \ i = 1, ..., n.$$

Under this assumption we simulate the risk vector F, getting the simulations

$$\mathsf{F}^k = \left(\mathsf{F}_1^k, ..., \mathsf{F}_m^k\right) \text{ with probability } p_k > 0, \ k = 1, ..., K.$$

Inserting these simulations, we get the simulations for the returns and future values

$$R_i^k(t_0, T) := R_i\left(\mathsf{F}^k, t_0, T\right) \text{ and } V_i^k(t_0, T) := V_i\left(\mathsf{F}^k, t_0, T\right), \ k = 1, ..., K,$$

for each $i = 1, .., n$. For any portfolio $\varphi = (\varphi_1, ..., \varphi_n)$ of the derivatives, which we consider to be fixed from t_0 to the end of the planning horizon T, the return $R(\varphi, t, T)$ and future value $V(\varphi, t, T)$ of this portfolio at the end of the planning horizon are given by the random variables

$$R(\varphi, t_0, T) = \sum_{i=1}^{n} \varphi_i \cdot R_i(t_0, T) \text{ and } V(\varphi, t_0, T) = \sum_{i=1}^{n} \varphi_i \cdot V_i(t_0, T),$$

with

$$V(\varphi, t_0) = \sum_{i=1}^{n} \varphi_i \cdot V_i(t_0)$$

denoting the portfolio value at time t_0. Using our simulations, the portfolio return $R(\varphi, t_0, T)$ and the future portfolio value $V(\varphi, t_0, T)$ are simulated by

$$R^k(\varphi, t_0, T) = \sum_{i=1}^{n} \varphi_i \cdot R_i^k(t_0, T)$$

and

$$V^k(\varphi, t_0, T) = \sum_{i=1}^{n} \varphi_i \cdot V_i^k(t_0, T), \ k = 1, ..., K.$$

As we have seen in Section 6.2.1, the discrete version of the lower partial moment of order $l \in I\!N$ and corresponding to an investor-specific benchmark or threshold return $b(t_0, T) \in I\!R$ is defined by

$$LPM_l(\varphi, R, b, t_0, T) = \sum_{\substack{k=1,...,K \\ R^k(\varphi, t_0, T) < b(t_0, T)}} p_k \cdot \left(b(t_0, T) - R^k(\varphi, t_0, T)\right)^l.$$

For ease of exposition we assume that the simulated lower partial moment is equal to the real one. This is true if the probability distribution of the portfolio return is discrete. If this is not the case the equality has to be interpreted as an approximation, where we suppose that the number and quality of the simulations is high enough to write the equality with good conscience.

If the portfolio is composed of different securities $i = 1, ..., n$ with relative weightings $\tilde{\varphi} = (\tilde{\varphi}_1, ..., \tilde{\varphi}_n)$, i.e. $\sum_{i=1}^{n} \tilde{\varphi}_i = 1$, the portfolio rate of return and therefore the lower partial moment of order l will depend on $\tilde{\varphi} = (\tilde{\varphi}_1, ..., \tilde{\varphi}_n)$, and will be denoted by $LPM_l(\tilde{\varphi}, R, b, t_0, T)$. Let $\mu(\tilde{\varphi}) := \sum_{i=1}^{n} \tilde{\varphi}_i \cdot \mu_i(t_0, T)$ with $\mu_i(t_0, T) := E_Q[R_i(t_0, T)]$ denote the expected return of the portfolio dependent on the weighting $\tilde{\varphi}$ of its assets and let $LPM_l(\mu, R, b, t_0, T)$ be the minimum value of $LPM_l(\tilde{\varphi}, R, b, t_0, T)$ for all $\tilde{\varphi}$ with $\mu(\tilde{\varphi}) = \mu$ and $\sum_{i=1}^{n} \tilde{\varphi}_i = 1$. As was pointed out by Bawa and Lindenberg [BL77] and by Harlow and Rao [HR89], $LPM_l(\tilde{\varphi}, R, b, t_0, T)$ is a convex functional in $\tilde{\varphi}$ for $l = 1$ and $l = 2$. Furthermore, $\mu \mapsto LPM_l(\mu, R, b, t_0, T)$ is an increasing and convex function for all $\mu \geq \mu(l, b)$, with $\mu(l, b)$ denoting the portfolio mean leading to the minimum of the function $\mu \mapsto LPM_l(\mu, R, b, t_0, T)$. In an empirical study Harlow [Har91] showed that his method of downside risk minimization at a given level of expected return significantly outperformed the corresponding mean-variance method. He considered a global asset allocation problem with fully currency-hedged equity and fixed income markets in 11 countries and returns spanning the period from January 1980 to December 1990. At a benchmark of 0% the lower partial moment (order 2) method provided more downside protection at the same level of expected return than the mean-variance decision rule, due to some skewness or asymmetry of the return distribution. This result was realized allocating a higher percentage of coupon bonds (67.17%) than the mean-variance approach (59.47%). He states that l should be at least of order 1 to consider any form of risk aversion. To be more precise, $l = 1$ is consistent with finite, increasing and concave utility functions while $l = 2$ is consistent with finite, increasing and concave utility functions displaying absolute risk aversion (see Harlow and Rao [HR89]). For $l = 0$ the lower partial moment does not especially weight the distance between the possible returns and the benchmark. But it still makes sense to incorporate the lower partial moment of order 0 in the asset allocation process to isolate the segment of the investment opportunity set most relevant to the investor (see Harlow [Har91]).

Following this idea we consider the portfolio optimization problem of maximizing the expected return at a given maximum risk level $\widetilde{A}_{\nu}^{l} \in I\!\!R$, $\nu = 1, ..., \Upsilon$, $l \in \{0, 1, 2\}$, where risk is measured by the lower partial moment of order l at several benchmark returns $b_{\nu}(t_0, T)$, quoted for discrete compounding. Additionally, there may be lower ($\widetilde{s}_i$) or upper ($\widetilde{S}_i$) constraints on the level of investment for each possible asset $i = 1, ..., n$ to consider the investors' strategic view of the optimal portfolio structure. So we are facing the following optimization problem:

$$
\left(\widetilde{P}\right)
\begin{cases}
\mu\left(\widetilde{\varphi}\right) \longrightarrow \max \\[2mm]
LPM_l\left(\widetilde{\varphi}, R, b_{\nu}, t_0, T\right) \leq \widetilde{A}_{\nu}^{l},\ \nu = 1, ..., \Upsilon, l \in \{0, 1, 2\} \\[2mm]
\widetilde{s}_i \leq \widetilde{\varphi}_i \leq \widetilde{S}_i,\ i = 1, ..., n \\[2mm]
\displaystyle\sum_{i=1}^{n} \widetilde{\varphi}_i = 1.
\end{cases}
$$

This formulation generalizes those in Leibowitz and Henriksson [LH89] and Leibowitz and Kogelman [LK91] in two ways. First, it is not based on the assumption of normally distributed returns, and second it is not restricted to lower partial moments of order $l = 0$. Especially the first generalization in our case is very important as we want to deal with rather asymmetric return distributions as a result of portfolios which are composed of interest-rate derivatives.

Using the future value simulations $V_i^k(t_0, T)$, $k = 1, ..., K$, of the assets $i = 1, ..., n$, and the absolute asset weights $\varphi = (\varphi_1, ..., \varphi_n)$, the absolute benchmarks $B_{\nu}(t_0, T)$ to be compared with the simulated portfolio prices $V^k(\varphi, t_0, T)$ are given by

$$
B_{\nu}(t_0, T) = V(\varphi, t_0) \cdot (1 + b_{\nu}(t_0, T)),\ \nu = 1, ..., \Upsilon.
$$

As we have seen in Section 6.2.1, the corresponding absolute lower partial moment of order $l \in \{0, 1, 2\}$ for the future portfolio value can be written as

$$LPM_l(\varphi, V, B_\nu, t_0, T) = \sum_{\substack{k=1,\dots,K \\ V^k(\varphi,t_0,T)<B_\nu(t_0,T)}} p_k \cdot \left(B_\nu(t_0,T) - V^k(\varphi, t_0, T)\right)^l.$$

(7.9)

A portfolio manager or trader may be restricted to specific trading limits. So we introduce the absolute lower (s_i) and upper bounds (S_i) for the amount φ_i of security $i = 1, ..., n$ in the portfolio corresponding to the relative investment restrictions of problem $\left(\widetilde{P}\right)$, and claim that

$$s_i \leq \varphi_i \leq S_i, \ i = 1, ..., n.$$

Beside these bounds for the volume of the securities, we also assume that there is a special *budget* for the portfolio manager to be invested resulting in the restriction[6]

$$V(\varphi, t_0) = budget.$$

For $A_\nu^l \in \mathbb{R}$, $\nu = 1, ..., \Upsilon$, and $l \in \{0, 1, 2\}$, let us consider the following optimization problem:

$$(P) \begin{cases} \displaystyle\sum_{i=1}^n \varphi_i \cdot E_Q[V_i(t_0, T)] \to \max \\[2mm] LPM_l(\varphi, V, B_\nu, t_0, T) \leq A_\nu^l, \ \nu = 1, ..., \Upsilon, l \in \{0, 1, 2\} \\[2mm] s_i \leq \varphi_i \leq S_i, \ i = 1, ..., n \\[2mm] \displaystyle\sum_{i=1}^n \varphi_i \cdot V_i(t_0) = budget. \end{cases}$$

The relation between $\left(\widetilde{P}\right)$ and (P) is summarized in the following lemma.

Lemma 7.3 *Let budget $\neq 0$ and $V_i(t_0) \neq 0$ for all $i = 1, ..., n$. Furthermore, let*

$$\widetilde{s}_i := s_i \cdot \frac{V_i(t_0)}{budget}, \ \widetilde{S}_i := S_i \cdot \frac{V_i(t_0)}{budget}, \ and \ \widetilde{A}_\nu^l := \frac{A_\nu^l}{budget^l}$$

for $i = 1, ..., n$, $\nu = 1, ..., \Upsilon$, and $l \in \{0, 1, 2\}$. Then the optimization problems (P) and $\left(\widetilde{P}\right)$ are equivalent in the following sense: $\varphi^* = (\varphi_1^*, ..., \varphi_n^*)$ is an optimal solution for (P) if and only if $\widetilde{\varphi}^* = (\widetilde{\varphi}_1^*, ..., \widetilde{\varphi}_n^*)$ with $\widetilde{\varphi}_i^* := \varphi_i^* \cdot V_i(t_0)/budget$, $i = 1, ..., n$, is an optimal solution for $\left(\widetilde{P}\right)$.

---

[6] We assume that there is always a cash account on which we can put our money as a riskless investment. Thus, it doesn't make sense to leave any money uninvested.

Proof. For the goal functions we have

$$
\begin{aligned}
\sum_{i=1}^{n} \widetilde{\varphi}_i^* \cdot \mu_i\left(t_0, T\right) &= \sum_{i=1}^{n} \widetilde{\varphi}_i^* \cdot E_Q\left[R_i\left(t_0, T\right)\right] \\
&= \sum_{i=1}^{n} \varphi_i^* \cdot \frac{V_i\left(t_0\right)}{budget} \cdot E_Q\left[\frac{V_i\left(t_0, T\right) - V_i\left(t_0\right)}{V_i\left(t_0\right)}\right] \\
&= \frac{1}{budget} \cdot \left(\sum_{i=1i}^{n} \varphi_i^* \cdot E_Q\left[V_i\left(t_0, T\right)\right]\right) - 1,
\end{aligned}
$$

which means that the maximizations of the goal functions in (P) and $\left(\widetilde{P}\right)$ are equivalent. Furthermore,

$$
\begin{aligned}
V\left(\varphi^*, t_0, T\right) &= \sum_{i=1}^{n} \varphi_i^* \cdot V_i\left(t_0, T\right) \\
&= \sum_{i=1}^{n} \varphi_i^* \cdot V_i\left(t_0\right) \cdot \left(1 + R_i\left(t_0, T\right)\right) \\
&= budget \cdot \sum_{i=1}^{n} \widetilde{\varphi}_i^* \cdot \left(1 + R_i\left(t_0, T\right)\right) \\
&= budget \cdot \left(1 + \sum_{i=1}^{n} \widetilde{\varphi}_i^* \cdot R_i\left(t_0, T\right)\right) \\
&= budget \cdot \left(1 + R\left(\widetilde{\varphi}^*, t_0, T\right)\right).
\end{aligned}
$$

Hence, since $V\left(\varphi^*, t_0\right) = budget$,

$$
V\left(\varphi^*, t_0, T\right) < B_\nu\left(t_0, T\right) = V\left(\varphi^*, t_0\right) \cdot \left(1 + b_\nu\left(t_0, T\right)\right)
$$

is equivalent to

$$
R\left(\widetilde{\varphi}^*, t_0, T\right) < b_\nu\left(t_0, T\right)
$$

and

$$
B_\nu\left(t_0, T\right) - V\left(\varphi^*, t_0, T\right) = V\left(\varphi^*, t_0\right) \cdot \left(b_\nu\left(t_0, T\right) - R\left(\widetilde{\varphi}^*, t_0, T\right)\right).
$$

Using these two results, we get

$$
LPM_l\left(\varphi^*, V, B_\nu, t_0, T\right) = \sum_{\substack{k=1,\ldots,K \\ V^k\left(\varphi^*, t_0, T\right) < B_\nu\left(t_0, T\right)}} p_k \cdot \left(\begin{array}{c} B_\nu\left(t_0, T\right) \\ -V^k\left(\varphi^*, t_0, T\right) \end{array}\right)^l
$$

$$
= \left[V\left(\varphi^*, t_0\right)\right]^l \cdot \sum_{\substack{k=1,\ldots,K \\ R\left(\widetilde{\varphi}^*, t_0, T\right) < b_\nu\left(t_0, T\right)}} p_k \cdot \left(b_\nu\left(t_0, T\right) - R^k\left(\widetilde{\varphi}^*, t_0, T\right)\right)^l
$$

$$
= \left[V\left(\varphi^*, t_0\right)\right]^l \cdot LPM_l\left(\widetilde{\varphi}^*, R, b_\nu, t_0, T\right).
$$

So using again that $V\left(\varphi^*, t_0\right) = budget$,

$$
budget^l \cdot LPM_l\left(\widetilde{\varphi}^*, R, b_\nu, t_0, T\right) = LPM_l\left(\varphi^*, V, B_\nu, t_0, T\right) \le A_\nu^l
$$

is equivalent to

$$LPM_l\left(\widetilde{\varphi}^*, R, b_\nu, t_0, T\right) \le \frac{A_\nu^l}{budget^l} = \widetilde{A}_\nu^l.$$

For the volume restrictions, we have

$$\widetilde{s}_i = s_i \cdot \frac{V_i\left(t_0\right)}{budget} \le \widetilde{\varphi}_i^* = \varphi_i^* \cdot \frac{V_i\left(t_0\right)}{budget} \le \widetilde{S}_i = S_i \cdot \frac{V_i\left(t_0\right)}{budget},$$

which is equivalent to

$$s_i \le \varphi_i^* \le S_i.$$

Furthermore, the budget constraint

$$\sum_{i=1}^{n} \varphi_i \cdot V_i\left(t_0\right) = budget$$

is equivalent to

$$\sum_{i=1}^{n} \widetilde{\varphi}_i = \sum_{i=1}^{n} \varphi_i \cdot \frac{V_i\left(t_0\right)}{budget} = 1,$$

which completes the proof.   □

Because the value of assets such as futures or forward agreements may be zero the returns used in optimization problem $\left(\widetilde{P}\right)$ could be misleading. Furthermore, optimization problem (P) is consistent with the absolute formulation of the value at risk. So we prefer to use optimization problem (P) in the sequel. To implement this problem, we choose the numbers $m_{\nu k} < 0$ to be sufficiently small and the numbers $M_{\nu k} > 0$, $m_{\nu k}^l > 0$ and $M_{\nu k}^l > 0$ to be sufficiently large and define the constraints

$$M_{\nu k} \cdot y_{\nu k} + V^k\left(\varphi, t_0, T\right) \ge B_\nu\left(t_0, T\right), \tag{A}$$

$$m_{\nu k} \cdot \left(1 - y_{\nu k}\right) + V^k\left(\varphi, t_0, T\right) < B_\nu\left(t_0, T\right), \tag{B}$$

$$0 \le \left(V^k\left(\varphi, t_0, T\right) - B_\nu\left(t_0, T\right)\right)^l + (-1)^{l-1} \cdot w_{\nu k}^l \le M_{\nu k}^l \cdot \left(1 - y_{\nu k}\right), \tag{C}$$

$$0 \le w_{\nu k}^l \le m_{\nu k}^l \cdot y_{\nu k}, \tag{D}$$

with $w_{\nu k}^l \in \mathbb{R}$, $y_{\nu k} \in \{0, 1\}$ for all $k = 1, ..., K$, where we consider $l \in \{0, 1, 2\}$ and $\nu \in \{1, ..., \Upsilon\}$ as arbitrary but fixed.

Lemma 7.4 *Let $\nu \in \{1, ..., \Upsilon\}$ and $l \in \{0, 1, 2\}$ be arbitrary but fixed and $y_{\nu k} \in \{0, 1\}$ for all $k = 1, ..., K$.*

a) *Under condition (A), we have for all $k = 1, ..., K$:*

$$y_{\nu k} = 1 \quad if \quad V^k(\varphi, t_0, T) < B_\nu(t_0, T).$$

Under condition (B), we have for all $k = 1, ..., K$:

$$y_{\nu k} = 0 \quad if \quad V^k(\varphi, t_0, T) \geq B_\nu(t_0, T).$$

b) *Under conditions (A) and (B), we have for all $k = 1, ..., K$:*

$$V^k(\varphi, t_0, T) < B_\nu(t_0, T) \quad if \ and \ only \ if \ y_{\nu k} = 1.$$

c) *Under conditions (A), (C), and (D), we have[7]:*

$$LPM_l(\varphi, V, B_\nu, t_0, T) \leq \sum_{k=1}^{K} p_k \cdot w_{\nu k}^l.$$

d) *Under conditions (A), (B), (C), and (D), we have:*

$$LPM_l(\varphi, V, B_\nu, t_0, T) = \sum_{k=1}^{K} p_k \cdot w_{\nu k}^l.$$

Proof.

a) is straightforward, since $y_{\nu k} = 1$ is the only way for condition (A) to hold if $V^k(\varphi, t_0, T) < B_\nu(t_0, T)$ and $y_{\nu k} = 0$ is the only way for condition (B) to hold if $V^k(\varphi, t_0, T) \geq B_\nu(t_0, T)$.

b) Let conditions (A) and (B) be satisfied. If $V^k(\varphi, t_0, T) < B_\nu(t_0, T)$, then $y_{\nu k} = 1$ because of (A). On the other hand, if $V^k(\varphi, t_0, T) \geq B_\nu(t_0, T)$, then $y_{\nu k} = 0$ because of (B). Since $y_{\nu k} = 1$ directly implies $V^k(\varphi, t_0, T) < B_\nu(t_0, T)$ because of (B) and $y_{\nu k} = 0$ directly implies $V^k(\varphi, t_0, T) \geq B_\nu(t_0, T)$ because of (A), we conclude

$$V^k(\varphi, t_0, T) < B_\nu(t_0, T) \quad if \ and \ only \ if \ y_{\nu k} = 1.$$

c),d) Let conditions (C) and (D) be satisfied in addition to (A). If $V^k(\varphi, t_0, T) < B_\nu(t_0, T)$ we conclude that $y_{\nu k} = 1$ because of (A) and thus

$$w_{\nu k}^l = \left(B_\nu(t_0, T) - V^k(\varphi, t_0, T) \right)^l$$

---

[7] To be precise, we only need the first inequality of (D) in addition to (A) and (C) to hold this statement. Furthermore, we only need the first inequalities of (C) and (D) if $l = 1$.

using (C). Hence,

$$
\sum_{k=1}^{K} p_k \cdot w_{\nu k}^{l} = \sum_{\substack{k=1,\ldots,K \\ V^k(\varphi,t_0,T)<B_\nu(t_0,T)}} p_k \cdot w_{\nu k}^{l} + \sum_{\substack{k=1,\ldots,K \\ V^k(\varphi,t_0,T)\geq B_\nu(t_0,T)}} p_k \cdot w_{\nu k}^{l}
$$

$$
\geq \sum_{\substack{k=1,\ldots,K \\ V^k(\varphi,t_0,T)<B_\nu(t_0,T)}} p_k \cdot w_{\nu k}^{l} \quad \text{using } w_{\nu k}^{l} \geq 0 \; (D)
$$

$$
= \sum_{\substack{k=1,\ldots,K \\ V^k(\varphi,t_0,T)<B_\nu(t_0,T)}} p_k \cdot \Big(B_\nu(t_0,T) - V^k(\varphi,t_0,T)\Big)^{l}.
$$

Using only the left inequality in (C) it can be easily seen that we can still write "$\geq$" instead of "$=$" in the last equation if $l = 1$. Also we can change the inequality in the second line of the previous statement to an equality under conditions (B) and (D) because $V^k(\varphi,t_0,T) \geq B_\nu(t_0,T)$ implies $y_{\nu k} = 0$ under (B) and thus we get $w_{\nu k}^{l} = 0$ using (D). $\qquad\qquad\Box$

Note that for the special case $l = 0$, we can conclude that conditions (C) and (D) are equivalent to $w_{\nu k}^{0} = y_{\nu k}$ giving us

$$
LPM_0\,(\varphi, V, B_\nu, t_0, T) = \sum_{k=1}^{K} p_k \cdot y_{\nu k} \qquad (7.10)
$$

under the additional conditions (A) and (B) with "$\leq$" instead of "$=$" if only condition (A) is satisfied in addition to (C) and (D).

Using Lemma 7.4, we can replace the lower partial moment constraint

$$
LPM_l\,(\varphi, V, B_\nu, t_0, T) \leq A_\nu^{l} \qquad (LPM)
$$

by the constraint

$$
\sum_{k=1}^{K} p_k \cdot w_{\nu k}^{l} \leq A_\nu^{l} \qquad (E)
$$

for $A_\nu^{l} \in \mathbb{R}$, $\nu \in \{1,\ldots,\Upsilon\}$ and $l \in \{0,1,2\}$, if conditions (A), (B), (C), and (D) are satisfied. Hence, instead of using constraint (LPM), we can use inequality (E) if we add conditions (A), (B), (C), and (D) to the optimization problem. For $l = 0$ condition (LPM) is called a *shortfall constraint* and the corresponding A_ν^{0}, in this case chosen to be an element of $(0,1)$, is called the *shortfall probability*.

Usually all commercial optimization tools use inequalities of the form $\leq$ or $\geq$ and a precision or tolerance level, expressed by the smallest absolute number that can be recognized numerically within the tool and which will be denoted by $\varepsilon > 0$ here. For this reason we rewrite equation (B) in the following form:

$$m_{\nu k} \cdot (1 - y_{\nu k}) + V^k (\varphi, t_0, T) \leq B_\nu (t_0, T) - \varepsilon. \qquad (B')$$

This gives us the following optimization problem (P_1), which is (approximately) equivalent to (P):

(P_1)
$$
\begin{cases}
\displaystyle\sum_{k=1}^{K} p_k \cdot V^k (\varphi, t_0, T) \to \max \\[2mm]
M_{\nu k} \cdot y_{\nu k} + V^k (\varphi, t_0, T) \geq B_\nu (t_0, T), \, \nu = 1, ..., \Upsilon, \, k = 1, ..., K \\[1mm]
m_{\nu k} \cdot (1 - y_{\nu k}) + V^k (\varphi, t_0, T) \leq B_\nu (t_0, T) - \varepsilon, \, \nu = 1, ..., \Upsilon, k = 1, ..., K \\[1mm]
0 \leq \left(V^k (\varphi, t_0, T) - B_\nu (t_0, T)\right)^l + (-1)^{l-1} \cdot w_{\nu k}^l \leq M_{\nu k}^l \cdot (1 - y_{\nu k}) \text{ and} \\[1mm]
0 \leq w_{\nu k}^l \leq m_{\nu k}^l \cdot y_{\nu k}, \, \nu = 1, ..., \Upsilon, \, k = 1, ..., K, \, l \in \{0, 1, 2\} \\[1mm]
y_{\nu k} \in \{0, 1\}, \, \nu = 1, ..., \Upsilon, \, k = 1, ..., K. \\[1mm]
\displaystyle\sum_{k=1}^{K} p_k \cdot w_{\nu k}^l \leq A_\nu^l, \, \nu = 1, ..., \Upsilon, l \in \{0, 1, 2\} \\[1mm]
s_i \leq \varphi_i \leq S_i, \, i = 1, ..., n \\[1mm]
\displaystyle\sum_{i=1}^{n} \varphi_i \cdot V_i (t_0) = budget.
\end{cases}
$$

As mentioned above, some of the restrictions can be omitted or simplified depending on the specific choice of l. Especially, we can omit condition (B) for a sufficient set of constraints. The variables of (P_1) to be optimized are $\varphi_1, ..., \varphi_n$, $w_{\nu k}^l$ and $y_{\nu k} \in \{0, 1\}$, $\nu = 1, ..., \Upsilon$, $k = 1, ..., K$, $l \in \{0, 1, 2\}$. Hence we have to solve a mixed-integer program, which can be rather easily done by commercial optimization tools[8].

---

[8] If we define

$$
a_{ik} := \begin{cases} S_i, & \text{if } V_i^k (t_0, T) < 0 \\ s_i, & \text{if } V_i^k (t_0, T) \geq 0 \end{cases}
\quad \text{and} \quad
b_{ik} := \begin{cases} s_i, & \text{if } V_i^k (t_0, T) < 0 \\ S_i, & \text{if } V_i^k (t_0, T) \geq 0 \end{cases}
$$

for $i = 1, ..., n$ and $k = 1, ..., K$, we can choose

$$
M_{\nu k} = B_\nu - \sum_{i=1}^{n} a_{ik} \cdot V_i^k (t_0, T), \, m_{\nu k} = B_\nu - \varepsilon - \sum_{i=1}^{n} b_{ik} \cdot V_i^k (t_0, T),
$$

$$
M_{\nu k}^l = \left[\sum_{i=1}^{n} b_{ik} \cdot V_i^k (t_0, T) - B_\nu\right]^l, \text{ and } m_{\nu k}^l = \left[B_\nu - \sum_{i=1}^{n} a_{ik} \cdot V_i^k (t_0, T)\right]^l,
$$

$k = 1, ..., K$, $\nu = 1, ..., \Upsilon$, $l \in \{0, 1, 2\}$, for a practical application.

Case Study (Optimal Portfolio Protection).

We now apply the previous results to optimize portfolios with asymmetric return distributions, i.e. portfolios including interest-rate derivatives. For this case study we use derivatives to protect the portfolio from falling below a given benchmark. We use the Hull-White model for an evaluation of the portfolio price. The yield curve is calculated by a quadratic spline procedure using the coupon-bond prices as quoted on the Reuters chain DETSY= at 11/27/97 and is shown in figure 7.18. The volatility structure is derived by fitting the model prices (volatility curve) to cap prices quoted as Black volatilities by Intercapital Brokers ltd. on their Reuters page VCAP with maturities of 2, 3, 5 and 7 years. This is done using the optimizazion problem $\left(P_{a,\sigma}^{Cap}\right)$ of Section 5.8.3. The resulting parameter values at 11/27/97 were $a = 0.0328$ and $\sigma = 0.7702\%$. The corresponding volatility structure is plotted in figure 7.19.

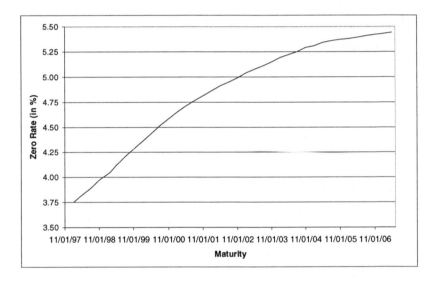

FIGURE 7.18. Zero rate curve at November 27, 1997

As a simulation model we use the method explained in the appendix based on the Hull-White model where, for ease of exposition, the market price of risk is set to zero, i.e. $\gamma \equiv 0$. Choosing $K = 7$ for this method, we calculated $2^K - 1 = 127$ different yield curve scenarios, resulting in a maximum approximation error of $\alpha_{error} = 0.78\%$. Note that compared to the shortfall probability α of any shortfall constraint, we should always have $\alpha_{error} < \alpha$. The trading limits for the portfolio manager are no short sales, i.e. $s_i = 0$, an upper trading limit of $S_i = \max\left\{\frac{budget}{D_i(t_0)}, -\frac{budget}{D_i(t_0)}\right\}$, if the price of the derivative at time t_0 is $D_i(t_0) \neq 0$, and $S_i = \frac{budget}{notional}$, if $D_i(t_0) = 0$, where *notional* denotes the notional amount of derivative

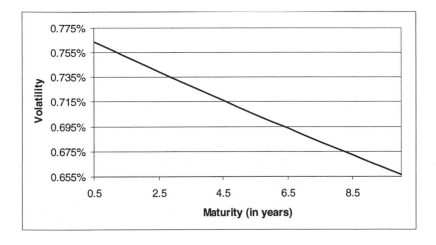

FIGURE 7.19. Volatility curve of the continuous zero bond yields at 11/27/97 calculated out of cap prices

$i = 1, ..., n$, and *budget* denotes the budget which we set to 10 Mio. The instruments used are a 6% coupon bond with a notional amount of 100, annual coupon payments and maturity 11/27/03, and a cap with maturity 11/27/03, annual payments based on a 6% cap rate and a notional amount of 100. The actual prices at 11/27/97 are $D_1(t_0) = 103.94$ for the coupon bond and $D_2(t_0) = 1.43$ for the cap, giving us $S_1 = \frac{10\ Mio.}{103.94} = 96,209.31$ for the coupon bond and $S_2 = \frac{10\ Mio.}{1.43} = 6,993,006.99$ for the cap. The time horizon or planning period is 1 year. The precision used for the case studies is $\varepsilon = 0.01\%$. The aim of the portfolio manager in this example is to find an optimal allocation or insurance strategy, where the downside risk, measured by the lower partial moment of order $l = 0$ of his portfolio at a shortfall probability of $\alpha = 5\%$, is restricted by an adequate number of caps. The results for different benchmarks show a typical shift to a full protection for increasing benchmarks $B := B(0, 1)$ leading to more and more asymmetric return distributions. The distributions are plotted in figures 7.20-7.22.

□

7.2.2 Risk Management Based on Value at Risk

In the last section, we discussed the optimization of a portfolio under limited downside risk. The downside risk was measured by the lower partial moment of order $l \in \{0, 1, 2\}$, which is the expectation of all possible returns or future values of a portfolio below a given benchmark, measured to the power of l. At least one benchmark, by definition, may be considered to be destined as a candidate to be applied for an optimal asset allocation

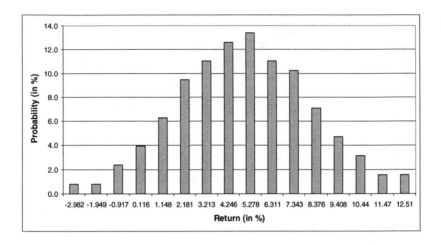

FIGURE 7.20. Return distribution for $\alpha = 5\%$ and $B = -1\%$ resulting in 100% bonds and 0% caps

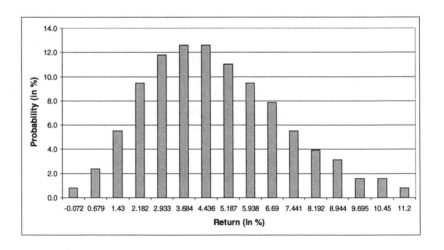

FIGURE 7.21. Return distribution for $\alpha = 5\%$ and $B = 1\%$ resulting in 51.75% bonds and 48.25% caps

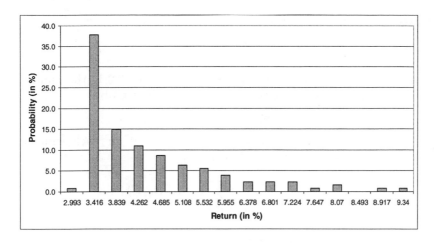

FIGURE 7.22. Return distribution for $\alpha = 5\%$ and $B = 3\%$ resulting in 30.32% bonds and 69.68% caps

under shortfall constraints, the value at risk (VaR). As defined in Section 6.2.2, the VaR is the maximum possible loss for a portfolio or trading book with respect to a given time horizon from t_0 to T, a given significance level $1 - \alpha \in (0,1)$ and measured by a change of the portfolios market value $\Delta V (\varphi, t_0, T) := V (\varphi, t_0, T) - V (\varphi, t_0)$ over this period. Here, $V (\varphi, t_0, T)$ denotes the future value of a portfolio $\varphi = (\varphi_1, ..., \varphi_n)$ of interest-rate derivatives with values of $V_i (t_0, T)$, $i = 1, ..., n$, at the end of the planning horizon T measured at time t_0. The change $\Delta V (\varphi, t_0, T)$ of the portfolio value from time t_0 to time T is given by

$$\Delta V (\varphi, t_0, T) = V (\varphi, t_0) \cdot R (\varphi, t_0, T),$$

with $R (\varphi, t_0, T)$ denoting the portfolio return over the planning horizon from t_0 to T. Following Definition 6.2.2, the value at risk is defined by

$$VaR (\alpha, \varphi, t_0, T) := E_Q [\Delta V (\varphi, t_0, T)] - \sup \left\{ x \in I\!R : F_{\Delta V(\varphi, t_0, T)} (x-) \le \alpha \right\},$$

where $F_{\Delta V(\varphi, t_0, T)}$ is the distribution function of $\Delta V (\varphi, t_0, T)$. So we have to find

$$\sup \left\{ x \in I\!R : F_{\Delta V(\varphi, t_0, T)} (x-) \le \alpha \right\} = \sup \left\{ x \in I\!R : Q (\Delta V (\varphi, t_0, T) < x) \le \alpha \right\},$$

with Q denoting the probability measure of our market model. Since we often do not know the distribution of $\Delta V (\varphi, t_0, T)$, and since we want to find the portfolio φ due to a given optimization problem, we simulate the distribution starting with simulations of the individual derivative values, in the previous section. To do this, let $R_i (t_0, T)$, $i = 1, ..., n$, denote the random return of an investment in derivative i, which is usually quoted for discrete compounding and defined by

$$R_i(t_0, T) := \frac{V_i(t_0, T) - V_i(t_0)}{V_i(t_0)}, \ i = 1, ..., n.$$

Under the assumptions and definitions of Section 7.2.1, the simulated returns and future values of the derivatives are denoted by

$$R_i^k(t_0, T) \text{ and } V_i^k(t_0, T), \ k = 1, ..., K,$$

for each $i = 1, .., n$, with a probability $p_k > 0$ for each simulation $k = 1, ..., K$. This results in the simulations

$$R^k(\varphi, t_0, T) = \sum_{i=1}^{n} \varphi_i \cdot R_i^k(t_0, T)$$

and

$$V^k(\varphi, t_0, T) = \sum_{i=1}^{n} \varphi_i \cdot V_i^k(t_0, T), \ k = 1, ..., K$$

of the portfolio return $R(\varphi, t_0, T)$ and the future portfolio value $V(\varphi, t_0, T)$. Thus, $Q(\Delta V(\varphi, t_0, T) < x)$ can be simulated by

$$Q(\Delta V(\varphi, t_0, T) < x) = F_{\Delta V(\varphi, t_0, T)}(x-) = \sum_{\substack{k=1,...,K \\ \Delta V^k(\varphi, t, T) < x}} p_k.$$

Choosing the numbers $m_k < 0$ sufficiently small and the numbers $M_k > 0$ sufficiently large, we define the constraints

$$M_k \cdot y_k + \Delta V^k(\varphi, t_0, T) \geq x, \qquad (\Delta A)$$

$$m_k \cdot (1 - y_k) + \Delta V^k(\varphi, t_0, T) < x \qquad (\Delta B)$$

for $x \in \mathbb{R}$. As in Lemma 7.4 and equation (7.10), we can show that

$$Q(\Delta V(\varphi, t_0, T) < x) = LPM_0(\varphi, x, t_0, T) \leq \sum_{k=1}^{K} p_k \cdot y_{1k}$$

under condition (ΔA), with equality if condition (ΔB) holds in addition to (ΔA). We can therefore conclude that, for a fixed portfolio $\varphi = (\varphi_1, ..., \varphi_n)$ of interest-rate derivatives, $\sup\{x \in \mathbb{R} : Q(\Delta V(\varphi, t_0, T) < x) \leq \alpha\}$, $\alpha \in (0, 1)$, is given as the element x^* of the solution $(x^*, y_1^*, ..., y_K^*)$ of the mixed-integer optimization problem

$$(P_2) \begin{cases} x \to \max \\ M_k \cdot y_k + \Delta V^k(\varphi, t_0, T) \geq x, \ k = 1, ..., K \\ m_k \cdot (1 - y_k) + \Delta V^k(\varphi, t_0, T) < x, \ k = 1, ..., K \\ \sum_{k=1}^{K} p_k \cdot y_k \leq \alpha \\ y_k \in \{0, 1\}, \ k = 1, ..., K. \end{cases}$$

Now let

$$(P_3) \begin{cases} x \to \max \\ M_k \cdot y_k + \Delta V^k \left(\varphi, t_0, T \right) \geq x, \ k = 1, ..., K \\ \sum_{k=1}^{K} p_k \cdot y_k \leq \alpha \\ y_k \in \{0, 1\}, \ k = 1, ..., K. \end{cases}$$

Then the following lemma holds.

Lemma 7.5 *Let* $\varphi = (\varphi_1, ..., \varphi_n)$ *be a fixed portfolio of interest-rate derivatives and* $\alpha \in (0, 1)$. *Furthermore, let* $(x^{**}, y_1^{**}, ..., y_K^{**})$ *be a solution of the mixed-integer optimization problem* (P_3). *Then the value at risk with respect to the time-period from* t_0 *to* T *and a significance level of* $1 - \alpha$ *is given by*

$$VaR \left(\alpha, \varphi, t_0, T \right) := \sum_{k=1}^{K} p_k \cdot \Delta V^k \left(\varphi, t_0, T \right) - x^{**}.$$

Proof. Let $(x^*, y_1^*, ..., y_K^*)$ denote the solution of (P_2) and $(x^{**}, y_1^{**}, ..., y_K^{**})$ the solution of (P_3). As we have already seen, x^{**} satisfies the condition

$$Q \left(\Delta V \left(\varphi, t_0, T \right) < x^{**} \right) \leq \sum_{k=1}^{K} p_k \cdot y_k \leq \alpha.$$

Hence, $x^{**} \leq x^* = \sup \{ x \in I\!R : Q \left(\Delta V \left(\varphi, t_0, T \right) < x \right) \leq \alpha \}$. On the other side, optimization problem (P_2) is identical to (P_3) plus the addition of one restriction. So the set of all possible solutions of (P_2) is a subset of those of (P_3). Hence, $x^* \leq x^{**}$ which completes the proof. □

The first extension of (P_3) is dedicated to the problem of deriving an asset allocation under the restrictions of getting a minimum expected change $\mu_{\min}^{\Delta V} (t_0, T)$ of the future portfolio value from t_0 to T and of ensuring that the VaR for the given time horizon and a significance level is limited by an upper bound $VaR_{\max}$. For the optimization problem, this gives us the additional restrictions

$$\sum_{k=1}^{K} p_k \cdot \Delta V^k \left(\varphi, t_0, T \right) \geq \mu_{\min}^{\Delta V} (t_0, T)$$

and

$$\sum_{k=1}^{K} p_k \cdot \Delta V^k \left(\varphi, t_0, T \right) - x \leq VaR_{\max}.$$

Additionally, the portfolio manager or trader may be restricted to the absolute lower (s_i) and upper trading limits (S_i) for the amount φ_i of asset $i = 1, ..., n$, i.e.

$$s_i \leq \varphi_i \leq S_i, \, i = 1, ..., n,$$

and to a special *budget* to be invested resulting in the restriction

$$V(\varphi, t_0) = budget.$$

On the whole, we are facing the following extension of the optimization problem (P_3):

$$(P_4) \begin{cases} x \to \max \\[2mm] M_k \cdot y_k + \Delta V^k(\varphi, t_0, T) \geq x, \, k = 1, ..., K \\[2mm] \sum_{k=1}^{K} p_k \cdot y_k \leq \alpha \\[2mm] y_k \in \{0, 1\}, \, k = 1, ..., K \\[2mm] \sum_{k=1}^{K} p_k \cdot \Delta V^k(\varphi, t_0, T) \geq \mu_{\min}^{\Delta V}(t_0, T) \\[2mm] x \geq \sum_{k=1}^{K} p_k \cdot \Delta V^k(\varphi, t_0, T) - VaR_{\max} \\[2mm] s_i \leq \varphi_i \leq S_i, \, i = 1, ..., n \\[2mm] \sum_{i=1}^{n} \varphi_i \cdot V_i(t_0) = budget. \end{cases}$$

Again, this is a mixed-integer linear program which can be rather easily solved by commercial optimization tools[9]. The difficulty a portfolio

---

[9] Let $\Delta\mu_i(t_0, T) := \sum_{k=1}^{K} p_k \cdot \Delta V_i^k(t_0, T)$, $i = 1, ..., n$. Furthermore, let

$$a_{ik} := \begin{cases} S_i, & \text{if } \Delta V_i^k(t_0, T) < 0 \\ s_i, & \text{if } \Delta V_i^k(t_0, T) \geq 0 \end{cases}, \, b_{ik} := \begin{cases} s_i, & \text{if } \Delta V_i^k(t_0, T) < 0 \\ S_i, & \text{if } \Delta V_i^k(t_0, T) \geq 0 \end{cases},$$

and

$$c_i := \begin{cases} S_i, & \text{if } \Delta\mu_i(t_0, T) < 0 \\ s_i, & \text{if } \Delta\mu_i(t_0, T) \geq 0 \end{cases}$$

for $i = 1, ..., n$ and $k = 1, ..., K$, as well as

$$M := \max\left\{ \sum_{i=1}^{n} b_{ik} \cdot \Delta V_i^k(t_0, T), \, k = 1, ..., K \right\}$$

and

$$m := \sum_{i=1}^{n} c_i \cdot \Delta\mu_i(t_0, T) - VaR_{\max}.$$

manager may have using problem (P_4) is that he has no freedom in choosing a goal function such as the expected future value to be maximized or the tracking error to be minimized. Also, it may quite well be possible that there is more than one possible solution for (P_4). So we extend the set of restrictions to introduce the freedom of choosing a specific goal function in the second extension of problem (P_3). For sufficiently large numbers $M_{kj}^1 > 0$ and $M_{\nu k}^2 > 0$, we define the constraints

$$M_{kj}^1 \cdot z_{kj} + \Delta V^k\left(\varphi, t_0, T\right) > \Delta V^j\left(\varphi, t_0, T\right), \qquad (\Delta A')$$

$$\Delta V^k\left(\varphi, t_0, T\right) \le \Delta V^j\left(\varphi, t_0, T\right) + M_{kj}^2 \cdot (1 - z_{kj}), \qquad (\Delta B'')$$

$$(1 + \varepsilon) \cdot w_j + \sum_{k=1}^{K} p_k \cdot z_{kj} > \alpha, \qquad (C1)$$

$$\sum_{k=1}^{K} p_k \cdot z_{kj} - p_j \le \alpha + w_j, \qquad (C2)$$

$$\sum_{j=1}^{K} w_j = K - 1 \qquad (C3)$$

for $z_{kj}, w_j \in \{0, 1\}$, $k, j = 1, ..., K$, and $\alpha \in (0, 1)$.

Lemma 7.6 *Let $z_{kj} \in \{0, 1\}$ for all $k, j = 1, ..., K$, and $\alpha \in (0, 1)$.*

a) *Under conditions $(\Delta A')$ and $(\Delta B'')$, we have for all $k, j = 1, ..., K$:*

$$\Delta V^k\left(\varphi, t_0, T\right) \le \Delta V^j\left(\varphi, t_0, T\right) \quad \text{if and only if } z_{kj} = 1.$$

Especially,

$$Q\left(\Delta V\left(\varphi, t_0, T\right) \le \Delta V^j\left(\varphi, t_0, T\right)\right) = F_{\Delta V(\varphi, t_0, T)}\left(\Delta V^j\left(\varphi, t_0, T\right)\right)$$

$$= \sum_{k=1}^{K} p_k \cdot z_{kj}.$$

b) *Let conditions $(\Delta A')$, $(\Delta B'')$, $(C1)$, $(C2)$, and $(C3)$ be satisfied. Furthermore, let $w_{j^*} = 0$, $j^* \in \{1, ..., K\}$. Then we have*

$$\sup\{x \in \mathbb{R} : Q\left(\Delta V\left(\varphi, t_0, T\right) < x\right) \le \alpha\} = \Delta V^{j^*}\left(\varphi, t_0, T\right).$$

Then, we can choose

$$M_k = M - \sum_{i=1}^{n} a_{ik} \cdot \Delta V_i^k\left(t_0, T\right), \ k = 1, ..., K,$$

for a practical application.

Especially,

$$VaR\left(\varphi, t_0, T\right) = \sum_{k=1}^{K} p_k \cdot \Delta V^k\left(\varphi, t_0, T\right) - \Delta V^{j*}\left(\varphi, t_0, T\right).$$

Proof.

a) Let conditions $(\Delta A')$ and $(\Delta B'')$ be satisfied. Furthermore, let $j = 1, ..., K$ be arbitrary but fixed. If $\Delta V^k\left(\varphi, t_0, T\right) \le \Delta V^j\left(\varphi, t_0, T\right)$, then $z_{kj} = 1$ because of $(\Delta A')$. On the other hand, if $\Delta V^k\left(\varphi, t_0, T\right) > \Delta V^j\left(\varphi, t_0, T\right)$, then $z_{kj} = 0$ because of $(\Delta B'')$. Since $z_{kj} = 1$ directly implies $\Delta V^k\left(\varphi, t_0, T\right) \le \Delta V^j\left(\varphi, t_0, T\right)$ because of $(\Delta B'')$ and $z_{kj} = 0$ directly implies $\Delta V^k\left(\varphi, t_0, T\right) > \Delta V^j\left(\varphi, t_0, T\right)$ because of $(\Delta A')$, we conclude

$$\Delta V^k\left(\varphi, t_0, T\right) \le \Delta V^j\left(\varphi, t_0, T\right) \quad \text{if and only if } z_{kj} = 1.$$

So

$$F_{\Delta V(\varphi, t_0, T)}\left(\Delta V^j\left(\varphi, t_0, T\right)\right) = \sum_{\substack{k=1,...,K \\ \Delta V^k(\varphi, t_0, T) \le \Delta V^j(\varphi, t_0, T)}} p_k$$

$$= \sum_{k=1}^{K} p_k \cdot z_{kj}.$$

b) Let conditions $(C1), (C2)$ and $(C3)$ be satisfied in addition to $(\Delta A')$ and $(\Delta B'')$. Furthermore, let j^* be the unique (because of $(C3)$) element of $\{1, ..., K\}$ with $w_{j*} = 0$. Then

$$\sum_{k=1}^{K} p_k \cdot z_{kj*} - p_{j*} \le \alpha \quad \text{and} \quad \sum_{k=1}^{K} p_k \cdot z_{kj*} > \alpha.$$

Using a), this means that

$$F_{\Delta V(\varphi, t_0, T)}\left(\Delta V^{j*}\left(\varphi, t_0, T\right) -\right) \le \alpha$$

and

$$F_{\Delta V(\varphi, t_0, T)}\left(\Delta V^{j*}\left(\varphi, t_0, T\right)\right) > \alpha,$$

which is equivalent to

$$\sup\left\{x \in \mathbb{R} : Q\left(\Delta V\left(\varphi, t_0, T\right) < x\right) \le \alpha\right\} = \Delta V^{j*}\left(\varphi, t_0, T\right).$$

Using Lemma 7.5, the equation for the VaR is an immediate consequence of this result. □

Note that constraints $(C1)$ and $(C2)$ are always satisfied if $w_{j^*} = 1$. Condition $(C3)$ ensures that there is exactly one $j^* \in \{1, ..., K\}$ with $w_{j^*} = 0$. If we choose M_j^3 to be sufficiently large, constraint

$$\Delta V^j (\varphi, t_0, T) \geq \sum_{k=1}^{K} p_k \cdot \Delta V^k (\varphi, t_0, T) - VaR_{\max} - M_j^3 \cdot w_j, \; j = 1, ..., K$$

is always satisfied for $w_j = 1$, $j = 1, ..., K$, and therefore ensures that, for $j^* \in \{1, ..., K\}$ with $w_{j^*} = 0$, the constraint realizes an upper bound for the VaR. As already mentioned in Section 7.2.1, commercial optimization tools usually use inequalities of the form $\leq$ or $\geq$ and a precision expressed by the smallest absolute number that can be recognized numerically within the tool and which we will denote by $\varepsilon > 0$. For this reason we rewrite equations $(\Delta A')$ and $(C1)$ in the following form:

$$M_{kj}^1 \cdot z_{kj} + \Delta V^k (\varphi, t_0, T) \geq \Delta V^j (\varphi, t_0, T) + \varepsilon, \qquad (\Delta A'')$$

$$(1 + \varepsilon) \cdot w_j + \sum_{k=1}^{K} p_k \cdot z_{kj} \geq \alpha + \varepsilon. \qquad (C1')$$

This gives us the following optimization problem (P_5) for maximizing the expected change of the portfolio value under a limited VaR:

$$(P_5) \begin{cases} \displaystyle\sum_{k=1}^{K} p_k \cdot \Delta V^k (\varphi, t_0, T) \to \max \\[2mm] M_{kj}^1 \cdot z_{kj} + \Delta V^k (\varphi, t_0, T) \geq \Delta V^j (\varphi, t_0, T) + \varepsilon, \; k, j = 1, ..., K \\[2mm] \Delta V^k (\varphi, t_0, T) \leq \Delta V^j (\varphi, t_0, T) + M_{kj}^2 \cdot (1 - z_{kj}), \; k, j = 1, ..., K \\[2mm] (1 + \varepsilon) \cdot w_j + \displaystyle\sum_{k=1}^{K} p_k \cdot z_{kj} \geq \alpha + \varepsilon, \; j = 1, ..., K \\[2mm] \displaystyle\sum_{k=1}^{K} p_k \cdot z_{kj} - p_j \leq \alpha + w_j, \; j = 1, ..., K \\[2mm] \displaystyle\sum_{j=1}^{K} w_j = K - 1 \\[2mm] \Delta V^j (\varphi, t_0, T) \geq \displaystyle\sum_{k=1}^{K} p_k \cdot \Delta V^k (\varphi, t_0, T) - VaR_{\max} - M_j^3 \cdot w_j \\[2mm] \text{and } z_{kj}, w_j \in \{0, 1\}, \; j, k = 1, ..., K \\[2mm] s_i \leq \varphi_i \leq S_i, \; i = 1, ..., n \\[2mm] \displaystyle\sum_{i=1}^{n} \varphi_i \cdot V_i (t_0) = budget. \end{cases}$$

It should be mentioned that the optimization problem (P_5) is of large scale and may lead to numerical problems. In that case we propose to solve problem (P_4) for an increasing sequence of minimum expected changes of the future portfolio value $\mu_{\min}^{\Delta V}(t_0, T) + m \cdot s$, with $s > 0$ denoting the step size, $m = 0, 1, 2, ...$, and $\mu_{\min}^{\Delta V}(t_0, T)$ the starting value. Let $m_{\max}$ be the first m in this increasing sequence where the problem (P_4) is unsolvable. Then we choose the solution of the problem (P_4) with $\mu_{\min}^{\Delta V}(t_0, T) + (m_{\max} - 1) \cdot s$ as a lower bound for the expected change of the future value to get an approximation for the optimal portfolio[10].

Case Study (Portfolio Optimization Under Limited VaR).

The previous results are now applied to optimize portfolios with asymmetric return distributions, i.e. portfolios including interest-rate derivatives under the constraint of a limited VaR. Again, we use the Hull-White model for an evaluation of the portfolio price, setting the market price of risk equal to zero for ease of exposition. We use the same yield curve and volatility structure as in the case study of Section 7.2.1 and shown in figures 7.18 and 7.19. The procedure for setting the trading limits for the portfolio manager, the budget, the time horizon, and the scenario generating procedure (in this case using $K = 6$) is also the same. We claim that there are no short sales, i.e. $s_i = 0$, $i = 1, ..., n$. The instruments used are a 6% coupon bond with a notional of 100, annual coupon payments and maturity 11/27/03, and a put option on this coupon bond with maturity 11/27/98 and strike price of

---

[10] If we define

$$d_{ikj} := \begin{cases} S_i, & \text{if } \Delta V_i^j(t_0, T) - \Delta V_i^k(t_0, T) \geq 0 \\ s_i, & \text{if } \Delta V_i^j(t_0, T) - \Delta V_i^k(t_0, T) < 0 \end{cases}$$

and

$$e_{ikj} := \begin{cases} S_i, & \text{if } \Delta V_i^k(t_0, T) - \Delta V_i^j(t_0, T) \geq 0 \\ s_i, & \text{if } \Delta V_i^k(t_0, T) - \Delta V_i^j(t_0, T) < 0 \end{cases},$$

for $i = 1, ..., n$ and $j, k = 1, ..., K$, we can choose

$$M_{kj}^1 := \varepsilon + \sum_{i=1}^{n} d_{ikj} \cdot \left[\Delta V_i^j(t_0, T) - \Delta V_i^k(t_0, T) \right]$$

and

$$M_{kj}^2 := \sum_{i=1}^{n} e_{ikj} \cdot \left[\Delta V_i^k(t_0, T) - \Delta V_i^j(t_0, T) \right]$$

for $k, j = 1, ..., K$, as well as

$$M_j^3 = \sum_{i=1}^{n} f_{ij} \cdot \left[\Delta \mu_i(t_0, T) - \Delta V_i^j(t_0, T) \right] - VaR_{\max}, \quad j = 1, ..., K,$$

with

$$f_{ij} := \begin{cases} S_i, & \text{if } \Delta \mu_i(t_0, T) - \Delta V_i^j(t_0, T) \geq 0 \\ s_i, & \text{if } \Delta \mu_i(t_0, T) - \Delta V_i^j(t_0, T) < 0 \end{cases},$$

$i = 1, ..., n$, $j = 1, ..., K$, for a practical application.

108.16. The actual prices at $11/27/97$ are $D_1(t_0) = 103.94$ for the coupon bond and $D_2(t_0) = 5.81$ for the option, giving us $S_1 = \frac{10\ Mio.}{103.94} = 96,209.31$ for the coupon bond and $S_2 = \frac{10\ Mio.}{5.81} = 1,721,170.40$ for the option. The precision used for the case studies is $\varepsilon = 0.01\%$. In our first application, the aim of the portfolio manager is to find an allocation or insurance strategy by using problem (P_4), where the downside risk, measured by the VaR of his portfolio at a significance level of $1-\alpha = 95\%$, is restricted by the upper bound $VaR_{\max} = 300,000$ and the minimum expected change of the future portfolio value is set to $\mu_{\min}^{\Delta V} = 100,000$ or 1.0%. The solution of problem (P_4) is a portfolio with 63.59% of the notional invested in the coupon bond and 36.41% of the notional invested in the option. The portfolio has an expected future portfolio value of $10,199,476.74$, a VaR of $300,000$ and shows a rather asymmetric return distribution, which is plotted in figure 7.23.

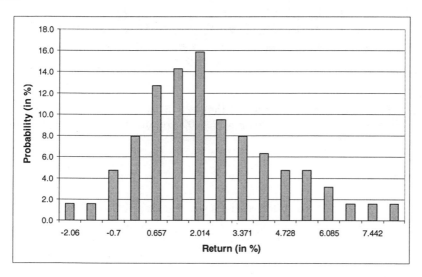

FIGURE 7.23. Return distribution for $VaR_{\max} = 300,000$ and $\mu_{\min}^{\Delta V} = 100,000$ or 1.00% resulting in 63.59% bonds and 36.41% put options

In our second case study, the portfolio manager wants to find an optimal insurance strategy where his aim is to maximize the expected future portfolio value while restricting the downside risk, measured by the VaR of his portfolio at a significance level of $1 - \alpha = 95\%$, by the upper bound $VaR_{\max} = 200,000$. The solution of the corresponding problem (P_5) is a portfolio with 52.53% of the notional invested in the coupon bond and 47.47% of the notional invested in the option. The portfolio has an expected future value of $10,061,958.72$ and a VaR of $200,000$. Due to the lower VaR allowed for the optimal portfolio, we get an even more asymmetric return

distribution compared to the previous example which is shown in figure
7.24.   □

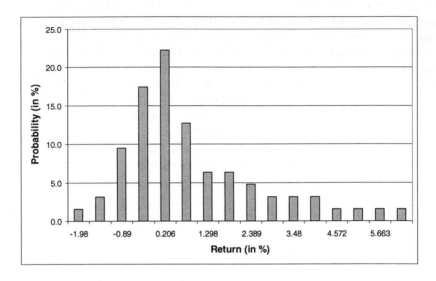

FIGURE 7.24. Return distribution of the portfolio with maximum expected future portfolio value under $VaR_{max} = 200,000$ resulting in 52.53% bonds and
47.47% put options

8
Appendix

The standard normal distribution function can be inverted using the function

$$Q(x) := \frac{1}{\sqrt{2\pi}} \int_x^\infty e^{-\frac{1}{2} \cdot z^2} \, dz = 1 - N(x),$$

where $N(x)$ denotes the value of the standard normal distribution function at point $x \in \mathbb{R}$. Following Abramowitz and Stegun [AS65], the inverse $Q^{-1}(y)$ of Q at point $y \in \left(0, \frac{1}{2}\right]$, i.e. that $x \in \mathbb{R}$ with $Q(x) = y$, can be approximated by

$$x = Q^{-1}(y) = t - \frac{c_0 + c_1 \cdot t + c_2 \cdot t^2}{1 + d_1 \cdot t + d_2 \cdot t^2 + d_3 \cdot t^3}, \quad t := \sqrt{\ln\left(\frac{1}{y^2}\right)},$$

with

$$\begin{array}{lll} c_0 = 2.515517, & c_1 = 0.802853, & c_2 = 0.010328, \\ d_1 = 1.432788, & d_2 = 0.189269, & d_3 = 0.001308, \end{array}$$

and an approximation error of less than $4.5 \cdot 10^{-4}$. Hence, the inverse N^{-1} of the standard normal distribution is given by

$$N^{-1}(y) = \begin{cases} -Q^{-1}(y), & y \in \left(0, \frac{1}{2}\right] \\ Q^{-1}(1-y), & y \in \left(\frac{1}{2}, 1\right). \end{cases}$$

To approximate the standard normal distribution function, we use the points $y^k = \frac{k}{2^n}$, $k = 0, ..., 2^n$, to divide the interval $[0, 1]$ into 2^n subintervals of length 2^{-n}. Then, following Theorem 17.9 in Hinderer [Hin85], the function

$$F_n(x) = \sum_{k=0}^{2^n-1} k \cdot 2^{-n} \cdot 1_{[k \cdot 2^{-n},(k+1) \cdot 2^{-n})}(N(x)),$$

with

$$1_{[a,b)}(N(x)) = \begin{cases} 1, & a \leq N(x) < b \\ 0, & \text{else.} \end{cases}$$

for $a, b \in \mathbb{R}$, satisfies

$$N(x) = \lim_{n \to \infty} F_n(x) \text{ for all } x \in \mathbb{R}.$$

Let us define the function N_n by

$$N_n(x) := \sum_{k=0}^{2^n-1} \frac{k}{2^n - 1} \cdot 1_{[k \cdot 2^{-n},(k+1) \cdot 2^{-n})}(N(x)), \ x \in \mathbb{R}.$$

Then for all $x \in \mathbb{R}$,

$$\begin{aligned}
\lim_{n \to \infty} N_n(x) &= \lim_{n \to \infty} \frac{2^n}{2^n - 1} \cdot F_n(x) \\
&= \lim_{n \to \infty} \frac{2^n}{2^n - 1} \cdot \lim_{n \to \infty} F_n(x) \\
&= N(x).
\end{aligned}$$

Furthermore, since

$$\frac{k}{2^n} \leq \frac{k}{2^n - 1} \leq \frac{k+1}{2^n} \text{ for all } k = 0, 1, ..., 2^n - 1,$$

we get $|N(x) - N_n(x)| \leq 1/2^n$ for all $x \in \mathbb{R}$, i.e.

$$\|N - N_n\|_{\max} := \sup_{x \in \mathbb{R}} |N(x) - N_n(x)| \leq \frac{1}{2^n}.$$

The maximum error approximating the standard normal distribution function N by the function N_n therefore is $1/2^n$. If we require a maximum error of $\alpha_{error} \in (0,1)$ we have to set

$$\frac{1}{2^n} \leq \alpha_{error} \text{ or equivalently } n \geq -\frac{\ln \alpha_{error}}{\ln 2}.$$

Using this accuracy with respect to the approximation of the distribution function, the standard normally distributed random variable X can be approximated by the discrete random variable X_n with possible realizations

$$x^k = N^{-1}\left(\frac{k}{2^n}\right), \ k = 1, ..., 2^n - 1, \tag{8.1}$$

each with probability $p_k = 1/(2^n - 1)$ where X_n converges in distribution to the random variable X.

Following Lemma 4.23, where we assume that the market price of risk is a deterministic function, the short rate $r(t)$ at time $t \in (t_0, T^*]$ is normally distributed (under Q) with expected value[1]

$$E_Q[r(t)|\mathcal{F}_{t_0}] = f(t_0, t) + \frac{1}{2} \cdot v^2(t_0, t) + \sigma \cdot e^{-at} \int_{t_0}^{t} \gamma(s) \cdot e^{as}\, ds$$

and variance

$$s_r^2(t_0, t) := Var_Q[r(t)|\mathcal{F}_{t_0}] = \frac{\sigma^2}{2a}\left(1 - e^{-2a\cdot(t-t_0)}\right),$$

with all parameters defined as in Section 4.5.3, especially

$$v(t_0, t) = \frac{\sigma}{a} \cdot \left(1 - e^{-a(t-t_0)}\right).$$

Hence, under Q, the short rate $r(t)$ at time t can be simulated by a discrete random variable with possible realizations

$$r^k := E_Q[r(t)|\mathcal{F}_{t_0}] + x^k \cdot s_r(t_0, t),\ k = 1, ..., 2^n - 1,$$

each with probability $p_k = \frac{1}{2^n-1}$, where x^k is as in equation (8.1). Furthermore, by Lemma 4.24 the zero-coupon bond prices in the Hull-White model are given by

$$P(r, t, T) = A(t, T) \cdot e^{-B(t,T)\cdot r}\ \text{for all}\ (r, t) \in \mathbb{R} \times [t_0, T],$$

with

$$B(t, T) = \frac{1}{a} \cdot \left(1 - e^{-a\cdot(T-t)}\right)$$

and

$$\ln A(t, T) = \ln\left(\frac{P(t_0, T)}{P(t_0, t)}\right) + B(t, T) \cdot f(t_0, t) - \frac{1}{2} \cdot s_P^2(t_0, t, T),$$

where

$$s_P(t_0, t, T) := B(t, T) \cdot s_r(t_0, t).$$

Inserting the simulations for the short rate $r(t)$ under Q, we can simulate the zero-coupon bond prices by

$$P^k(t, T) := P(r^k, t, T) = A(t, T) \cdot e^{-B(t,T)\cdot r^k} = P(t_0, T) \cdot e^{h(x^k, t_0, t, T)},$$

---

[1] For ease of exposition, especially if $t - t_0$ is small as it is for overnight exposures the term $\frac{1}{2} \cdot v^2(t_0, t)$ may be neglected, since

$$v^2(t_0, t) = \frac{\sigma^2}{a^2} \cdot \left(1 - e^{-a(t-t_0)}\right)^2 \approx \sigma^2 \cdot (t - t_0)^2 \approx 0$$

in this case.

$k = 1, ..., 2^n - 1$, with[2]

$$
\begin{aligned}
h\left(x^k, t_0, t, T\right) \ :=\ & R\left(t_0, t\right) \cdot \left(t - t_0\right) \\
& -\tfrac{1}{2} \cdot \left(s_P^2\left(t_0, t, T\right) + B\left(t, T\right) \cdot v^2\left(t_0, t\right)\right) \\
& -B\left(t, T\right) \cdot \sigma \cdot e^{-at} \int_{t_0}^{t} \gamma(s) \cdot e^{as}\, ds - x^k \cdot s_P\left(t_0, t, T\right).
\end{aligned}
$$

It should be noted that h does not depend on the forward rate $f(t_0, t)$ as $r(t)$ does, i.e. the simulation of the zero-coupon bond prices does not depend on the forward rate. The zero-rate curve is consequently simulated by

$$
R^k\left(t, T\right) := -\frac{1}{T - t} \cdot \ln P^k\left(t, T\right), \ k = 1, ..., 2^n - 1,
$$

which does also not depend on the forward rates. These simulations can now be used to simulate the prices of any interest-rate derivative with respect to changes of the short rate or, within the Hull-White model, with respect to changes of the zero-rate curve.

---

[2] As above, we may neglect the term $B\left(t, T\right) \cdot v^2\left(t_0, t\right)$ for ease of exposition if $t - t_0$ is small.

References

[ADEH99] P. Artzner, F. Delbaen, J.-M. Eber, and D. Heath. Coherent Measures of Risk. *Math. Finance*, 9(3):203–228, 1999.

[Arr71] K.J. Arrow. *Theory of Risk Aversion*. Markham, 1971.

[AS65] M. Abramowitz and I.A. Stegun. *Handbook of Mathematical Functions*. Dover Publications, 1965.

[Bac00] L. Bachelier. Théorie de la Spéculation. *Ann. Sci. École Norm. Sup.*, 17:21–86, 1900.

[Baw75] V.S. Bawa. Optimal Rules for Ordering Uncertain Prospects. *J. Finan. Econom.*, 2:95–121, 1975.

[BC73] S.P. Bradley and D.B. Crane. Management of Commercial Bank Government Security Portfolios: An Optimization Approach under Uncertainty. *J. Bank Res.*, pages 18–30, 1973.

[BC81] R. Bookstaber and R. Clarke. Options Can Alter Portfolio Return Distributions. *J. Portfolio Manag.*, 7:63–70, 1981.

[BC83] R. Bookstaber and R. Clarke. An Algorithm to Calculate the Return Distribution of Portfolios with Option Positions. *Manag. Sci.*, 29(4):419–429, 1983.

[BD86] S. Brown and P.H. Dybvig. The Empirical Implications of the Cox, Ingersoll, Ross Theory of the Term Structure of Interest Rates. *J. Finance*, 41:617–630, 1986.

[BDT90] F. Black, E. Derman, and W. Toy. A One-Factor Model of Interest Rates and its Application to Treasury Bond Options. *Finan. Analysts J.*, 46(1):33–39, 1990.

326   References

[BGM97]   A. Brace, D. Gątarek, and M. Musiela. The Market Model of Interest
          Rate Dynamics. *Math. Finance*, 7(2):127–155, 1997.

[BK91]   F. Black and P. Karasinski. Bond and Option Pricing When Short
          Rates Are Lognormal. *Finan. Analysts J.*, 46(4):52–59, 1991.

[BK98]   N.H. Bingham and R. Kiesel. *Risk-Neutral Valuation: Pricing and
          Hedging Financial Derivatives.* Springer, 1998.

[BL77]   V.S. Bawa and E.B. Lindenberg. Capital Market Equilibrium in a
          Mean-Lower Partial Moment Framework. *J. Finan. Econom.*, 5:189–
          200, 1977.

[Bla76]   F. Black. The Pricing of Commodity Contracts. *J. Finan. Econom.*,
          31:167–179, 1976.

[BM01]   D. Brigo and F. Mercurio. *Interest-Rate Models: Theory and Practice.*
          Springer, 2001.

[Boy77]   P.P. Boyle. Options: A Monte Carlo Approach. *J. Finan. Econom.*,
          4:323–338, 1977.

[BR96]   M. Baxter and A. Rennie. *Financial Calculus.* Cambridge University
          Press, 1996.

[BS73]   F. Black and M. Scholes. The Pricing of Options and Corporate
          Liabilities. *J. Political Econom.*, 81:637–659, 1973.

[BS82]   M.J. Brennan and E.S. Schwartz. An Equilibrium Model of Bond
          Pricing and a Test of Market Efficiency. *J. Finan. Quant. Anal.*,
          17:301–329, 1982.

[CEJ98]   L. Cathcart and L. El-Jahel. Valuation of Defaultable Bonds. *J.
          Fixed Income*, 8(1):65–78, 1998.

[CIR85]   J.C. Cox, J.E. Ingersoll, and S.A. Ross. A Theory of the Term Struc-
          ture of Interest Rates. *Econometrica*, 53:385–407, 1985.

[CJR79]   S. Ross Cox J. and M. Rubinstein. Option Pricing: A Simplified
          Approach. *J. Finan. Econom.*, 7:229–264, 1979.

[CKLS92] K.C. Chan, G.A. Karolyi, F. Longstaff, and A.B. Sanders. An Em-
          pirical Comparison of Alternative Models for the Short-Term Interest
          Rates. *J. Finance*, 47:1209–1228, 1992.

[CS93]   R. Chen and L. Scott. Maximum Likelihood Estimation for a Multi-
          Factor Equilibrium Model of the Term Structure of Interest Rates. *J.
          Fixed Income*, 4:14–31, 1993.

[Dah89]   H. Dahl. Variationer I Rentestrukturen Og Styring Af Renterisiko.
          Finans/Invest, 1, 1989.

[Dah93]   H. Dahl. *A Flexible Approach to Interest-Rate Risk Management.*
          Cambridge University Press, 1993.

[DK94]   D. Duffie and R. Kan. Multi-Factor Interest-Rate Models. *Phil.
          Trans. Roy. Soc. London Ser. A*, 347:577–586, 1994.

[DK96]   D. Duffie and R. Kan. A Yield-Factor Model of Interest Rates. *Math.
          Finance*, 6:379–406, 1996.

[DS94a]   F. Delbaen and W. Schachermayer. A General Version of the Funda-
          mental Theorem of Asset Pricing. *Math. Ann.*, 300:463–520, 1994.

[DS94b]   D. Duffie and K. Singleton. Modeling Term Structures of Defaultable
          Bonds. *Working Paper*, 1994.

[Duf92]   D. Duffie. *Dynamic Asset Pricing Theory*. Princeton University
          Press, 1992.

[EE76]   M.E. Echols and J.W. Elliot. A Quantitative Yield Curve Model for
          Estimating the Term Structure of Interest Rates. *J. Finan. Quant.
          Anal.*, 11:87–114, 1976.

[EG91]   E.J. Elton and M.J. Gruber. *Modern Portfolio Theory and Invest-
          ment Analysis*. John Wiley & Sons, 1991.

[EKM97]   P. Embrechts, C. Klüppelberg, and T. Mikosch. *Modelling Extremal
          Events*. Springer, 1997.

[FFP91]   F.J. Fabozzi, T.D. Fabozzi, and I.M. Pollack, editors. *The Handbook
          of Fixed Income Securities*. Business One Irwin, third edition, 1991.

[FH96]   B. Flesaker and L. Hughston. Positive Interest. *Risk*, 9(1):46–49,
          1996.

[Fle90]   R. Fletcher. *Practical Methods of Optimization*. John Wiley & Sons,
          1990.

[FLT93]   M.D. Fitzgerald, C. Lubochinsky, and P. Thomas. *Financial Futures*.
          Euromoney Books, second edition, 1993.

[Fri64]   A. Friedman. *Partial Differential Equations of the Parabolic Type*.
          Prentice-Hall, 1964.

[Fri75]   A. Friedman. *Stochastic Differential Equations and Applications*, vol-
          ume I. Academic Press, 1975.

[Gar86]   K. Garbade. Modes of Fluctuations in Bond Yields - an Analysis
          of Principal Components. Bankers Trust Company, Monea Market
          Center, New York, June 1986.

[Gar89]   K. Garbade. Polynomial Representations of the Yield Curve and
          its Modes of Fluctuation. Bankers Trust Company, Money Market
          Center, New York, July 1989.

[Ges77]   R. Geske. The Valuation of Corporate Liabilities as Compound Op-
          tions. *J. Finan. Quant. Anal.*, 12:541–552, 1977.

[GR93]   M. Gibbons and K. Ramaswamy. A Test of the Cox, Ingersoll, Ross
          Theory of the Term Structure. *Rev. Finan. Studies*, 6:619–658, 1993.

[Gre93]   W.H. Greene. *Econometric Analysis*. Prentice-Hall, 1993.

[Har91]   W.V. Harlow. Asset Allocation in a Downside-Risk Framework. *Fi-
          nan. Analysts J.*, 47(5):28–40, 1991.

[HH64]   J.M. Hammersley and D.C. Handscomb. *Monte Carlo Methods*. John
          Wiley & Sons, 1964.

[Hic62]   J.R. Hicks. Liquidity. *Econom. J.*, 72:787–802, 1962.

[Hin85]   K. Hinderer. *Grundbegriffe der Wahrscheinlichkeitstheorie*. Springer, 1985.

[HJM92]   D. Heath, R. Jarrow, and A. Morton. Bond Pricing and the Term Structure of Interest Rates: A New Methodology for Contingent Claim Valuation. *Econometrica*, 60:77–105, 1992.

[HK84]   R. Hogg and S. Klugmann. *Loss Distributions*. John Wiley & Sons, 1984.

[HL86]   T.S.Y. Ho and S.-B. Lee. Term Structure Movements and Pricing Interest-Rate Contingent Claims. *J. Finance*, 41:1011–1029, 1986.

[HP83]   J.M. Harrison and S.R. Pliska. A Stochastic Calculus Model of Continuous Trading: Complete Markets. *Stochastic Process. Appl.*, 15:313–316, 1983.

[HR89]   W.V. Harlow and K.S. Rao. Asset Pricing in a Generalized Mean-Lower Partial Moment Framework: Theory and Evidence. *J. Finan. Quant. Anal.*, 24:285–311, 1989.

[HS84]   T. Ho and R. Singer. The Value of Corporate Debt with a Sinking Fund Provision. *J. Business*, 57:315–336, 1984.

[Hub81]   P.J. Huber. *Robust Statistics*. John Wiley & Sons, 1981.

[Hul00]   J.C. Hull. *Options, Futures, and Other Derivatives*. Prentice-Hall, fourth edition, 2000.

[HW90]   J. Hull and A. White. Pricing Interest-Rate Derivative Securities. *Rev. Fin. Stud.*, 3(4):573–592, 1990.

[HW93]   J. Hull and A. White. One-Factor Interest-Rate Models and the Valuation of Interest-Rate Securities. *J. Finan. Quant. Anal.*, 28:235–254, 1993.

[HW94a]   J. Hull and A. White. Numerical Procedures for Implementing Term-Structure Models II: Two-Factor Models. *J. Derivatives*, 4:37–48, 1994.

[HW94b]   J. Hull and A. White. Numerical Procedures for Implementing Term-Structure Models I: Single-Factor Models. *J. Derivatives*, 3:7–16, 1994.

[HW96]   A. Hull and J. White. Using Hull-White Interest-Rate Trees. *J. Derivatives*, 1:26–36, 1996.

[HW00]   J. Hull and A. White. Forward Rate Volatilities, Swap Rate Volatilities, and Implementation of the LIBOR Market Model. *J. Fixed Income*, 3:46–62, 2000.

[Inc44]   E.L. Ince. *Ordinary Differential Equations*. Dover, 1944.

[Ing87]   J.E. Ingersoll. *Theory of Financial Decision Making*. Rowman & Littlefield, fifth edition, 1987.

[Itô44]   K. Itô. Stochastic Integral. *Proc. Imperial Acad. Tokyo*, 20:519–524, 1944.

[Jam89]   F. Jamshidian. An Exact Bond Option Pricing Formula. *J. Finance*, 44:205–209, 1989.

[Jam91]   F. Jamshidian. Bond and Option Evaluation in the Gaussian Interest Rate Model. *Res. Finance*, 9:131–170, 1991.

[Jam93]   F. Jamshidian. Options and Futures Evaluation with Deterministic Volatilities. *Math. Finance*, 3(2):149–159, 1993.

[Jam96]   F. Jamshidian. Bond, Futures and Options Evaluation in the Quadratic Interest Rate Model. *Appl. Math. Finance*, 3:93–115, 1996.

[Jam98]   F. Jamshidian. LIBOR and Swap Market Models and Measures. *Finance Stochast.*, 1(4):293–330, 1998.

[Jor84]   J.V. Jordan. Tax Effects in Term Structure Estimation. *J. Finance*, 39:393–406, 1984.

[JT95]   R.A. Jarrow and S.M. Turnbull. Pricing Options on Derivative Securities Subject to Credit Risk. *J. Finance*, 50:53–85, 1995.

[JT00]   R.A. Jarrow and S.M. Turnbull. *Derivative Securities*. South-Western College Publishing, Cincinnati, 2000.

[JZ97]   F. Jamshidian and Y. Zhu. Scenario Simulation: Theory and Methodology. *Finance Stochast.*, 1:43–67, 1997.

[KK99]   R. Korn and E. Korn. *Optionsbewertung und Portfolio-Optimierung*. Vieweg, 1999.

[KN94]   M. Kijima and I. Nagayama. Efficient Numerical Procedures for the Hull-White Extended Vasicek Model. *J. Finan. Engrg.*, September-December:275–292, 1994.

[KN96]   M. Kijima and I. Nagayama. A Numerical Procedure for the General One-Factor Interest Rate Model. *J. Finan. Engrg.*, December:317–337, 1996.

[KRS93]   I.J. Kim, K. Ramaswamy, and S. Sundaresan. Does Default Risk in Coupons Affect the Valuation of Corporate Bonds? A Contingent Claim Model. *Finan. Manag.*, 22:117–131, 1993.

[Kry80]   N. Krylov. *Controlled Diffusion Processes*. Springer, 1980.

[KS91]   I. Karatzas and S.E. Shreve. *Brownian Motion and Stochastic Calculus*. Springer, second edition, 1991.

[Lan94]   D. Lando. Three Essays on Contingent Claims Pricing. *Working Paper*, 1994.

[Lan96]   D. Lando. Modelling Bonds and Derivatives with Default Risk. *Working Paper*, 1996.

[Lan98]   D. Lando. On Cox Processes and Credit Risky Securities. *Working Paper*, 1998.

[Leo96]   K. Leong. *The Right Approach*. Risk Publications, 1996.

[LH89]   M.L. Leibowitz and R.D. Henriksson. Portfolio Optimization with Shortfall Constraints: A Confidence-Limit Approach to Managing Downside Risk. *Finan. Analysts J.*, 45(2):34–41, 1989.

[Lin72]   J. Lintner. Equilibrium in a Random Walk and Lognormal Securities Market. *Discussion Paper 235*, 1972.

[LK91]   M.L. Leibowitz and S. Kogelman. Asset Allocation under Shortfall Constraints. *J. Portfolio Manag.*, 4:18–23, 1991.

[LL97]   D. Lamberton and B. Lapeyre. *Introduction to Stochastic Calculus Applied to Finance.* Chapman & Hall, 1997.

[LR84]   R.H. Litzenberger and J. Rolfo. An International Study of Tax Effects on Government Bonds. *J. Finance*, 39:1–22, 1984.

[LRS95]   A. Li, P. Ritchken, and L. Sankarasubramanian. Lattice Models for Pricing American Interest-Rate Claims. *J. Finance*, 50(2):719–737, 1995.

[LS88]   R. Litterman and J. Scheinkman. Common Factors Affecting Bond Returns. Goldman, Sachs & Co., Financial Strategies Group, New York, September 1988.

[LS92]   F.A. Longstaff and E.S. Schwartz. Interest-Rate Volatility and the Term Structure: A Two-Factor General Equilibrium Model. *J. Finance*, 47:1259–1282, 1992.

[LS95]   F.A. Longstaff and E.S. Schwartz. A Simple Approach to Valuing Risky Fixed and Floating Rate Debt. *J. Finance*, 50:789–819, 1995.

[Mar52]   H.M. Markowitz. Portfolio Selection. *J. Finance*, 7:77–91, 1952.

[Mar70]   H.M. Markowitz. *Portfolio Selection: Efficient Diversification of Investments.* John Wiley & Sons, 1970.

[Mar91]   H.M. Markowitz. *Portfolio Selection.* Blackwell, 1991.

[May98]   S.R. Mayer. *Bewertung Exotischer Zinsderivate.* IFA-Verlag, 1998.

[MC95a]   W. Morokoff and R.E. Caflish. Quasi-Monte Carlo Integration. *J. Comput. Physics*, 122(2):218–230, 1995.

[MC95b]   B. Moskowitz and R.E. Caflish. Smoothness and Dimension Reduction in Quasi-Monte Carlo Methods. *Mathematical and Computer Modelling*, 23(8/9):37–54, 1995.

[McC71]   J.H. McCulloch. Measuring the Term Structure of Interest Rates. *J. Business*, 44:19–31, 1971.

[McC75]   J.H. McCulloch. The Tax-Adjusted Yield Curve. *J. Finance*, 30:811–830, 1975.

[MCM88]   J. Martin, S. Cox, and R. MacMinn. *The Theory of Finance, Evidence and Applications.* Dryden, 1988.

[Mer74]   R.C. Merton. On the Pricing of Corporate Debt: The Risk Structure of Interest Rates. *J. Finance*, 2:449–470, 1974.

[MR97]   M. Musiela and M. Rutkowski. *Martingale Methods in Financial Modelling.* Springer, 1997.

[MSS97]   K.R. Miltersen, K. Sandmann, and D. Sondermann. Closed Form Solutions for the Term Structure Derivatives with Log-Normal Interest Rates. *J. Finance*, 52(1):409–430, 1997.

[NC91]   B.K. Newton and P.B. Chau. Valuation and Risk Analysis of International Bonds. In T.D. Fabozzi F. J. Fabozzi and I.M. Pollack, editors, *The Handbook of Fixed Income Securities*, pages 1320–1334. Business One Irwin, 1991.

[Ncl96]   I. Nelken. *The Handbook of Exotic Options: Instruments, Analysis, and Applications.* Irwin Professional Publishing, Chicago, 1996.

[NM65]   J.A. Nelder and R. Mead. A Simplex Method for Function Minimization. *Computer J.*, 7:308–313, 1965.

[Øks98]   B. Øksendal. *Stochastic Differential Equations: An Introduction with Applications.* Springer, fifth edition, 1998.

[Pel00]   A. Pelsser. *Efficient Methods for Valuing Interest Rate Derivatives.* Springer, 2000.

[Pro92]   P. Protter. *Stochastic Integration and Differential Equations.* Springer, second edition, 1992.

[Pro99]   F. Proske. Feynman-Kac for the Hull-White Process. In *Risklab Technical Reports No. 9901.* RiskLab GmbH, Munich, 1999.

[PS00]   M.B. Pedersen and A. Schumacher. *LIBOR-Marktmodelle.* Uhlenbruch, Bad Soden, 2000.

[PV95]   A. Pelsser and T. Vorst. Optimal Optioned Portfolios with Confidence Limits on Shortfall Constraints. *Adv. Quant. Anal. Finance Accounting*, 3(Part A):205–220, 1995.

[Reb96]   R. Rebonato. *Interest-Rate Option Models.* John Wiley & Sons, 1996.

[RW87]   L.C.G. Rogers and D. Williams. *Diffusions, Markov Processes, and Martingales: Itô Calculus*, volume 2. John Wiley & Sons, 1987.

[Sch96]   P.J. Schönbucher. The Term Structure of Defaultable Bond Prices. *Working Paper*, 1996.

[Sha64]   W.F. Sharpe. Capital Asset Prices: A Theory of Market Equilibrium under Conditions of Risk. *J. Finance*, 29:425–442, 1964.

[SM96]   C. Smithson and L. Minton. Value-at-Risk. *Risk*, 9(1):25–27, 1996.

[SMC78]   D.B. Suits, A. Mason, and L. Chan. Spline Functions Fitted by Standard Regression Methods. *Rev. Econom. Statist.*, 60:132–139, 1978.

[SML94]   M. Steiner, F. Meyer, and K. Luttermann. *Die preisliche Be-wertung von Zins-Futures*, volume 8. 1994.

[STD93]   D. Shimko, N. Tejima, and D. Van Deventer. The Pricing of Risky Debt When Interest Rates are Stochastic. *J. Fixed Income*, 3(3):58–66, 1993.

[SZ96]   G. Scheuenstuhl and R. Zagst. Optimal Optioned Portfolios with Limited Downside Risk. In P. Albrecht, editor, *Aktuarielle Ansätze für Finanz-Risiken*, volume II, pages 1497–1517. VVW Karlsruhe, 1996.

[SZ97]   G. Scheuenstuhl and R. Zagst. Asymmetrische Renditestrukturen und ihre Optimierung im Portfoliomanagement mit Optionen. In C. Kutscher and G. Schwarz, editors, *Aktives Portfolio Management*, volume 2, pages 153–174. Verlag Neue Zürcher Zeitung, 1997.

[SZ00]   B. Schmid and R. Zagst. A Three-Factor Defaultable Term-Structure Model. *J. Fixed Income*, 10(2):63–79, 2000.

[Tur95]   S.M. Turnbull. Interest Rate Digital Options. *J. Derivatives*, 3:92–101, 1995.

[Vas77]   O. Vasicek. An Equilibrium Characterization of the Term Structure. *J. Finan. Econom.*, 5:177–188, 1977.

[WDH93]   P. Wilmott, J. Dewynne, and S. Howison. *Option Pricing: Mathematical Models and Computation*. Oxford Financial Press, Oxford, 1993.

[Wie23]   N. Wiener. Differential Space. *J. Math. and Phys.*, 2:131–174, 1923.

[Wil96]   T.C. Wilson. Calculating Risk Capital. In C. Alexander, editor, *The Handbook of Risk Management and Analysis*, pages 195–232. John Wileys & Sons, 1996.

[YW71]   T. Yamada and S. Watanabe. On the Uniqueness of Solutions of Stochastic Differential Equations. *Journal of Mathematics of Kyoto University*, 11:155–167, 1971.

[Zag97a]   R. Zagst. Efficient Value-at-Risk Calculation for Interest Rate Portfolios. *Financial Markets and Portfolio Management*, 11(2):165–178, 1997.

[Zag97b]   R. Zagst. Value at Risk (VaR) - Part I: Methods with Full Valuation. *Solutions*, 1(1):11–15, 1997.

[Zag97c]   R. Zagst. Value at Risk (VaR) - Part II: Methods with Approximate Valuation. *Solutions*, 1(2):13–21, 1997.

Index

Printing: Saladruck, Berlin
Binding: H. Stürtz AG, Würzburg